Recent Titles in This Series

(See the AMS catalog for earlier titles)

Photograph courtesy of Claude Doeblin.

Wolfgang Doeblin
1915–1940

CONTEMPORARY MATHEMATICS

149

Doeblin and Modern Probability

Proceedings of the Doeblin Conference
"50 Years after Doeblin: Development in the Theory
of Markov Chains, Markov Processes,
and Sums of Random Variables"
held November 2–7, 1991, with support
from the Applied Probability Trust

Harry Cohn
Editor

American Mathematical Society
Providence, Rhode Island

Proceedings of the Doeblin Conference "50 Years after Doeblin: Development in the Theory of Markov Chains, Markov Processes, and Sums of Random Variables" held November 2–7, 1991 at the University of Tubingen's Heinrich Fabri Institut, Blaubeuren, Germany, through the efforts of Professors H. Hering and J. Gani, and with support from the Applied Probability Trust.

1991 *Mathematics Subject Classification*. Primary 60–06, 60–02, 60–03.

Library of Congress Cataloging-in-Publication Data

Doeblin and modern probability/Harry Cohn, editor.
    p. cm.—(Contemporary mathematics, ISSN 0271-4132; v. 149)
    "Proceedings of the Doeblin conference '50 years after Doeblin: development in the theory of Markov chains, Markov processes, and sums of random variables' held November 2–7, 1991, through the efforts of Professors H. Hering and J. Gani, and with support from the Applied Probability Trust."
    ISBN 0-8218-5149-7
    1. Probabilities—Congresses. 2. Stochastic processes—Congresses. I. Doeblin, Wolfgang. II. Cohn, Harry, 1940– . III. Series: Contemporary mathematics (American Mathematical Society); v. 149.
QA273.A1D64   1993   93-17268
519.2—dc20   CIP

This volume was printed directly from author-prepared copy.
Portions of the volume were typeset by the authors using $\mathcal{A}\mathcal{M}\mathcal{S}$-TEX and $\mathcal{A}\mathcal{M}\mathcal{S}$-LATEX, the American Mathematical Society's TEX macro system.

10 9 8 7 6 5 4 3 2 1   98 97 96 95 94 93

Contents

PREFACE

J.L. DOOB

My first contact with Doeblin was a letter from him, protesting (justifiably) that I should have written him about a slip in his 1938 paper instead of simply noting the lapse in a later paper of my own. I was making so many errors in my own work, that the error in Doeblin's paper seemed no big deal to me! To do him justice, Doeblin held no grudge about my reference and sent me a card showing how to fix his proof. Thereafter, we exchanged reprints and letters; in particular he sent me his thesis. This thesis was published in 1937 in a rather inaccessible journal, and I thought so highly of it that I printed some of its results, with few changes, in my book sl "Stochastic Processes" in 1953. When the news spread that I had a copy of the thesis, I received so many requests for it in those pre Xerox days that I finally had it microfilmed to facilitate distribution.

In the late thirties, probability theory was in the process of modernization and was gradually being incorporated into the body of standard mathematics. Deep researchers like Doeblin delayed the acceptance of probability as mathematics by their virtuosity in dealing not only with probabilities, but also with the subtleties of sample functions. The latter seemed alien to classical analysts who while experiencing no difficulty in accepting analytical studies of conditional probabilities which led to difference equations, partial differential equations, and integral equations, in which the probabilistic background was hardly hinted at, found sample function contexts incomprehensible and therefore considered them outside mathematics.

I found it extraordinary that Doeblin could still do mathematical research while he was in the French army. The paragraph which I quote below, translated from a letter to me in French dated 16 April 1939 illustrates how he managed to work even in the army.

> *I had been intending since October to write out the above proof the following Sunday, but for many reasons I always deferred it. For the moment I spend rather little time on mathematics as military service scarcely leaves me any spare time. I have returned to my research on the set of powers of a probability law, which I hope to finish before October. I shall send you some reprints one of these days.*

The following eloquent paragraph was written about Doeblin by his mentor Paul Lévy in his obituary; this is my translation:

> *One is always struck by the sureness and precision of his reasoning, and his extraordinary ability to solve the most varied difficulties either by frontal attack or by finding an indirect approach. I think I can say, to give an idea of the level at which he deserves to be placed, that one can count on the fingers of a single hand the mathematicians, who since Abel and Galois, died so young leaving such an important body of work. There is no doubt that his name should maintain an important place in the history of the calculus of probability.*

It is to be hoped that the young probabilists of today will learn from this book how much they owe not only to the results obtained by Doeblin, but to the new approach to probability that he was instrumental in creating.

101 WEST WINDSOR ROAD, #1104 URBANA, ILLINOIS, 61801, U.S.A.

INTRODUCTION

HARRY COHN

The present volume is based on the papers read at the conference "**50 years after Doeblin: Developments in the theory of Markov chains, Markov processes and sums of random variables**" held between 2-7 November 1991 at Blaubeuren.

The conference was dedicated to the memory of one of the greatest probabilists of this century, Wolfgang Doeblin, who died in action during World War II at the age of 25. Doeblin left behind several seminal contributions which have profoundly influenced many areas of probability and still continue to be a source of inspiration for modern developments.

A group of probabilists from a number of countries thought that a commemoration of his achievements by the community of scientists was in order 50 years after his death, and that the time was long overdue for a well documented account of his life and work. This led to a conference in Germany, where distinguished scientists from all over the world whose research has been influenced by Doeblin's papers presented lectures and engaged in discussions which were later to materialize in this book dedicated to Doeblin's memory.

An organizing committee was formed in 1989 consisting of A Blanc-Lapierre (France), KL Chung (USA), H Cohn (Australia), J Gani (Australia), H Hering (Germany) and M Iosifescu (Romania). In the final analysis, the conference became possible through the efforts of Professor H Hering who obtained funding for the accommodation of 35 participants in the modern building of the Heinrich Fabri Institute at Blaubeuren, a small picturesque town near Ulm. Some funds were also made available by the Applied Probability Trust through Professor J Gani. The participants, representing

16 countries, payed tribute to Doeblin's legacy in talks and discussions
which covered many of the directions Doeblin had worked on during his
short life. We were privileged to have Professor JL Doob (USA) attend
the conference as guest of honour. His book *"Stochastic Processes"* published in 1953 was the first to establish Doeblin as a basic contributor to
modern Probability Theory.

Although an article on Doeblin's life and work was written by Paul Lévy
in 1955, there has been rather scanty information available on his life and
only part of his 26 papers were known. There is however a well researched
life of Doeblin's father, the noted novelist Alfred Doeblin, whose best
known work *"Berlin, Alexanderplatz"* first published in 1929 has been
internationally acclaimed and translated into many languages. Later this
novel has become known to millions of people from Fassbinder's masterly
adaptation to a TV series. Much unknown and very relevant material was
also found in the Paris archives and brought to light through the efforts
of Professors Blanc-Lapierre and Bru. We also wish to acknowledge the
assistance of Doeblin's younger brother Claude, who lives in Nice, France,
and kindly provided some material on Doeblin from the family archives,
later to be incorporated in Bru's article.

Doeblin's life and work were illustrated by T Lindvall (Sweden) who gave
a talk entitled *Wolfgang Doeblin: Snapshots of his life and work*, B Bru
(France) who dealt with *W Doeblin: life and work from Parisian archives*
and KL Chung (USA) who described how he became acquainted with
Doeblin's work on Markov chains in *Discovering Doeblin's work*.

These talks were followed by a number of sessions where the participants discussed various contributions of Doeblin together with further
developments in the current literature. Among the contributors were
S Asmussen* (Denmark): *Light traffic theorems for random walks and
queues*, KB Athreya (USA): *Continuous time stochastic gambling problems* E Bolthausen (Switzerland) *Large deviations and problems with
long range interactions of the paths*, P Brémaud (France): *Maximal coupling and sensitivity analysis*, JE Cohen (USA): *Relative entropy under
mappings by stochastic matrices*, H Cohn (Australia): *On a paper by
Doeblin on finite non-homogeneous Markov chains*, M Csörgő (Canada):
*Almost sure summability of partial sums of independent random vari-
ables*, I Cuculescu (Romania): *Applications of a construction of Markov
processes*, Y Derriennic (France): *On the local limit theorem for some
Markov chains on some unilateral random walks in random environments*,

---

*Some of these talks have not appeared as papers in this book, while others have
had their titles altered when written up in their final forms.

M Gordin (Russia): *Homoclinic approach to the central limit theorem for dynamical systems*, J Hajnal (England): *A theory of shuffling*, A Hordijk (Netherland): *The Doeblin condition and generalized strong ergodicity*, M Iosifescu (Romania): *A basic tool in mathematical chaos theory: the 1937 ergodic theorem of Doeblin and Fortet*, P Jagers (Sweden): *The Markov renewal structure of stable populations*, A Joffe (Canada): *Perturbations of homogeneous Markov chains*, S Kalpazidou (Greece): *On Doeblin-Fortet chains occurring in the number expansion theory*, JFK Kingman (England): *Poisson processes and random sets*, TM Liggett (USA): *The coupling technique in interacting particle systems*, S Meyn (USA): *Stability criteria for continuous time Markov processes*, A Mukherjea (USA): *Random walks on matrices and attractors*, P Ney (USA): *Regeneration for non-Markovian chains*, P Nüesch (Switzerland): *A multivariate look at ES Andersen's equivalence principle*, E Nummelin (Finland): *Duality for Markov operators with applications to interval maps*, M Peligrad (USA): *Limit theorems for dependent sequences*, P Révész (Austria): *Path properties of an infinite system of Wiener processes*, M Rosenblatt (USA): *The central limit theorem for Markov sequences*, H Thorisson (Iceland): *Shift-coupling*, R Tweedie (USA): *Conditions for Doeblin decomposition of general Markov chains*, E Seneta (Australia): *Applications of ergodicity coefficients to homogeneous Markov chains*.

Discussions followed each talk and often continued informally as the participants were accommodated in the same building and had their meals together. The talks revealed an amazing number of important contributions made by Doeblin; this was a surprise even for those participants, like Chung and Doob, who have written comprehensive books incorporating Doeblin's contributions. A number of Doeblin's articles published in obscure journals were known only to some of the participants; it turned out that very new developments such as chaos theory and random mappings transformations benefit from crucial contributions made by Doeblin more than 50 years ago.

Although a thorough analysis of Doeblin's contributions to probability has yet to be written, this book endeavours to highlight Doeblin's importance in the subject by emphasising the influence of his ideas on a number of contemporary research directions.

On completing this project, I feel gratitude to the organizing committee and the contributors to the Blaubeuren conference as well as to other colleagues who could not attend the conference but offered their support and advice. Special thanks are due to Kai Lai Chung who was enthusiastic about the project when it was first suggested and constantly offered very valuable advice, and my old friend Marius Iosifescu—one of the best connoisseurs of Doeblin's work. Last but not least, I am enormously grateful to Joe Gani whose help during all the stages of this project was essential in bringing it to fruition.

DEPARTMENT OF STATISTICS, UNIVERSITY OF MELBOURNE, PARKVILLE, VICTORIA 3052, AUSTRALIA.

Contemporary Mathematics
Volume **149**, 1993

DOEBLIN'S LIFE AND WORK
FROM HIS CORRESPONDENCE

BERNARD BRU

ABSTRACT. We examine W. Doeblin's life and work through his correspondence, deposited in different archives in Paris and in the Marbach Schiller Museum

1. Doeblin's Life and Work from his Correspondence

Let me begin by thanking the organizers of the Doeblin Conference who invited me to speak at Blaubeuren. It is A. Blanc-Lapierre who should have talked about the life and work of Wolfgang Doeblin (W.D.) in Paris; he asked me to take his place since he could not come to the Conference. Though he has stated that he had no direct association with W.D., Professor Blanc-Lapierre is obviously much more qualified than I to speak here, since he was born in 1915 like W.D., and knew many of the same people and the same events.

Instead of personal souvenirs, I shall try to examine with you some of the sources available in Paris, such as the National Archives where you can find records of the Faculty of Sciences of Paris, the "Académie des Sciences" archives where Maurice Fréchet's family has deposited the greatest part of his personal papers, the "Bulletin des Sciences Mathématiques" archives, the War Archives and other archival collections. These archives contain letters, manuscripts, and annotations which give details of some aspects of W.D.'s life and work and especially his scientific training, the chronology of his works and the conditions under which he brought them to a succesful conclusion between his arrival inParis

---

AMS 1991 subject classifications. Primary 01A60, 01A70; secondary 60-03
This paper is in final form and no version of it will be submitted for publication elsewhere.

in July 1933 until his death in June 1940, (1)[1].

There are other archival sources which I have examined, especially in the Marbach Schiller Museum where W.D.'s personal papers are deposited. I hope that other research workers will complete the present work, whose only ambition is to render homage to Wolfgang Doeblin.

1.1. Studies: October 1933 - June 1935. In July 1933 the Döblin family established itself in Paris, first in a hotel, then in Maisons-Laffitte in the western suburbs, where they lived six months, and finally in Paris, 5 Square Delormel, near Place Denfert-Rochereau in December 1934 where they were to stay until November 1939.

At the beginning of the University term in October 1933, W. Doeblin, who had started his scientific studies in Zürich in May 1933 (2), applied to the Paris Faculty of Sciences for a degree in mathematics.

In order to understand what follows, let us outline the peculiarities of the French University system. After their lycée studies, the best pupils in mathematics do not generally go to the University; they continue their studies in the best Paris lycées, preparing for entrance examinations to the higher state schools such as the "Ecole Normale Supérieure" or the "Ecole Polytechnique". After two or three years of preparation, those who succeed in the entrance examination to the "Ecole Normale Supérieure" must enrol in the University and obtain a degree equivalent to the American Master's, called at that time "licence". The Mathematics degree was composed of three compulsory and encyclopedic courses, Integral and Differential Calculus, Rational Mechanics and General Physics. For each one of these courses there were two examination sessions per year. The "Ecole Normale Supérieure" students did not attend the University lectures and this explains why A. Blanc-Lapierre, G. Choquet and L. Schwartz, for example, do not remember ever having met W.D., though they were all born in 1915 and passed the same examinations at the same time. Except for some French students who had given up the lycée preparations for the "Ecole Normale Supérieure", the courses were essentially attended by foreign students, who in the thirties represented more than half of the registered students; they came essentially from Central and Eastern Europe: Bulgaria, Hungary, Poland, Rumania, Czechoslovakia and Germany after 1933. So W.D. was not an isolated case.

The lectures were given at the new Henri Poincaré Institute. The buildings, erected with the help of the Baron E. de Rothschild and Rockefeller Foundation donations, had been inaugurated in 1928; at that time they accepted students in mathematics and theoretical physics, in accordance with the wish of the promoters of the Institute, Emile Borel for France and George Birkhoff for the Foundation, who wanted the newest theories in physics to be enriched and perhaps clarified by contact with the most advanced mathematics.

The examinations register of the Faculty of Sciences of Paris for the year 1934[2] indicates that W.D. succeeded in three examinations during his first year in Paris.

1° "Mathématiques générales" (general mathematics), a first year course created in 1902. It was not compulsory for the "licence" but was recommended for students who had not attended the "mathématiques spéciales" classes of lycées.

---

[1]Numbers in these brackets refer to the Notes at the end of the paper
[2]National Archives AJ/16/5508

In 1933-1934 this course was given by G. Valiron (1884-1955), author of a well known treatise in analysis, assisted by two lecturers, E. Leroy (1870-1954), a disciple of Bergson and philosophy professor at the "Collège de France", and E. Cahen (1865-1941), professor of a "mathématiques spéciales" class in a Paris lycée, who in 1899 had defended a thesis on the ζ function.

In June 1934, out of 246 candidates for the "Mathématiques Générales", 61 succeeded in the finals. W.D. obtained 33 out of 60 and 34 out of 40 in the two written examinations, 12 out of 20 in the application test and 81 out of 100 at the oral examination; he passed with distinction ("assez bien").

These are not exceptional marks; for example Michel Loève (1907-1979) had passed "mathématiques générales" in Paris in 1929 with the distinction: "très bien" and the highest marks. One can understand the reasons for these results by comparing the perfect correctness and clarity of Loève's style with the liberties taken by W.D., who was extraordinarily intuitive and so had difficulty in observing academic standards of rigour (3), which did not make him easy to read.

2° Rational Mechanics, was, as we have said, a compulsory course of the "licence de mathématiques"; the French students attended it four years after the baccalauréat and it was given by J. Chazy (1882-1955). In June 1934, out of 242 candidates 54 succeeded. Here again W.D. obtained a distinction ("assez bien") with 26 out of 40 in the written examination, 8 out of 20 in the practical test and 26 out of 40 in the oral examination.

3° Probabilities and Mathematical Physics, Statistics option. This was a graduate course, for the French "diplôme d'études supérieures". It was intended to give advanced students the opportunity to come into contact with less traditional material and to open up research subjects.

This course had been created in 1930 by George Darmois (1888-1960). It included a very general probability course given by E. Borel (1871-1956) and a Statistics course given by Darmois who that year had also taught "les probabilités en chaînes" instead of Fréchet (1878-1973).

Though G. Darmois was a remarquable scientist, his work is now somewhat forgotten. In 1914 he had defended a very brilliant thesis on geometry. After the First World War, he had been appointed Professor at the University of Nancy where he took an interest in Relativity Theory and Statistics. He was the one who, in France, at that period, developed Fisher's ideas. E. Borel, who wanted to create a statistics course, and was aware of his strength, had summoned him to Paris.

Unlike Borel, who before the First World War was a great scientist but who now was no longer much interested in mathematics, Darmois was in 1934 a very enthusiastic man, openminded and welcoming to young people; his lectures were very modern, (e.g. [96][3] p. 224).

Examinations in statistics were held in March and October. W.D. did not take his in March 1934, but succeeded in October with distinction ("assez bien"); his marks were 6 out of 20 and 16 out of 20 for the written examination, 13 out of 20 for the practical test; 11 out of 20 and 15 out of 20 for the oral examination.

So in his very first year at the Sorbonne, W.D., hardly 19 years old, had already completed the full cycle of University studies, from the first to the last

---

[3]Numbers in square brackets refer to the general references at the end of the paper

year and had found a subject for research: the theory of probability. Obviously
one may wonder about the reasons for this unusual route and for this choice.
Lacking positive indications we must content ourselves with assumptions. One
of them might be explained this way.

It is fairly certain that W.D. had never received any special advice from any
of the scholarly Faculty Professors when he arrived in Paris. He had started his
mathematical studies in Zürich, and so naturally applied to study mathematics
in Paris, obviously for the preparatory course "mathématiques générales" which
was attended by all the foreign students. If he registered in Rational Mechanics it
must have been because the title of the course made him think it might be easier
than Integral and Differential Calculus which covered all the classical analysis of
Jordan's, Picard's or Hadamard's big treatises. On the other hand, enrolling in
the Probability course is not natural at all and presupposes a deeply considered
choice. It is conceivable that W.D. decided to apply for Statistics at some time
in the year, since he was not present at the examination in March 1934.

If only to improve his French (4), W.D. must have attended the general mathematics and mechanics lectures and have met other students, probably foreigners
like himself. One of them may have motivated his orientation. For that same year
1933-1934, Ervin Feldheim, a Hungarian student born in 1912, was attending
the mechanics course given by Chazy. Feldheim had undertaken mathematical
studies in Paris in 1931 and while working for his degree in 1934 he attended Darmois's lectures which had the reputation of being the easiest of all the graduate
courses. Indeed he was not the only one to do so. Another Hungarian student,
Geza Kunetz, attended Darmois's lectures that same year and in March 1937 he
defended a University thesis under Darmois's direction, [90].

Darmois's course also provided an opportunity to follow a career as either a
statistician or a professional accountant, careers which were already at that time
more accessible in France and all over the world. Foreign students in mathematics were particularly attracted to these possibilities as they could not hope to
obtain full positions in the French state Lycées or Universities. Thus, M. Loève,
who was stateless, and had to teach in private schools in order to support his
family, had studied actuarial statistics hoping in this way to find a better job.
He also found, like W.D., a research program and we have quite a number of
other instances in Europe and in the United States of such strong probabilistic
vocations attributable to the necessities of life.

One can imagine that W.D.'s choice may have been due to the same reasons,
as he certainly wished at that time to help his father financially, since his father
was already advanced in age and did not have a regular income.

It is possible that Feldheim who, as well as Kunetz, was successful in Borel's
and Darmois's examinations in March 1934, might have suggested that Doeblin
apply for the same course. Feldheim failed in the Mechanics examination in
June 1934 and succeeded only in October. Let us note that Feldheim then
prepared a thesis dedicated to Darmois and P. Lévy, on stable distributions in
the plane, which he defended in February 1937. Back in Hungary, in October
1934, Feldheim obtained a job in an insurance company and went on publishing
papers on analysis up to 1942. He was murdered in 1944 (63).

In his correspondence W.D. speaks of Feldheim in friendly terms, so it is at
least likely that they met.

Indeed G. Darmois seems to have been the first person to take an interest in W.D. and to have helped his decision to work on the theory of probabilities (5). Mme Schwartz, P. Lévy's daughter, remembers having heard W.D. say to her father "that he did not know yet if he was more interested in probability than in other parts of mathematics, but that he had chosen to work in that direction because he felt that it would become one of the great branches in future mathematics".

We now know that the thirties was the golden age for probability theory but this was not at all obvious at the time, particularly in Paris, where probability calculus was not generally considered to be real mathematics. Traditionally the most brilliant students took the path of geometry or analysis and more recently of number theory; those who took up probability did so after having succeeded in other fields. Apart from the exceptional case of Bachelier, it was only after the arrival in Paris of Fréchet in 1929 and Darmois in 1933 that probability theses were defended at the Sorbonne, and even then they were not readily accepted. Thus E. Borel, though one of the promoters of the modern theory of probability, who presided over nearly all the probability theses, once said to J. Ville, author of a very remarquable thesis on martingales: "when are you going to make up your mind to work on analysis ?"

However, W.D.'s choice was not an unreasonable one, as in 1934 Paul Lévy, professor at the "Ecole Polytechnique", was to publish his well known memoir on infinitely divisible distributions and to demonstrate Feller-Lévy's theorem, which at last explained the Laplace-Liapounoff theorem. Furthermore, Maurice Fréchet, who before the First World War had started analysis on abstract spaces and who, after Hadamard, had been interested, on his arrival in Paris, in the theory of chain events, henceforth devoted himself almost exclusively to that theory. Fréchet maintained a copious correspondence with a large number of analysts and closely followed the literature of his time which he received from everywhere. In particular Fréchet was in contact with all the mathematicians interested in the theory of Markov chains, Bernstein, Hostinský, Kolmogorov, von Mises, Onicescu and others, inviting them to lecture in Paris, thanks to the generosity of the Rockefeller Foundation.

So in 1934 the Henri Poincaré Institute was, no doubt without fully realizing it, one of the most propitious places in the world for studies in probability theory. Probably W.D. was not conscious of it either. We must not, however, idealize the situation. Research in mathematics in Paris at that time was not an easy task. There was an insuperable barrier between young scientists and full professors due to the great difference in age (a professor was seldom appointed in Paris before the age of 50), and to a tradition of absolute individualism. Young people were left to work on their own, particularly those who, unlike the "Ecole Normale Supérieure" students, could not take advantage of the surroundings favorable to fruitful scientific exchanges. Moscow, Göttingen, or Berlin before 1933, for example, would certainly have welcomed W.D. better. He however, as we shall see, does not seem to have suffered too much from this situation.

Let us come back to 1934. After his successes in June and October, W.D. applied on the 17th of October for the two year-long courses which he needed to complete his degree in mathematics, Integral and Differential Caculus, and General Physics[4]. The 1935 examinations register unfortunately disappeared in

---

[4]National Archives AJ/16/5182 n°73

the fire which destroyed part of the Sorbonne archives in June 1968, and so we do not have the marks obtained by W.D. It is certain however that he temporarily withdrew from General Physics in which he would succeed only in June 1938 (6); however he passed Integral and Differential Calculus in June 1935. This course was divided into three parts; A. Denjoy (1884-1974), who held the chair, lectured on ordinary and partial differential equations, M. Fréchet on Fourier analysis, and R. Garnier (1887-1984) on differential geometry. An examination was held for each course and W.D. must have been bright enough since A. Denjoy, who did not have the reputation of being easily impressed, agreed together with G. Darmois, to introduce him to the "Société Mathématique de France" on the 13th of November 1935. W.D. was then 20 years old [20].

It is likely that W.D. followed other courses in 1934-1935 as well as lectures at the Henri Poincaré Institute that same year. W.D. has followed "Analyse Supérieure", a very original but rather awkward graduate course, taught by J. Drach (1874-1949), who lectured, that year, on abelian integrals and algebraic functions. We do not know if he passed this exam in June 1935, (7). W.D. could also have followed Elie Cartan's geometry course which attracted the most brilliant students. He could even have followed the "Séminaire Julia" meetings held every week in the Darboux amphitheatre of the Henri Poincaré Institute, which were organised by the founders of the Bourbaki group, H. Cartan, C. Chevalley, J. Delsarte, R. de Possel, A. Weil and others. Throughout the year 1934-1935 the subject for seminar study was "Hilbert space". R. de Possel lectured on set measure and integration theory (Bourbaki had not yet adopted the Bourbaki integration theory), J. Leray lectured on the Carleman theory, A. Weil on Haar measure and J. von Neumann, then Rockefeller visiting lecturer in Paris, explained his operator theory. In 1935-1936 this same "Séminaire Julia" was to study "topology", then the "Elie Cartan works" in 1936-1937, and "algebraic functions" in 1937-1938 before the publication in 1939 of the first Bourbaki volume. W.D. could also have attended the prestigious Hadamard seminar at the Collège de France where the latest results in all mathematical fields were presented. Finally he could have followed the "Cours Peccot" at the Collège de France which was then shared by J. Leray and R. de Possel, who lectured on "abstract measure and integration theory".

1.2. The research years at the Henri Poincaré Institute: October 1935 - October 1938. At the beginning of the University term in October 1935, W.D. did not enrol immediately in graduate studies; maybe he was still hesitating about his final direction. He applied as thesis candidate only on the 22nd of January, 1936[5]. Perhaps Darmois advised him to wait for Fréchet who was in Russia in October and November 1935. He surely started working by himself on the theory of Markov chains, which was at that time complete only under rather strong hypotheses. Possible generalizations could only be foreseen; but methods were too specialized to be significantly extended. However W.D. certainly obtained his first results at the end of 1935 and submitted them to Fréchet as soon as he was back. W.D.'s correspondence with Fréchet, deposited at the Académie des Sciences archives, and which we have faithfully transcribed, shows that this period is extraordinarily productive for W.D. He presented new results to Fréchet every week, and the latter was soon overwhelmed, all the more

---

[5]National Archives AJ/16/5183 n°2611

so because W.D.'s reports were extremely concise, most of the intermediate steps seeming obvious to him, and because they sometimes exhibited inaccuracies. As a precaution, he provided several very different proofs for each of his results.

In June 1936, W.D. was in possession of the essential results of his thesis, as they are presented in the two long memoirs on chains published in Bucharest and in Athens in 1937 and 1938, respectively. There are very few examples in the history of mathematics of such an achievement. W.D., who was ignorant of Hadamard's work on shuffling cards, and (of course) of the development Kolmogorov would give to it, extended Hadamard's direct method to the case of simple chains and deduced their complete study, including the law of the iterated logarithm. It was only on the 29th of June, 1936, that Fréchet agreed to present Doeblin's first note on chains, after having tried to obtain a precise and clear draft, in vain as his paper still contains a few errors. Meanwhile W.D. presented his results on chains at the Société Mathématique de France meetings on the 13th of May and the 10th of June, 1936; he was the only speaker at these meetings and he was barely 21 years old.

Throughout this period W.D. worked every day at the Henri Poincaré Institute library which was then on the first floor of the building. It was a large room supplied with mural shelves, with two or three long tables for the readers, of whom there were very few at that time. There W.D. met mathematics and theoretical physics graduate students, and in particular the Fréchet students, such as Robert Fortet and Jean Ville, and later on Michel Loève who was interested in 1935 in theoretical physics. All of them remember with admiration his intelligence and his extraordinary acuteness of mind. However he does not seem to have become very intimate with his fellows. He also met some of Louis de Broglie's students, such as Assène Datzeff who in 1937 was to defend a thesis on the Schrödinger equation and who would become professor of Theoretical Physics at the University of Sofia, and particularly Marie-Antoinette Baudot (1912-1980) for whom he would become more than a mere friend. According to his youngest brother Steve who shared his room he must have felt bitterly disappointed when she became engaged to Jacques Tonnelat, one of her University friends, son of Ernest Tonnelat (1877-1948) German scholar and professor at the Collège de France, and Alfred Döblin's friend and commentator. In 1941 Marie-Antoinette Tonnelat-Baudot was to defend a thesis on photon theory in a Riemann space; later she was to succeed Louis de Broglie's in the theoretical physics chair at the Sorbonne. Unfortunately W.D.'s letters to Marie-Antoinette Baudot seem to have disappeared, (see [133], [134]). It is she who would search for W.D. after his disappearance in June 1940 and would, in 1944, inform his family of his death.

Incidentally, in the Spring of 1936, W.D. solved a problem set by P. Lévy on the increase of the dispersion of sums of independent random variables. Lévy who had just solved the oldest problem of the theory, had presented his results at the Henri Poincaré Institute and at the Société Mathématique de France, leaving one difficulty unsolved. W.D. brought him the desired solution, which Lévy considered to be rather difficult, [96] p.156, [99] p.112, and which W.D. described as "rather easy" in a remark later suppressed on the draft of his memoir "on the sums of a large number of random variables", {9}[6].

---

[6]Numbers in curly brackets refer to the Doeblin bibliography at the end of the paper

Generally speaking W.D. seemed to consider all his 1936 results as "rather easy" and he already considered previous results as rather obvious. Furthermore, at that time he was in constant fear of seeing other mathematicians publish one or other of his results before he could find time to write them down, as they all seemed so simple and natural to him.

Lévy's deep admiration and esteem for W.D. stemmed from these days. According to Laurent Schwartz, P. Lévy's son-in-law, W.D. was "one of the only young people in whom P. Lévy had taken an interest at that time". As early as June 1936 he spoke of W.D. not as a student but as a partner. The correspondence between W.D. and P. Lévy was destroyed during the war when P. Lévy's flat was sacked by the Nazis, (8), but we know that P. Lévy welcomed W.D. into his home and enjoyed discussing their research work with him, [102] p. 116. From October 1936, W.D. and Ervin Feldheim helped P. Lévy to read the proofs of his well-known book "Théorie de l'addition des variables aléatoires" which was to be published in January 1937, [96] p. XVII; so it is from that time that W.D.'s first thoughts on the sums of independent random variables can be dated. They were to develop into the two great 1939 and 1940-1947 memoirs. The first, which again takes up and proves a number of Lévy's results through original and direct methods, was to be received by the "Bulletin des sciences mathématiques" on the 15th of February (1938), sent to the printers on the 26th of July, and published in January and February 1939. We shall discuss the second one later.

At the end of that same year, 1936, W.D. worked out his general theory of Markov chains, and presented his results to Fréchet in the Spring of 1937[7]. It was, according to W.D., the "most difficult" problem he had ever solved[8]. The corresponding memoir was to be written only later by W.D., probably between February and March 1938, (9), after he gave up working on the sums of independent random variables and all his other current projects. The manuscript was handed to P. Gauja, archivist of the Academy of Sciences in charge of publications on the 21st of July and published in the Ecole Normale Supérieure's Annals only in 1940. The "Note aux Comptes Rendus" announcing the general theory is dated 5th of July, 1937. It completes W.D.'s thoughts on Markov chains.

Because of a lack of new documents, I shall not have anything to say about the collaboration between Doeblin and Fortet. During the Summer of 1937, W. Doeblin corrected proofs of his thesis together with part of the proofs of Fréchet's book on the "Théorie des événements en chaîne dans le cas d'un nombre fini d'états possibles". These simultaneous readings forced W.D. to define his original views [20 to 24] on Markov processes in continuous time with a finite number of states. His correspondence with Fréchet in September 1937 indicates how W.D. had modified yet again the final chapter of his thesis on the Smoluchowski equation in the finite case. The Smoluchowski equation and more generally the Chapman-Kolmogorov equation would henceforth be one of his constant subjects of research until the time of his death.

In this connection let us recall that his first work on the Chapman equation in the case of jump processes, which dates from 1938 and was to be published in 1939 (10), was in part inspired by a work of B. Pospišil, [119], published in 1936. We can find a reprint annotated by W.D. in the Fréchet archives of the

---

[7]Box Fréchet n°8, Academy of Sciences Archives
[8]Letter to Fréchet of the 28/10/39

"Laboratoire de Probabilités" of the Paris 6 University. It may be opportune to recall here the memory of B. Pospišil (1912-1944) who also died before giving all he had to give. Pospišil studied at the Masaryk University of Brno from 1931 to 1935. Influenced by Hostinský, he proved in 1935 the first theorem on the existence and uniqueness of solutions of a Chapman-Kolmogorov equation under precise and explicit conditions. B. Hostinský (1884-1951), professor of theoretical physics in Brno, is a scientist somewhat neglected nowadays. Even Czech scholars do not consider him to have been profound and do not forgive him for having been from the outset a firm opponent of Einstein's ideas. However, Hostinský was the first in 1928, even before Hadamard, to extend the ergodic principle for finite Markov chains to the continuous case and to propose in 1932 the application of Volterra's multiplicative integral to the study of the Chapman equation, a method which Pospišil took up again and extended. B. Hostinský had corresponded with Fréchet from 1919 to his death in 1951 and it was Hostinský who popularized the study of Markov processes in France through his numerous memoirs and two series of lectures at the Henri Poincaré Institute in 1930 and 1936. See, e.g., [4], [63], [64].

After this first memoir on the Chapman equation, Pospišil abandonned probability theory. In 1936 he participated in the Brno topology colloquium founded by the great geometer E. Čech, and in 1939 defended a thesis on topology [9]. Arrested by the Gestapo in 1941, he died in 1944 after three years in a concentration camp.

W.D. had an opportunity to meet Hostinský again at the conference on probability theory held in Geneva from the 11th to the 16th of October 1937. In Geneva W.D. also met H. Cramér, W. Feller, B. de Finetti, H. Steinhaus, A. Wald and other researchers. It was the only international congress W.D. ever attended.

Before clarifying his early ideas on continuous Markov processes, at the beginning of 1938 W.D. wrote his memoir on the sum of independent random variables, which he finished in February 1938. Following Lévy and Khinchine, he also began to study the set of limit distributions of the powers of a given type of distribution. He wrote to Fréchet in October 1939 that he had "slaved away" in vain all through the Spring of 1938 to discover a necessary and sufficient condition so that a set of infinitely divisible distributions should be precisely the set of the limit distributions of the powers of a certain type of distribution, a problem on which he was not to work again until February 1939 after enrolling in the French army.

In January 1937 W.D.'s personal financial situation improved. Thanks to Fréchet's recommendation, he obtained an Arconati-Visconti grant of 6,750 francs for the three first terms of 1937, (11). This grant was renewed for the 1937-1938 academic year. The application filled in by W.D. in May 1937 can be consulted at the National Archives in Paris[9]. It contains a half burnt identification photograph, a certificate of nationality signed by two witnesses, Jean Dubuffet (probably the painter) and Victor Detouche and a "reseach program" for the year, which is half destroyed but can easily be reconstructed. W.D., wrote "I intend to carry on my research work on constant chains, on the continuous case, and on a set of infinite measure. I wish to go on studying variable chains.

"Apart from these two questions on which I have been working for a long

---

[9] AJ/16/5796

time, I wish to study the probability of large values of the sums of independent random variables and some questions relating to parabolic [partial differential] equations".

We should note that in 1937-1938 W.D. with J. Ville organised in Paris a study group on probability theory, at which the most important new results were presented, (12). One of the main topics in 1938 was continuous Markov processes. The group was soon established by E. Borel, who held the Chair of "Calculus of Probability and Mathematical Physics". It developed into the probability seminar of the Paris Faculty of Sciences. This seminar is still functioning and welcomes many scientists from all over the world.

In 1938 W.D. seemed to have reached maturity. He was interested in everything (13). For instance, in a conversation with Paul Lévy, reported by L. Schwartz, he announced his intention to study stochastic integrals and differential equations; the subject had hardly been touched by Bernstein. His first note on the Kolmogorov equation was presented at the Academy on the 24th of October, 1938, just before he left for military service.

1.3. In the French Army: November 1938 - June 1940. As soon as he had defended his thesis in March 1938, W.D. who, with his parents and his two younger brothers, had obtained French nationality on the 16th of October, 1936[10], renounced deferral of his eligibility for conscription.

Delayed because of the Czech crisis, he enlisted only on the 3rd of November 1938, in the 91st Infantry Regiment (I.R.) stationed in the Ardennes at Givet. W.D., who refused to have military training, was assigned to the rank of private.

It was the first time W.D. had left his family. He was to see them again only on the occasion of short leaves, which were infrequent. From the letters addressed to his friends, which we have had the opportunity to consult, it appears that, for the first time since his arrival in France, W.D. had neither the time nor the wish to pursue his work. Barrack-room life weighed heavy on his spirits and he became home-sick. All his research was suspended between November 1938 and February 1939, (14).

In order to escape his melancholy, in February 1939 he took up the problem he had been working on so hard and had failed to solve in the Spring of 1938. It concerned the general characterization of domains of partial attraction of a distribution. In his free time he began writing down the results he had already obtained and at the end of July 1939 he was in possession of the characterization he had sought. In October 1939 he wrote to Fréchet that, next to the general theory of chains, it was the most difficult problem he had ever solved. Unfortunately he was short of time and could not complete the writing of his demonstration. Indeed, at the beginning of August 1939 he had to attend a training squad for corporals, and the exercise became so strenuous that he could not go on with his work. He wanted to finalize his manuscript before the end of August, but the Polish crisis broke out and his regiment was dissolved on the 22nd of August and reorganised with mobilized reservists. War was declared on the 2nd of September 1939, and he just had time to send his unfinished manuscript to his elder brother Peter in Philadelphia, and all copies of his proof to his parents in Paris.

---

[10]Ordinance n°26035/36

Nobody since then seems to have understood what it was that W.D. had proved in July 1939, nor how he had reached his result. In his last note on Doeblin's work, P. Lévy even doubts that W.D. had really proved anything. The Philadelphia draft brought back to Paris by his parents has been examined by several experts such as Lévy, Fortet, Loève, Pollaczek and Chung ,who were only able to state that it is an incomplete paper of 20 pages containing formulae devoid of explanations. This manuscript is deposited at the Academy of Sciences Archives and is easily accessible to any interested scholar. W.D.'s last theorem on the characterization of domains of partial attraction remains unfathomable today. Yet W.D. seemed to hold fast to his result. Perhaps he was right?

Once war was declared, W.D. was assigned to the 291st Infantry Regiment commanded by lieutenant-colonel Modot, which belonged to the 52nd Infantry Division (général Echard), a division which at the end of 1939, was to be sent to Lorraine in the Sarre group (général Hubert) of the 3rd army (général Condé).

W.D. was a telephone operator of the third Battalion (commandant Charles). He became homesick again and was unable to work. At the end of October 1939, he started writing demonstrations of his notes on the Kolmogorov equation which had been published in the Comptes Rendus in October 1938 and January 1939, and which he soon completed with a new note, ready in January 1940.

Meanwhile M. Fréchet was in charge of the statistics section of the Henri Poincaré Institute assigned to war service under the direction of Emile Borel, who thus returned to a post at the highest level that he had taken on in the First World War. M. Fréchet contacted W.D. again to ask for his collaboration, but W.D. does not seem to have been very excited by the problem submitted by Fréchet, which was: to find numerical approximation rules for the binomial distribution, valid in all cases.

The winter of 1939-1940 was severe in Lorraine and no preparation had been made to provide for such a large number of soldiers. In January, W.D. "slept in an attic with no heat where the snow entered". In spite of these conditions, and thanks to the telephone booth where it was warm and where he could be alone, W.D. finished his memoir on the Kolmogorov equation in February 1940 and sent it to the Academy of Sciences in a sealed letter, where it was registered as number 11668.

This sealed letter has never been opened. M. Fréchet forgot it, though he had been informed of it. It is still in the Academy of Sciences' Archives. The statutes of the Academy specify that it cannot be opened until the heirs make the request or until one hundred years have passed. It is supposed to consist of a complete memoir, so when opened it should provide interesting information on Doeblin's methods in studying the Kolmogorov equation.

Now that he was back at work, and as the phoney war seemed to be lasting, W.D. agreed to join a cadet training squad, which should have given him the chance to work under better conditions, but the opportunity never arose.

He was then working on the Chapman equation in the general case; he wrote to Fréchet that he was "mainly studying cases in which the Hostinský solutions happen to be the general solutions". On the 10th of April 1940 he sent Fréchet an initial note "on mixed movements" and the same day "another note which followed the previous one but was not intended to be published for several months". This second note, entitled "On Hostinský's solution of the Chapman equation", has never been published. Doeblin undertook a general theory of Markov processes based directly on Chapman's equation; it would be interesting to be able to judge his contribution in that field.

In a letter of the 13th of April, W.D. communicated other results to Fréchet, completing the two notes of the 10th of April. He was optimistic and thought the war would soon be over. He sent Fréchet an application for a grant for the following year.

On the 21st of April, 1940, W.D. wrote a last postcard to Fréchet in which he thanked him for a university grant which would help him in his work when he could return to it. He thanked him equally for sending reprints and added: "Feller's work contains a solution of the Chapman equation which is absolutely equivalent to Hostinský's solution. Obviously Feller did not notice this". It was the last of W.D.'s scientific assertions that I know of.

On the first of May, W.D.'s Regiment was sent to the front line. On the 10th of May the German army attacked Belgium and rapidly outflanked the allied armies. The Sarre district remained relatively quiet. Just after the offensive, W.D. received his first citation, and he was decorated with the "Croix de Guerre", (15).

The German offensive ended at Dunkirk on the 4th of June, 1940. The British expeditionary force and half of the French army were either no longer able to fight or were in retreat.

The second German offensive started on the 5th of June. The Von Boch army group attacked on the Somme river and reached the Seine at Rouen on the 9th of June; it would be in Paris by the 14th. Also, on the 9th of June the Von Rundstedt army group attacked in the Argonne. In spite of the desperate resistance of the 2nd French army, the front was broken through and on Monday the 17th of June the Guderian armoured divisions reached the Swiss frontier, encircling the French eastern group of armies and cutting off 500,000 soldiers in Lorraine. However, an order for general withdrawal of the Lorraine troops had been given on the 12th, and the troops were to withdraw at night on the 13th to the Marne to Rhine canal. The Sarre group, to which W.D. belonged, was to withdraw on the night of the 14th.

But on the 14th at 7 a.m. the first German army of general Von Witzleben attacked in the Sarre. Their artillery attack was the heaviest so far in the war, 1,000 pieces of artillery being concentrated along a front of 28 km. Moreover, the Germans had total command of the air. However, the assault was repelled in the only real French victory at this stage of the war. During the night, after a whole day of hard fighting, the troops left their fortified positions, and withdrew on foot with all their gear. The retreat was very painful physically and morally, carried out by night to avoid air attacks. In the day the fighting went on without interruption.

From the 15th to the 18th of June the 291st Infantry Regiment was in charge of protecting the retreat of the troops. It fought against crushing odds to delay the German offensive and suffered heavy losses (16). Totally exhausted, the third Batallion, to which W.D. belonged, crossed the canal on the Hénaménil bridge on the afternoon of the 17th ([7] 1, p. 312). On the morning of the 19th, after a deadly German assault, the command post of the 291st regiment was trapped near Thiébaumesnil, east of Lunéville ([7], 2, p. 197). However, on the same day the "Compagnie d'accompagnement" of the third Battalion, to which W.D. belonged, escaped and retreated to the south. On the 20th of

June, captain Renard, chief of the Doeblin's company, realized that they were surrounded by the German troops, and that there was no hope of escaping capture. Until that moment, according to captain Renard, (17), W.D. had been a constant model of bravery and self-sacrifice. But, knowing that the battle was lost, he decided to escape alone. Without informing his comrades, W.D. left his company and probably walked all the night. On the morning of the 21st he arrived at Housseras, a little village near Rambervillers, where he met soldiers from the 82nd Fortress Infantry Regiment who were ready to surrender. They, like himself, were exhausted. He was alone.

W.D. had always declared that he was ready to die for his ideals. He could not accept capture by the Germans who would have regarded him as a traitor and might have forced him to betray his father, wanted by the Gestapo. He burnt all his papers in a farm building and shot himself, electing to die. He was buried that night in a grave behind the church. His body was identified in April 1944 (18). Fighting ceased in Lorraine on the 22nd of June at 3 p.m..

W.D. was decorated posthumously with the "Croix de Guerre avec Palme" and the "Médaille Militaire", the highest distinction granted to a French soldier. Last year his name was placed on the war memorial at Housseras, the village where he is buried with his father and mother [113].

2. Appendices

2.1. CORRESPONDENCE BETWEEN
W. DOEBLIN AND M. FRÉCHET,
1936-1940,
DEPOSITED AT THE INSTITUT DE FRANCE.

The correspondence of W.D. with his thesis director, M. Fréchet, is classified in chronological order. The numbers (1), (2) ... refer to the notes (1), (2) ... below.

I

1-8-1936   (19)

Monsieur le Professeur

Vous trouverez dans ce paquet la fin du 1er chapitre (20). J'avais voulu y ajouter un § sur le battage des cartes (pour les variables aléatoires). J'avais voulu aussi simplifier et revoir la plupart des démonstrations de la 2ème partie. Mais je n'en ai plus eu le temps, ayant été obligé de m'aliter pendant quelques jours.

Toutefois je ne crois pas qu'il y ait de fautes importantes. J'ai en effet pour presque tous les théorèmes plusieurs démonstrations quelquefois assez différentes. Lors de la rédaction, je me suis aperçu que mon analyse du cas $\sigma = 0$ à l'intérieur d'un groupe final était fausse (21). Mais j'ai eu de la chance, l'énoncé de ma note

(22) est quand même vrai. Je n'ai eu qu'à baptiser cas non aléatoire le cas où la connaissance de l'état initial et de l'état final détermine complètement $S^{(n)}$.

Après les vacances je rédigerai le 2ème chapitre (23), comprenant la généralisation d'à peu près tous les résultats touvés au cas continu. Il y aura peut être quelques difficultés pour quelques points de détails, mais j'ai vu que l'essentiel reste vrai.

Permettez moi de vous souhaiter de bonnes vacances, les miennes s'annoncent pluvieuses.

Veuillez croire, Monsieur le Professeur, à mon plus profond respect.

W.D.

II

Paris, 7 septembre 1936   (24)

De retour à Paris, je viens de recevoir votre carte et je m'empresse de mettre tout de suite à la poste ces papiers en espèrant qu'ils vous reviennent encore à temps.

Contrairement à ce que je pensais, l'énoncé pour les fréquences dans le cas où l'écart type est borné n'est pas correct. Comme je constate que les errata que j'avais signalés avant les vacances n'ont pas encore paru aux Comptes Rendus je vais y encore ajouter les deux suivants (qui sont plutôt des modifications) (25):

p.25 ligne 14 d'en bas.

lire l'écart type de m_i^n est soit identiquement nul lorsque l'état initial est connu (cas non aléatoire) soit ...

au lieu de l'écart type ...

p.25 ligne 8 d'en bas.

lire si $X^{(1)} + ... + X^{(n)}$ reste encore aléatoire après la connaissance de l'état initial et de l'état final du système

au lieu de si nous sommes dans un cas aléatoire.

Les énoncés seront alors absolument corrects.

J'espère que vous passez de bonnes vacances.

Veuillez agréer, Monsieur le Professeur, l'expression de mon plus profond respect.

W.D.

III

Monsieur le Professeur   (26)

Chaque dimanche à peu près j'ai la frousse des lundi c'est à dire des Comptes Rendus, en craignant que des résultats que je n'ai communiqués à personne peuvent se trouver dans une note des Comptes Rendus. Cette fois ce que je vous communique n'est peut être pas très important. Ce ne sont même pas de nouveaux résultats mais avec Monsieur Paul Lévy j'estime qu'il faut tirer toutes les conclusions de ses travaux si on ne veut pas être taxé d'incapable.

J'ai revu votre mémoire: "Sur l'allure asymptotique des densités ... " (27). Les résultats de Monsieur Hadamard (28) sur les fonctions fondamentales correspondant à des constantes de module 1 s'expliquent de la façon suivante.

Les $\varphi(E)$ sont des combinaisons linéaires de ce que j'appelle des $Pr\left[E, \overline{l(\alpha)}\right]$ (généralisations des $Pr\left(i, \overline{l(\alpha)}\right)$ cas discontinu Ch.I §5 (29)). Il est possible de donner les coefficients de ces $Pr\left[E, \overline{l(\alpha)}\right]$ pour $\varphi(\mathrm{E})$.

Les résultats de Monsieur Hadamard en résultent immédiatement.

Veuillez croire, Monsieur le Professeur, à mon profond respect.

W.D.

IV

17 Octobre 1936   (30)

Monsieur le Professeur

Comme j'ignore si vous viendrez lundi à l'IHP je vous communique par lettre mes résultats de cette semaine. Vous savez que depuis longtemps j'ai étendu les résultats de ma note au cas continu sous l'hypothèse que

1) l'ensemble dans lequel se meut le point mobile est de mesure finie

2) et qu'il existe deux nombres η et ε et un entier N tels que

$$P^N(E, \mathcal{E}) < 1 - \eta \quad si \quad mes(\mathcal{E}) < \varepsilon \qquad (31)$$

(ce cas contient comme cas particulier le cas dénombrable régulier au moins si je me rappelle bien la terminologie de Fortet) (32).

Maintenant j'ai pu démontrer que les théorèmes sur la stabilité de Poisson, sur la tendance dans les ensembles finaux, la division dans des sous-ensembles cycliques, la périodicité asymptotique des probabilités se conservent si l'on suppose que l'entier N qui intervient dans l'hypothèse 2) dépend de E état initial.

A partir de là je suis (indépendamment de Kolmogoroff) (33) en mesure de donner dans le cas dénombrable la condition nécessaire et suffisante pour que les $p_{i,k}^n$ sont asymptotiquement périodiques mais ne tendent pas vers zéro. D'ailleurs Kolmogoroff n'a pas démontré que les probabilités de rester en dehors des ensembles finaux dans cette hypothèse est nulle (il ne l'a pas envisagé).

Je vous avais fait remarquer avant les vacances que ce que vous avez donné dans les Commentaires (34) n'est que la condition pour que les $P^n(E, F)$ tendent uniformément par rapport à E vers une limite avec $\int P(F)dF \neq 0$. Je pense à l'aide de ces derniers cas donner les conditions nécessaires et suffisantes pour que les $P^n(E, F)$ tendent uniformément vers $P(F)$.

Veuillez croire, Monsieur le Professeur, à mon profond respect.

W.D.

V

Paris le 26-12-36   (35)

Monsieur le Professeur

Je viens d'être averti que le Conseil de la Faculté m'a accordé une bourse Arconati-Visconti (36) de 6750 francs. J'ai aussitôt renoncé à l'allocation de l'Alliance Israélite et remboursé ce que j'avais obtenu.

Cette bourse, dont je n'oublierai pas que je ne la dois qu'à vous constitue les plus belles étrennes que je n'ai jamais eues et je vous en remercie beaucoup bien sincèrement.

Veuillez croire, Monsieur le Professeur, à mes sentiments très respectueux et très reconnaissants.

W.D.

VI

Paris 27 aout   (37)

Monsieur le Professeur

Je vous renvoie ci-joint les 2 placards et le manuscrit (38) car contrairement à votre lettre je n'ai trouvé qu'un seul manuscrit dans l'enveloppe.

Quant au placard 29, ma remarque était due au mot arriver. Je pensais que la définition était conforme à ce que vous vouliez dire, alors les formules n'étaient "pas tout à fait correctes".

Ma remarque sur le placard 32 était certainement occasionnée par une lecture trop rapide. Le texte est bien exact. Mais il me parait que toute la démonstration donnée en petits caractères (39) pourrait être remplacée par une démonstration plus courte utilisant la formule d'approximations

$$(1) \qquad P_{ik}(s,t) = P_{ik}(s,t_0) + (t - t_0)[\frac{\partial}{\partial t}P_{ik}(s,t_0) + \varepsilon_{ik}(s,t)]$$

$\psi(t_0) = P_{ik}(s,t_0)$ pour $i = i_1 \ldots, k = k_1 \ldots$, il résulte immédiatement que $\psi(t_0 + \Delta t)$ ne peut prendre pour Δt suffisamment petit qu'une des valeurs $P_{ik}(s,t_0 + \Delta t)$ où $i \in (i_1, \ldots), k \in (k_1, \ldots)$, et même que les indices i, k possibles doivent appartenir au sous ensemble de ces indices pour lequel $\frac{\partial}{\partial t}P_{ik}(s,t_0)$ est minimum $= \mu$. Alors on a évidemment

$$\psi(t) = \psi(t_0) + (t - t_0)[\mu + \varepsilon(t)]$$

ce qui démontre vos affirmations. Au fond à partir de (1) c'est tellement évident qu'il suffit de signaler cette formule et de laisser au lecteur le soin de la vérification.

L'unique manuscrit ne donne lieu à aucune observation. Je viens de vous envoyer un exemplaire des épreuves de ma thèse. Le premier chapitre contient énormément d'errata. Au second chapitre § 2 (placard 27) il y a une coquille très grave (elle se trouve déjà dans mon manuscrit). Il faut remplacer la 3^{ième}ligne par "que le point mobile reste dans son évolution future toujours dans" (40). Le sens est évidemment tout à fait différent. Le §1 du dernier chapitre (équation de Smoluchowski) a été complètement changé par moi, j'avais envoyé la nouvelle rédaction au commencement de juillet à Onicescu (41), il faut croire que l'impression a été déjà finie à cette époque là. Je vous enverrai au commencement de septembre une lettre où j'aurai rassemblé tous les points où je ne sais pas comment faire la correction. Cette lettre contiendra probablement (42) aussi un certain nombre de résultats sans grande importance sur l'équation de Smoluchowski à un nombre fini d'états, tels que la démonstration de l'existence d'une limite pour $t \to \infty$ de $P_{ik}(t)$ (continue), forme de la solution continue la plus générale, démonstration directe que toutes les solutions continues $P_{ik}(t) \geq 0$ avec $\sum_k P_{ik}(t) = 1$ de l'équation de Smoluchowski ont des dérivées continues, forme des $P_{ik}(t)$ ayant à l'origine un point de discontinuité de première espèce. Une partie de ces résultats a été obtenue par moi il y a plus d'une année, le reste je le porte dans mon cerveau depuis assez longtemps, et je l'aurais laissé murir encore plus longtemps, sans la lecture de votre livre. Mais comme je vais travailler sitôt ma thèse soutenue sur l'équation de Smoluchowski dans le cas général, ces choses là reviennent tellement souvent à mon esprit que j'entends m'en débarasser une fois pour toutes, même si ça va me couter une journée de travail.

J'espère que vous avez dans les Alpes un meilleur temps que celui que j'ai eu la dernière semaine de mon voyage dans les Vosges.

Veuillez croire, Monsieur le Professeur, à mes sentiments les plus respectueux.

(J'espère que vous avez reçu le reste des épreuves de votre livre que j'avais envoyé le 19 juillet. Je me méfie des employés de poste de Toul).

VII

Lettre arrivée le 7 septembre.   (43)

Monsieur le Professeur

Je vous remercie de votre carte.

Les résultats auxquels j'ai fait allusion dans ma lettre n'étaient pas du tout la démonstration de l'existence de $\lim_{t \to \infty} \Gamma_{ik}(t)$ pour les solutions continues de l'équation de Smoluchowski, résultat qui m'est connu depuis longtemps et qui résulte d'ailleurs immédiatement comme corollaire aussi bien des travaux sur le cas continu de Bogoliouboff (44) que des miens (45).

Si vous voulez ajouter ce point à votre livre, vous n'aurez pas besoin de le gonfler de beaucoup. Les démonstrations ne manquent pas.

I Si vous vous bornez au cas dans lequel vous démontrez que la forme des solutions est de la forme

$$\mu_{ik}(t) = \sum_\rho P_{ik}^{(\rho)}(t) \lambda_\rho^t$$

la démonstration est immédiate. La partie périodique s'écrit

$$\pi_{ik} + \sum a_{ik}^{(\rho)} \lambda_\rho^t$$

λ_ρ étant de module 1 et λ_ρ^t doit être pour chaque t racine de l'unité. Donc $\lambda_\rho = 1$.

II Supposez maintenant que $\mu_{ik}(t)$ est continue pour $t > 0$. Vous voyez immédiatement de l'équation de récurrence que $\mu_{ik}(t)$ est uniformément continue pour $0 < t < \infty$. Alors la démonstration la plus immédiate pour l'esprit - qui était ma première et que j'ai abandonnée lorsque j'ai constaté que le principe de la démonstration de Bogoliouboff revenait à cela aussi - est: une fonction uniformément continue pour $t < \infty$ ne peut avoir 2 périodes asymptotiques dont le rapport est irrationnel, sans tendre vers une limite. Or $p_{ik}(nt)$ a comme période si $n \to \infty$ un multiple entier de t. Il en résulte la proposition. (46)

III La période $\leq D_s$ de $p_{ik}(ns)$ est $\leq rs$ (r le nombre d'états). $p_{ik}(t)$ admet donc des périodes asymptotiques arbitrairement petites. Comme $p_{ik}(t)$ est uniformément continue pour $t < \infty$, il résulte aussi la proposition. C'est à peu près la démonstration à laquelle je me suis arrêtée pour la thèse (cas continu) (47).

IV Il y a enfin des démonstrations où l'on introduit les groupes finals. Maintenant si vous supposez $p_{ii}(t) \xrightarrow[t \to 1]{} 1$ alors la <u>raison</u> de l'existence de $\lim_{t \to \infty} p_{ik}(t)$ me parait être que $p_{ii}(t)$ est > 0 quelque soit i.

Je me proposais d'établir plutôt trois autres choses.

<u>A</u> De $p_{ii}(t) \to 1$ si $t \to 0$ résulte que $p_{ik}(t)$ a une dérivée continue pour $t \geq 0$. [Ceci pour montrer directement que vos résultats s'appliquent dès que les $p_{ik}(t)$ sont continus pour $t = 0$].

<u>B</u> Analyse du cas où $\lim_{t \to 0} p_{ik}(t)$ existe sans être nécessairement δ_{ik}.

<u>C</u> Démonstration directe que toute solution $a_{ik}(t)$ continue pour $t > 0$ de l'équation de Smoluchowski (sans se borner au cas des probabilités) est nécessairement de la forme que vous établissez dans votre livre.

Les 3 propositions surtout les 2 dernières m'étaient connues depuis assez longtemps de même que leurs démonstrations. Seulement j'avais laissé tout cela dans le vague et je l'avais refoulé autant que possible dans ma subconscience (je trouve d'ailleurs que ma subconscience travaille beaucoup mieux que ma conscience). J'ai maintenant précisé et rédigé sommairement les démonstrations. Je considère donc A,B,C comme bien établies maintenant.

Si A vous intéresse pour votre livre je vous enverrai la démonstration, seulement je crains que ça fera une bonne page de mon écriture.

Ayant analysé le cas où l'origine est un point de discontinuité de 1ère espèce - j'ai constaté qu'il faut introduire des "groupes à liaisons instantanées" - je me suis demandé ce qui arrive si l'origine est un point de discontinuité de 2ème espèce pour les $p_{ik}(t)$, j'ai trouvé ce résultat étonnant: c'est impossible, (48). Par conséquent toute solution $p_{ik}(t) \geq 0$ de l'équation de Smoluchowski est nécessairement de la forme

$$\sum_{\rho=1}^{k} P_{ik}^{(\rho)}(t) \lambda_\rho^t$$

donc continue pour $t > 0$.

Je crois sans pouvoir le démontrer pour le moment que cela soit vrai aussi pour toute solution $a_{ik}(t)$ bornée à l'origine (49).

Dans le cas dénombrable je n'ai pu obtenir que des résultats moins précis (50).

Dans le cas général il faudra probablement faire intervenir des notions de voisinage.

Est ce que je ne ferai qu'une seule correction des épreuves ?

Il n'y a pas mal d'errata. Je voudrais remercier dans l'introduction Monsieur Onicescu, seulement je ne sais pas très bien qu'écrire. Je ne pense pas mettre "je remercie Monsieur Onicescu de s'être occupé de l'impression de ma thèse" (51).

J'allais vous demander de m'envoyer la mise en page de votre livre. Quand est-ce que le livre paraîtra? Cette année encore?

Veuillez agréer, Monsieur le Professeur, mes sentiments très respectueux.

                                                                        W.D.

VIII

                                                    Paris le 10 septembre (1937)   (52)

J'ai reçu avant hier votre lettre. Je n'ai pu compléter la liste qu'assez partiellement, surtout parce que l'IHP est fermé et mon fichier y est aussi. Dans votre liste se trouve un mémoire de Perron (avec titre incertain). Je ne sais pas s'il est le même que celui qui est cité par Hostinský (et qui contient l'expression des itérés $a_{ik}^{(n)}$ d'une matrice en fonction de i, k et n). Si c'est un mémoire différent il faudrait mettre 2 (53). On ne m'a pas envoyé jusqu'à maintenant de nouveaux placards de votre livre.

J'ai vérifié hier: Bogoliouboff (54) ne parle pas du cas fini, mais c'est un corollaire tellement immédiat qu'il n'y a aucun doute que c'est lui qui a la priorité, priorité que je n'ai jamais contestée. Quant à Romanovsky (55), il est certain que c'est lui qui a le 1er publié une démonstration complète de la loi de Gauss dans le cas positif régulier, seulement les travaux de Mihoc (56) et Schulz (57) ont été donnés à l'imprimerie avant la publication de Romanovsky, quant à moi lorsque je vous ai apporté le théorème du logarithme itéré dans le cas régulier, théorème qui présuppose la démonstration de la loi de Gauss, le mémoire de Romanovsky n'était pas encore paru. Il faut dire que pour le cas régulier Mihoc et moi ont pensé, jusqu'à Romanovsky, que le théorème en question pour le cas régulier était déjà démontré par Bernstein (58), j'avais obtenu la première des 2 démonstrations pour la loi de Gauss que je donne dans le ch.I (59) en janvier 36, sans y attribuer la moindre importance, et il est probable qu'il en a été de même pour Mihoc. Je ne sais d'ailleurs pas si Schulz (60) n'a pas obtenu le théorème avant 32. Il faut dire qu'avant vos travaux (61) sur les moments des 2 premiers ordres, les démonstrations de Bernstein ne s'appliquaient pas immédiatement au cas régulier pour $r > 2$.

Je vous serais reconnaissant si vous pouvez m'envoyer l'adresse de Ville (62), je ne vois que lui pour me signaler les lacunes du §2, à moins que je m'adresse à Feldheim (63) mais qui est en Hongrie. Fortet (64) connait trop bien toutes ces choses. Maintenant ce "rappel ... " n'était pas dans mon intention un résumé de la théorie pour des gens qui ne la connaissent pas du tout, mais plutôt une collection des propositions sur lesquelles je m'appuie dans le ch.I. Evidemment ma situation un peu bête est créée par le fait que l'exposé (65) ne paraîtra que beaucoup plus tard, à moins qu'il ne paraisse pas du tout. (Dans ce dernier cas je suis décidé à ne pas rédiger le mémoire développant ma dernière note aux Comptes Rendus) (66). En tout cas je vais envoyer à l'exposé et à votre livre. Pour ce dernier je vous prie de bien vouloir m'écrire le titre exact.

Cela m'amène à la question de la bibliographie. La solution la plus commode pour moi serait évidemment de renvoyer à votre livre et à Hostinský (67). Seulement cela ne serait pas idéal pour le lecteur. Si je donne les indications bibliographiques pour les travaux cités, je peux soit les mettre à la fin du mémoire, soit à la fin de chaque chapitre (parce que le mémoire va être coupé en 2) soit enfin en bas des pages, mais là je ne sais pas comment marquer cela sur les épreuves.

Au placard 6bis je légitime le remplacement de $S_i^{(n)}$ par $S^{(n)}$ (68), ce que je dis n'est peut être pas tout à fait suffisant, mais dans le §4 je ne suppose pas que j'ai affaire à un $S^{(n)}$ et comme je donne 2 démonstrations... J'ai d'ailleurs ajouté au §3 en note que l'hypothèse de remplacement des $S_i^{(n)}$ par $S^{(n)}$ n'est plus faite au paragraphe 4.

Le §3 contient environ 60 errata, est ce que c'est beaucoup ou normal? Ces errata portent de préférence sur les formules et sur les indices, et sont tels que votre lecture sera peut être très difficile. Le ch. I contient d'ailleurs autant d'errata que tout le reste. Je me demande si je ne dois pas vous envoyer mes épreuves corrigées depuis longtemps.

En attendant je travaille surtout sur l'équation de Smoluchowski dans le cas continu (69). J'obtiens des résultats, mais j'avance encore à tatons, et je n'ai pas encore trouvé la méthode qui s'applique au cas général. Veuillez agréer, Monsieur le Professeur, l'expression de mon profond respect.

P.S. J'ai porté votre index bibliographique le 8 chez Gauthier-Villars.

IX

$\sim$ 16 septembre   (70)

Monsieur le Professeur

J'ai renvoyé hier vos épreuves. Je n'ai qu'une observation à faire en ce qui les concerne: placard 24, $2^{\grave{e}}$ page, vous dites que $lim_{n\to\infty}\frac{\mathcal{N}_{kn}^{(3)}}{n^2}$ est généralement $\neq 0$ dans le cas d'une variable semi-régulière (je crois qu'il est en effet assez utile d'introduire ce concept). Or cette limite est toujours nulle dans ce cas (71).

En ce qui concerne la mise en page de votre livre je ne l'avais demandée que pour l'index. Comme Fortet va en faire une correction, il m'a paru aussi inutile que je les corrige aussi. Ville parait être occupé avec le livre de Borel (72), il ne croit pas pouvoir aller à Genève (73). Je ne suis pas très étonné que A.Wintner (74) ne vient pas, je me demandais depuis longtemps ce qu'il comptait dire à ce colloque.

Lorsque je vous ai écrit ma dernière lettre, j'avais démontré pour les solutions de l'équation de Smoluchowski

$$a_{ik}(t+s) = \sum_{j=1}^{r} a_{ij}(s)a_{jk}(t)$$

bornées uniquement ceci: les solutions sont de la forme de votre livre (75)

$$a_{ik}(t) = \sum_{\rho} P_{ik}^{\rho}(t)\lambda_{\rho}^{t}$$

losqu'on se borne pour t aux valeurs rationnelles, donc les solutions continues sont pour tout t de cette forme.

Tout le reste supposait $a_{ik}(t) \geq 0$ et $\sum_{k} a_{ik}(t) = 1$

Depuis j'ai progressé un peu, j'ai pu montrer que si x est rationnel, τ un nombre > 0 quelconque, alors

$$lim_{\alpha\to 0}a_{ik}(\alpha\tau) = a_{ik}(0)$$

existe indépendamment de τ. Je pense alors mettre $a_{ik}(\alpha\tau)$ sous la forme

$$a_{ik}(\alpha\tau) = a_{ik}(+0) + \alpha\tau[\beta_{ik}(\tau) + O(\alpha)] \qquad (76)$$

là dedans $\beta_{ik}(\tau) = \beta_{ik}(\alpha\tau)$, qui est une sorte de dérivée si l'on se rapproche de l'origine par valeurs multiples rationnelles d'un nombre τ, est, multiplié par $\alpha\tau$, une fonction linéaire

$$(\tau_1 + \tau_2)\beta_{ik}(\tau_1 + \tau_2) = \tau_1\beta_{ik}(\tau_1) + \tau_2\beta_{ik}(\tau_2)$$

Malheureusement je n'ai pas encore réussi à montrer dans le cas général que $\beta_{ik}(\tau).\tau$ est borné, et je ne sais pas si c'est vrai.

En tout cas je crois qu'on pourra montrer que les solutions bornées de l'équation de Smoluchowski sont de la forme indiquée lorsqu'elles sont mesurables. Il suffirait certainement de montrer que les solutions de

$$f(u_1 + u_2) = f(u_1) + f(u_2)$$

qui ne sont pas de la forme $k(u_1 + u_2)$ sont non mesurables, ce qui me parait être bien connu (77).

Si l'on revient au cas des probabilités, équation de Smoluchowski, dans le cas continu, j'ai constaté que la tendance vers une limite a toujours lieu s'il n'y a pour chaque t qu'un nombre $< \infty$ d'ensembles finals pour $P(E, \mathcal{E}, t)$ et s'il n'y a pas d'ensembles essentiels impropres. Il résulte que dans le cas de Bogoliouboff ainsi que dans le cas de ma thèse, l'hypothèse de continuité est superflue (78).

Dans le cas général, avec des légères restrictions mais en supposant $P(E, \mathcal{E}, t)$ continue en t j'ai aussi obtenu la tendance à la limite.

Veuillez agréer, Monsieur le Professeur, l'expression de mon profond respect

W.D.

P.S. Je viens de recevoir votre lettre. Pour ce qui est des signes $\approx$ et $\rightarrow$ je ne sais guère comment écrire commodément un développement limité, sans préciser l'ordre de grandeur du reste. Placard 8, il faut lire S' et S'' (79) au lieu de $S_1^{(n)}$ et $S_2^{(n)}$, je l'avais déjà corrigé dans mes épreuves. Mais je ne vois pas où, placard 8, j'emploie A pour $|A|$, en tout cas si plusieurs fois j'ai employé les signes $\approx$ et $\rightarrow$ d'une façon sciemment fausse, A pour $|A|$ est une faute d'impression ou d'inattention.

Maintenant placard 7 et 8 je résume d'abord une démonstration de Bernstein, c'est un résumé de démonstration et non pas une démonstration complète (80).

De l'hypothèse $\quad \sum_{i=1}^{n} \dfrac{\tau_i}{B_n^{3/2}} \rightarrow 0 \quad$ résulte $\quad \dfrac{E\left[u_i^2\right]}{B_n'} \rightarrow 0,$

il n'y a qu'à passer au log pour évaluer $lim_{n\rightarrow\infty}E_n$. Ensuite au lieu de γ_k il vaudrait mieux $\gamma_k^{(n)}$, $\left(lim_{n\rightarrow\infty} \sum_1^n \gamma_k^{(n)} = 0\right)$, mais j'utilise les notations de Bernstein. $\sum_k P_k \mathcal{E}_k(y_i)$ (81) est l'espérance mathématique à priori parce que par hypothèse la probabilité à priori de l'état E_k à toutes les épreuves est P_k.

J'ai dit au commencement du §3 qu'il suffit de considèrer les $S^{(n)}$, c'est au §4 que je supprime cette hypothèse. J'espère que la démonstration du placard 9 sera plus claire dans le texte corrigé. On a bien $S' = y_1^{(n)} + ... + y_\rho^{(n)}$, Bernstein utilise dans son mémoire à plusieurs reprises son lemme (82) pour des grandeurs $y_1^{(n)}, ..., y_\rho^{(n)}$, il n'y a rien à changer dans la démonstration, car tout dépend uniquement de l'ordre de grandeur de $\sum \dfrac{\alpha_i}{\sqrt{B_n}}$ etc ...

Je me suis permis de faire comme Bernstein, mais si vous voulez je peux changer cela. Maintenant le mot positivement dans l'énoncé du théorème avait déjà été supprimé par moi. Je vais vous envoyer demain mes épreuves du ch. I.

W.D.

X

19 septembre 1937   (83)

Monsieur le Professeur

J'avais démontré que $\dfrac{N_{kn}^{(3)}}{n^2} \to 0$, sans utiliser vos formules, mais cela résulte aussi immédiatement de l'expression que vous donnez (84)

$$\sum_l \Pi_{kl} z_l \sigma_l^2$$

Remarquez d'abord que $\Pi_{kl} = 0$ si $E_l \bar{\in} \sum \mathcal{G}_\alpha$,   ($\mathcal{G}_\alpha$ groupe final) $\Pi_{kl} = Pr[k, \mathcal{G}_\alpha]\Pi_l^{(\alpha)}$ si $E_l \in \mathcal{G}_\alpha (\Pi_l^{(\alpha)}$ désignant la valeur commune de Π_{kl} lorsque $E_k \in \mathcal{G}_\alpha)$. Alors vous pouvez écrire

$$\sum_l \Pi_{kl} z_l \sigma_l^2 = \sum_\alpha Pr[k, \mathcal{G}_\alpha] \sum_{E_l \in \mathcal{G}_\alpha} \Pi_l^{(\alpha)} z_l \sigma_l^2$$

Maintenant σ_l^2 est indépendant de l'état initial dans $\mathcal{G}_\alpha = \sigma_\alpha^2$

$$\Pi_l^{(\alpha)} = \Pi_{l_1 l} \qquad \left(E_{l_1} \in \mathcal{G}_\alpha\right)$$

Donc

$$\sum_{E_l \in \mathcal{G}_\alpha} \Pi_l^{(\alpha)} z_l \sigma_l^2 = \sigma_\alpha^2 \sum_{E_l \in \mathcal{G}_\alpha} \Pi_{l_1 l} z_l = \sigma_\alpha^2 \sum \Pi_{l_1 l} z_l = 0$$

Il résulte

$$\sum_l \Pi_{kl} z_l \sigma_l^2 = 0$$

si

$$\sum \Pi_{kl} z_l = 0 \qquad \left(k = 1, \ldots, r\right) \qquad c.q.f.d.$$

Je vous remercie beaucoup de votre réponse relative à $f(x + y)$ (85). Je regarderai votre mémoire dès que l'IHP sera ouvert. De quelle année à peu près est-il ? De toute façon il me paraît en résulter (je pense le vérifier encore une fois avant de l'indiquer comme sûr) que les solutions bornées mesurables sont toutes de la forme de votre livre. Je vais essayer après de me débarrasser d'une de ces conditions, je crois que toutes les solutions bornées sont mesurables et réciproquement, enfin je laisse cela pour novembre. (86)

En ce qui concerne cette chose de Kolmogoroff (87), ce m'est connu depuis assez longtemps. Je ne crois pas que cela se trouve dans une publication. Pour le cas semi-régulier cela résulte comme cas particulier du théorème du logarithme itéré, dans le cas général j'ai dit dans le §10 (?) (88) (celui qui suit le §sur le théorème du log itéré) tout à la fin quelques mots sur une autre "loi des grands nombres" de Finetti etc ... ; je me rappelle que j'avais voulu ajouter qu'il y a aussi une certaine "loi forte des grands nombres" dans ce sens, mais je l'ai oublié. Tout cela est vrai pour les $S_k^{(n)}$, mais la démonstration tient en quelques lignes pour les $f_{hk}^{(n)}$. 2 cas sont possibles $E_k \bar{\in} \sum \mathcal{G}_\alpha$, ou $E_k \in \mathcal{G}_\alpha$, dans le premier cas c'est evident puisque E_k n'est obtenu avec proba 1 qu'un nombre fini de fois. Dans le second cas, si le système passe dans les autres groupes finals, E_k n'est pas obtenu du tout, si au contraire le système passe dans $\mathcal{G}_\alpha$, la fréquence converge presque sûrement vers Π_k, d'où vous déduisez immédiatement la propriété de Kolmogoroff. La démonstration peut donc être faite sans utiliser des résultats plus précis que ceux de vos épreuves, vous n'avez qu'à le démontrer comme vous l'écrivez dans le cas semi-régulier (donc aussi à l'intérieur d'un groupe final), et vous pouvez le déduire alors sans difficulté dans le cas général. (Ce qui précède est évidemment une <u>esquisse</u> de démonstration).

J'ai reçu le commencement de la mise en page et j'ai fait l'index pour ces parties. Il y a certains concepts comme "cas le plus régulier" (que j'ai rangé sous r) "principe ergodique" (que j'ai rangé sous e) que vous mettriez peut être sous d'autres initiales, mais vous aurez toujours la possibilité de le faire une fois l'index terminé (89). J'ai cru utile de faire à part un index des principaux symboles que vous introduisez s_{ik}, P_k, etc..., condition (P), système (E) etc... (90) où est ce qu'il faut envoyer les épreuves ?

J'ai reçu hier les placards de ma thèse que vous m'avez envoyés. En ce qui concerne l'introduction je l'avais envoyé à Ville, je crois maintenant comprendre que vous jugez que je peux la laisser comme elle est en modifiant tout au plus certaines phrases. Je veux d'ailleurs y ajouter ce que j'entends en disant que certaines connaissances donne un renseignement moindre sur quelque chose qu'un autre. J'attends les observations de Ville pour vous envoyer l'introduction. Chapitre 1 §1, je veux bien ajouter la définition de $\mathcal{E}$, et citer Schulz et Mihoc, seulement je voudrais faire cela comme annotation en bas de la page (91), et je ne sais pas comment le marquer sur les épreuves. §2 il y avait surtout la question des $\approx$ et $\rightarrow$, j'espère que sous la forme actuelle changée tout soit correct. A la fin du § il y avait quelques phrases qui avaient provoqué la désignation incompréhensible, je les ai rayées. §3 j'ai substitué les $S_k^{(n)}$ aux $S^{(n)}$. J'ai constaté effectivement qu'au placard 8 à 2 reprises il y avait $|A|$ pour A, ce qui m'avait échappé lors de la première correction. Placard 7 il y a "incompréhensible" à côté de l'énoncé du lemme fondamental. Cet énoncé est copié mot à mot du mémoire de Bernstein (92). Je me demande s'il ne serait pas préférable d'écrire dans l'énoncé au lieu de "pourvu que ... tendent vers 0 avec $1/n$", "si ... tendent vers 0".

Veuillez agréer, Monsieur le Professeur, mes sentiments respectueux.

W.Doeblin

P.S. J'enverrai demain les placards 5 - 8 corrigés.

XI

Rapport sur la thèse de Monsieur Doeblin
soutenue le 26 mars 1938. (93)
Par Monsieur Fréchet (jury Borel-Fréchet-Garnier)

Au congrès international des mathématiciens de 1932, le grand mathématicien Serge Bernstein (94) déclarait que la théorie des probabilités en chaîne avait atteint sa forme presque parfaite. Et l'auteur du présent rapport n'était pas loin de partager son opinion.

Pourtant, plusieurs travaux importants ont transformé cette théorie et la thèse de Monsieur Doeblin est un de ceux là et sans doute le plus digne d'attention.

Dans le cas le plus fouillé, celui de Markoff, défini analytiquement par le système

(I)
$$P_{ij}^{(n+1)} = \sum_k p_{ik} P_{kj}^{(n)}$$

(T)
$$\sum_k p_{jk} = 1$$

(P)
$$p_{ij} \geq 0$$

Monsieur Doeblin s'est rencontré avec Monsieur Kolmogoroff pour étendre au cas général la méthode directe, sans calculs, par laquelle Monsieur Hadamard établissait, dans le cas où $\sum_j p_{jk} = 1$, l'existence de sous-groupements cycliques (c'est au moment où Monsieur Kolmogoroff venait d'annoncer cette extension encore inédite au rapporteur, dans une conversation privée, que Monsieur Doeblin apportait au rapporteur ses premiers résultats écrits sur cette extension).

Dans ce même cas simple de Monsieur Markoff c'est aussi simultanément mais par des méthodes différentes que Monsieur Mihoc (de Bucarest) et Monsieur Doeblin établissaient la validité du passage à la loi de Gauss, (en général), à la limite, de la loi réduite de probabilité de la somme de plusieurs grandeurs enchaînées. Tous deux notaient aussi l'existence d'un cas d'exception. Enfin, Monsieur Doeblin montrait en outre que le théorème du logarithme itéré dû à Khintchine concernant une somme de variables indépendantes s'étendait aussi à une somme de variables en chaîne.

En passant au cas continu, Monsieur Doeblin retrouve par l'emploi (réalisé par lui le premier dans le cas continu) de la méthode directe, les résultats obtenus par le rapporteur par la méthode des équations intégrales et des noyaux principaux. En outre, il obtient des résultats plus étendus. Le problème consiste à étudier les solutions du système

(I)
$$\bar{\omega}^{(n+1)}(E, v) = \int_V \bar{\omega}(F, v) d\bar{\omega}^{(n)}(E, w_F)$$

(T)
$$\bar{\omega}(E, V) = 1$$

(P)
$$\bar{\omega}(E, v) \geq 0$$

où $\bar{\omega}(E, v)$ est une fonction additive de l'ensemble v sous ensemble de l'ensemble fondamental fixe V. Le rapporteur avait fait une étude assez complète du cas où V est borné et où de plus $\bar{\omega}$ est représentable sous forme d'une intégrale de Lebesgue $\bar{\omega}(E, v) = \int_V P(E, F) dF$ telle que $P(E, F)$ ou au moins l'un de ses itérés soit borné sur V. Il avait seulement donné quelques indications sur le cas où ces hypothèses ne sont pas réalisées. Ce sont ces cas exceptionnels et difficiles qui ont fait l'objet des recherches de Monsieur Doeblin ainsi que de Messieurs Fortet, Kryloff et Bogoliouboff, mais les résultats de Monsieur Doeblin englobent ceux des autres auteurs. Il arrive aussi à obtenir des conditions nécessaires et suffisantes simples pour que les résultats que nous avions obtenus - par exemple $\bar{\omega}(E, v)$ est une fonction asymptotiquement périodique de n - restent valables. Mais de plus Monsieur Doeblin arrive à ces résultats par une méthode tout à fait distincte de la nôtre, la méthode directe, qui ne fait pas appel à des théories algébriques et analytiques mais reste strictement dans le domaine du calcul des probabilités.

Enfin le dernier chapitre traite un cas analogue au précédent mais où le mouvement s'effectue de façon continue. L'auteur traite particulièrement de l'équation de Smoluchowski. Il a ainsi à étudier un système de la forme

$$(I) \qquad \bar{\omega}(E, s, v, t) = \int_V \bar{\omega}(F, u, v, t) d\bar{\omega}(E, s, w_F, u)$$

$$(T) \qquad \bar{\omega}(E, s, V, t) = 1$$

$$(P) \qquad \bar{\omega}(E, s, v, t) \geq 0$$

avec cette fois

$$\bar{\omega}(E, s, v, s) = \begin{cases} 1 \; si \; E \; appartient \; à \; V \\ 0 \; dans \; le \; cas \; contraire \end{cases}$$

Là encore, il arrive dans un champ déjà étudié (par Messieurs Paul Lévy (95), Kryloff, Bogoliouboff) et il réussit à étendre leurs résultats à des cas plus étendus.

Il est impossible de résumer ici les résultats extrêmement nombreux de Monsieur Doeblin. D'une part il a réussi à étendre le mode de raisonnement direct de Monsieur Hadamard à toute la variété des cas qu'on rencontre dans la théorie des probabilités en chaîne, et par conséquent à faire mieux comprendre le mécanisme des mouvements que ne pouvaient le faire les méthodes analytiques antérieures. D'autre part, il a montré une grande puissance démonstrative en traitant les cas les plus difficiles qu'avaient d'abord dû laisser de côté ses prédécesseurs. Il a aussi fait le grand effort sur lequel insistait le rapporteur, de rendre convaincantes pour tous, des démonstrations que l'esprit très vif et très intuitif de Monsieur Doeblin lui faisaient apparaître comme très évidentes.

La richesse des résultats de Monsieur Doeblin, la force qu'elle révèle, autorisent le rapporteur à juger que son travail est hautement digne d'être présenté comme Thèse de Doctorat ès Sciences Mathématiques.

Le rapporteur   Monsieur Fréchet.

La soutenance de la Thèse a entièrement confirmé les appréciations favorables de Monsieur Fréchet. Mention très honorable.

Emile Borel.

XII

Paris le 8 septembre 1938.   (96)

J'ai quelque scrupule à vous envoyer du travail à si peu de distance de votre service. Toutefois comme c'est vous et non Ville ou Fortet qui avez été aidé pour venir à Genève, il me semble plus équitable de m'adresser à vous. Je vous envoie donc sous pli séparé placards et manuscrits de Bernstein.

Vous trouverez ci-joint une lettre de lui, à me retourner sous urgence, à l'occasion d'une lettre à m'écrire. Le message souligné au crayon vous permettra d'obtenir des éclaircissements. Vous dites en effet que vous n'avez pas été satisfait de son mémoire sur le même sujet. Cependant, je vous prie de rédiger vos demandes non sur le ton d'une autorité indiquant des erreurs, mais à la manière d'un lecteur qui trouve des difficultés et s'attend à les voir résolues: "je ne vois pas comment complèter le raisonnement de la p. , ligne ", il me semble qu'il y a contradiction entre etc... D'autre part, pour aller vite, attendez pour écrire d'avoir porté les placards corrigés à L. Freymann (97) (librairie Hermann, 6 rue de la Sorbonne), gardez provisoirement le manuscrit et écrivez à l'auteur d'après ce manuscrit qu'il faudra finalement rendre à M. Freymann. Dites à celui-ci qu'il faut peut être envoyer ces placards corrigés à M. Wavre (98) avant de les envoyer à l'imprimerie.

J'aurais en outre besoin d'avoir des résumés (en français) des conférences de MM. Slutsky, Glivenko et Romanovsky (à joindre aux résumés par de Finetti qu'il faut traduire en français). Vous me parlez de Loève. Est il exact qu'il soit en train de travailler avec M. Darmois? En ce cas je préfère ne rien lui demander, il est dans ce cas plus naturel de ne pas le détourner des études, recherches ou travaux qui lui auraient été proposés par M. Darmois , (99).

Ces résumés seraient à peu près dans les mêmes proportions que ceux qui ont été faits et publiés en Italien par de Finetti. Les placards correspondants seraient fournis par M. Freymann (à qui ils devraient être rendus en les lui demandant de ma part). Par ordre d'urgence, ils ne viendraient qu'après la correction de Bernstein.

On ne peut mettre la main sur l'adresse actuelle de Feller. Peut être serai je obligé de corriger ses placards à sa place. Il ne manque plus en effet pour les placards que les corrections de Bernstein et Feller.

Je suis content qu'on vous ait attribué 5000: Je craignais qu'on ne diminue ce chiffre.

Au sujet de vos travaux, vous feriez mieux de vous contenter de rédiger ce qui est fait. Je crains un peu les erreurs dans des recherches faites sous la pression du temps et votre départ prochain.

Bien à vous

M. Fréchet

P.S. Si à un moment quelconque vous avez accès à un Zentralblatt, pourriez vous regarder si Ta-Li (100) a publié quelque chose sur les équations différentielles linéaires, où et quoi?

XIII

Lettre du 28 octobre 1939.   (101)

Permettez moi de vous donner de mes nouvelles. Vous savez que depuis février j'ai travaillé sur une question relative aux variables aléatoires indépendantes; l'ensemble de puissances. J'ai d'abord rédigé les résultats déjà obtenus avant mon service (102), puis j'ai continué mes recherches en les rédigeant en même temps. Fin juillet j'ai envoyé un premier mémoire de 25 pages à Steinhaus pour les Studia Mathematica (103). Je ne crois pas que ce mémoire y verra jamais le jour (si on peut exprimer cette expression). Quelle malencontreuse idée que j'avais aussi en envoyant ce mémoire en Pologne! Je voudrais bien savoir ce qu'ils sont devenus les mathématiciens polonais (104). J'ai enfin réussi fin juillet à résoudre le problème suivant, contre lequel je m'étais vainement acharné au printemps 38 et qui est certainement après la théorie générale des chaînes le problème le plus difficile que j'ai pu résoudre; $\mathcal{L}$ étant une loi de probabilité, $X_1, ..., X_n$ des variables aléatoires indépendantes dépendant de $\mathcal{L}$, n_ρ une suite d'entiers, a_ρ et b_ρ deux suites de constantes, il se peut que les lois de $\left(a_\rho^2 \sum_{i=1}^{n_\rho} X_i - b_\rho\right)$ convergent vers une loi de probabilité limite I n'attribuant pas toute la probabilité à une valeur unique. Soit $\mathcal{F}[\mathcal{L}]$ l'ensemble de toutes les lois I qu'on peut obtenir en faisant varier n_ρ, a_ρ et b_ρ (dans le cas le plus simple $\mathcal{F}[\mathcal{L}]$ comporte toutes les lois du type de Gauss). Quelles sont les conditions nécessaires et suffisantes pour qu'un ensemble de lois $\mathcal{G}$ donné est un $\mathcal{F}[\mathcal{L}]$?

Fin juillet presque toute la démonstration était rédigée, il fallait encore 2-3 jours de travail. J'espérais pouvoir en finir en aout. Le sort ne l'a pas voulu. Je suis parti de Givet (105) le manuscrit étant à peu près au même point qu'en juillet. Je l'ai envoyé ainsi que les doubles du manuscrit que j'avais envoyé à Steinhaus à un frère que j'ai à Philadelphie (106), je ne sais pas s'il l'a reçu (107), je crains fort qu'il ne l'ait pas reçu. J'ai fait une très grave bêtise en me départant ainsi de tous mes brouillons et doubles. Après la guerre il me faudra bien 2 mois pour rédiger tout cela (108). Pendant le mois d'aout le peloton d'élèves-caporaux a plus bavé que lorsque nous étions bleus et il m'était impossible de travailler. D'abord il y a eu l'examen du peloton, le 10 (dont je suis sorti $2^{\text{ème}}$), je reste toutefois $2^{\text{è}}$ classe, les nominations devant avoir lieu le 1er septembre et le bataillon ayant été dissous le 22 aout). Avant l'examen, étude obligatoire tous les soirs. Après l'examen pour changer en même temps qu'on nous faisait particulièrement baver pendant l'exercice, on nous utilisait tous les soirs qu'on était libre pour faire des corvées diverses, généralement pour décharger du sable pour la défense passive de $5^h 1/2$ à $8^h 1/2$, et le dimanche, la garde. Comme la garde (tous les 3 jours) était devenue aussi plus fatigante par suite de l'augmentation du nombre de sentinelles vous comprenez pourquoi je n'ai pas pu travailler en aout. Depuis j'ai cessé tout travail. Je suis téléphoniste ici (pour le cas où le secteur s'animerait) pratiquement pour le moment je fais de la lecture au son, du morse avec les radios et de temps en temps on fait des travaux de campagne. Je ne travaille pas, je n'en ai pas le courage, d'ailleurs je n'ai aucun local où je pourrais m'isoler au moins relativement. Toutefois le jour où vous voudriez faire appel à moi, je serais entièrement à votre disposition, même si le secteur s'animerait, j'aurais du temps à réfléchir, la nuit par exemple pendant les heures de garde au téléphone.

Le paquet du petit japonais (109) dont je ne me rappelle plus le nom contenait 2 mouchoirs très fins. J'espère que vous avez reçu les épreuves que je me suis permis de vous envoyer ne sachant pas comment les envoyer directement (110).

J'espère que vous êtes en bonne santé et je vous prie de bien vouloir agréer, Monsieur le Professeur, l'expression de mes sentiments très respectueux.

W.Doeblin.

XIV

Lettre du 12 novembre 1939.   (111)

J'ai reçu il y a quelques jours (le 9) le tirage à part que vous m'avez envoyé. Je vous en remercie beaucoup. Je vous avais envoyé, il y a peut être un mois, des épreuves de mon mémoire sur l'équation de Chapman (112) , destiné à la Skandinavisk aktuarietidskrift en vous priant de l'envoyer à la rédaction (moi même je ne le puis pas), je ne sais pas si vous l'avez reçu.

Je suis toujours au cantonnement et les jours s'écoulent ici tranquillement, les alertes quotidiennes n'arrivent pas à briser la monotonie. On ne s'aperçoit guère de la guerre. Il est vrai que ça peut changer d'un jour à l'autre. En attendant j'ai repris mon travail; oh, pas beaucoup, une heure par jour à peu près. Je rédige les démonstrations de ma note sur l'équation de Kolmogoroff. On ne parle plus de permissions pour nous, mais mon moral est bon quand même.

Veuillez agréer, Monsieur le Professeur, mes sentiments très respectueux.

W.Doeblin.

XV

Lettre du 19 novembre 1939.   (114)

Monsieur le Professeur

J'ai reçu votre lettre du 15 novembre. Je vous remercie des renseignements donnés et je vous suis particulièrement reconnaissant d'avoir bien voulu expédier mes épreuves au Skandinavisk aktuarietidskrift.

Vous me demandez comment je me tire de la situation au point de vue pécuniaire. Pour le moment je n'ai aucun souci matériel. Certes ma solde de $2^{è}$ classe me permet juste de payer un journal, mais je n'ai pas encore épuisé mes économies et il me reste encore une partie de la subvention de 5000 francs que j'ai eue grâce à vous avant mon service. D'autre part je reçois régulièrement des mandats de mes parents.

Toutefois comme la guerre peut durer 4-5 ans, je suis matériellement intéressé à monter en grade et j' ai bien l'impression qu'ici je resterai toujours $2^{è}$ classe. Je vais essayer de me renseigner si je ne peux pas suivre maintenant encore un peloton d'officiers de réserve duquel je ne sortirai probablement pas comme officier mais au moins comme sergent.

En attendant ce qui m'intéresse le plus ce sont les permissions. Jusqu'à hier j'ai été absolument convaincu que l'active partirait avant les réservistes, au moins dans notre régiment. Il y a des artilleurs avec nous, pour eux c'est l'inverse, mais ce n'est pas là une consolation. Comme nous ne sommes que 4 de l'active à l'état major (115), je ne pourrai partir qu'en février-mars, ce qui fera 10 mois que je ne serai pas retourné en permission. D'ailleurs il est bien possible qu'il y aura une offensive allemande en février, alors les permissions seront suspendues jusqu'à une date indéfinie.

Heureusement mon travail mathématique m'aide à lutter contre le cafard. L'alcool ne me disant rien je n'ai pas comme d'autres la ressource de m'enivrer.

Veuillez croire, Monsieur le Professeur, à mes sentiments respectueux.

W.Doeblin.

XVI

Monsieur le Professeur   (116)

Voici la réponse à la question que vous m'avez posée.

Faisons n expériences, l'événement E ayant la probabilité p de se produire dans chacune des expériences qui sont indépendantes. Soit S_n le nombre des réalisations de E dans ces n expériences. La loi de probabilité de S_n est égale à une loi de Poisson $\left[e^{-\lambda}\frac{\lambda^p}{p!}, \lambda = np\right]$ avec une erreur $O\left[np^2\right]$. Si $np(1-p) = \sigma^2 n$ est suffisamment grand, la loi de $\left[S_n - np\right]/\sigma\sqrt{n}$ est très voisine d'une loi de Gauss, mais si $np(1-p)$ n'est pas assez grand, la loi de S_n diffère assez sensiblement de la loi de Gauss, l'erreur étant si mes souvenirs sont exacts $O\left(\frac{1}{\sqrt{np(1-p)}}\right)$. Une condition dans le genre $\sqrt{np(1-p)} > K$ est donc absolument nécessaire pour l'application de certains critères supposant que la loi de S_n est très sensiblement la loi de Gauss-Laplace. Si $\sqrt{np(1-p)}$ est $< K$ et $n > K'$, on a nécessairement np^2 ou $n(1-p)^2 < \varepsilon$ et la loi de Poisson sera applicable à S_n ou $n-S_n$. Voici à peu près tout ce que je vois à dire sur la question que vous m'avez posée. Je m'excuse de ne vous répondre que maintenant, mais je l'avais oublié complètement.

Mon train de vie n'a pas changé beaucoup. On est toujours au même endroit. Je travaille toujours sur l'équation de Kolmogoroff, dans un mois je publierai une note complémentaire à ma note d'octobre 1938 (117). Je me suis aperçu d'une chose assez amusante. Soit $F(x,y,s,t)$ une solution de l'équation de Chapman continue par rapport à x et correspondant à des mouvements continus presque sûrement. Comme j'ai prouvé depuis longtemps $1 - F(x,y,s,t)$ est alors par rapport à x une loi de probabilité. Soit $G(y,x,t,s) = 1 - F(x,y,s,t)$. J'ai constaté que G satisfait à l'équation fonctionnelle

$$G(y,x,t,s) = \int_{-\infty}^{+\infty} G(z,x,u,s)d_z G(y,z,t,u) \qquad (s < u < t) \qquad (118)$$

qui devient l'équation de Chapman si on pose par exemple $t' = -t$, $s' = -s$.

Je vous avais dit que j'avais l'intention de faire une demande pour suivre un peloton d'E.O.R. Un tel peloton a été constitué pendant ma permission. Il faut que j'attende 3-4 mois pour essayer de suivre le second.

Lorsque vous aurez des problèmes pour la Défense Nationale intéressant le Calcul des Probabilités je serais toujours heureux si vous faisiez appel à moi.

Je vous souhaite une bonne année et vous prie de bien vouloir agréer mes sentiments très respectueux.

W.Doblin.

XVII

(Lettre de M.Fréchet à W.Doeblin)   (119)

Mon cher ami

La réponse que vous m'avez donnée et dont je vous remercie montre dans quelle direction il faudrait faire les calculs nécessaires pour aboutir à des règles numériques.

Ce sont ces règles numériques dont j'aurais besoin. Savoir par exemple pour quelles valeurs de k la condition que vous indiquez $\sqrt{npq} > k$ donne une erreur relative par la loi de Laplace-Gauss $< 1/100$. Idem pour l'erreur absolue en ce qui concerne non la répétition mais la fréquence. Question analogue pour la loi de Poisson; question analogue pour les frontières où Poisson doit être préféré à Laplace-Gauss et réciproquement?

Je sais en outre que c'est là un genre de questions qui bien que très utile, est difficile à traiter pour vous qui ne disposez pas de tables de la fonction de Laplace ou peut être même de table de logarithme.

XVIII

Lettre de Doeblin du 12 mars 1940   (120)

Monsieur le Professeur

Permettez moi de vous donner de mes nouvelles. Depuis janvier on a un peu voyagé, j'ai vu trois-quatre nouveaux cantonnements, on n'a pas toujours été bien installé, nous avons couché pendant plusieurs semaines après l'alerte sur la Belgique en janvier dans un grenier sans poêle dans lequel il neigeait et dans lequel on gelait drôlement. Mais maintenant depuis plus d'un mois on est assez bien installé dans des baraques spécialement construites pour la troupe. On se trouve dans un petit bled de 150 habitants environ quelque part en Lorraine. Il y a ici autant de fumiers que dans le village des Ardennes où on a passé les premiers mois de la guerre, même plus.

J'ai fini mon mémoire sur l'équation de Kolmogoroff il y a à peu près un mois, plutôt, à un moment donné j'en ai eu assez de l'équation de Kolmogoroff et j'ai arrêté mon mémoire. J'ai renvoyé un brouillon à la maison où il est arrivé et le manuscrit à l'Académie de Sciences comme pli cacheté, mais je crains qu'ils ne l'ont pas reçu (121).

Depuis je me suis occupé surtout de l'équation de Chapman, dans le cas des probabilités. J'ai obtenu un certain nombre de formules valables dans le cas général et j'ai étudié surtout certains cas dans lesquels des solutions de la forme indiquée par Hostinský (122) se trouvent être les solutions générales. Lorsque je viendrai en permission je vous demanderai de me prêter pendant 8 jours ceux des tirages à part de Hostinský concernant l'équation de Chapman qui ne se trouvent pas à l'IHP.

J'ai eu pendant plus d'un mois beaucoup de temps libre pour travailler et surtout je pouvais m'isoler à la cabine téléphonique. Mais ce temps là a l'air d'être à peu près terminé.

J'ai fait une demande d'admission au peloton d'élèves-aspirants qui va commencer en juin. Je ne sais pas encore quand je viendrai en permission, soit dans quelques jours, soit dans 2 mois.

J'espère que vous allez bien et je vous prie, Monsieur le Professeur, de bien vouloir agréer mes sentiments très respectueux.

W.Doblin.

XIX

Lettre de Doeblin du 10 avril 1940   (123)

Monsieur le Professeur

Je vous envoie ci-joint une Note destinée aux Comptes Rendus intitulée "Sur des mouvements mixtes" (124). Il y a une référence bibliographique que je n'ai pas pu donner exactement, c'est celui du travail de Feller des Math. Ann. quelque chose comme Existenz und Eindentigkeitsätz für stochastische Prozesse vers 1936 (125). Je vous prierai d'y faire les changements que vous jugez nécessaires de faire et de la donner à Hadamard ou Borel. Quant aux épreuves, Fortet pourrait sans doute se charger de leur correction.

Dans une lettre à part je vous envoie aussi une autre note faisant suite à la précédente mais qui n'est pas destinée à être publiée avant quelques mois. Je vous avais fait remettre sur votre bureau les tirages à part que je vous avais empruntés, ainsi que le résumé du travail de Wold (126) et la Notice individuelle pour ma demande de bourse et la demande. J'espère que vous les avez trouvés. En ce qui concerne la liste de mes travaux , elle groupe tous mes travaux publiés à l'exception d'à peu près toutes les Notes aux Comptes Rendus. Je pense avoir pour mes demandes ultérieures une réserve de sécurité suffisante d'abord pour l'année prochaine il y aura mon mémoire sur la théorie générale des chaînes (dont je viens de finir la correction des premières épreuves) ainsi que celui sur la théorie métrique des fractions continues. Cela suffira bien pour une année et après il y aura mes travaux sur l'ensemble des puissances et ceux sur les équations de Kolmogoroff et Chapman.

Il paraît qu'il faut joindre à la demande une notice détaillée sur les travaux en cours. En principe les travaux en cours sont ceux qui ne sont pas encore publiés. Est ce que c'est la Notice sur les travaux que j'ai indiqués pour ma demande de bourse et que vous m'avez dit de faire ? En tout cas je vais pour ne pas perdre de temps commencer dès demain la rédaction de cette note. Il ne me reste en effet que quelques jours pour le faire.

Veuillez agréer, Monsieur le Professeur, l'expression de mes sentiments respectueux.

W. Doblin

XX

Lettre de Fréchet à Doeblin, Paris le 12 avril 1940   (127)

J'avais pensé d'abord opérer comme vous me l'indiquiez pour votre première note. Réflexion faite, j'aime mieux vous renvoyer la note pour deux raisons, 1° pour que vous voyiez vous même les modifications (de forme) que j'y propose, 2° pour que vous la recopiez en modifiant (ou supprimant) à votre gré mes modifications qui ne sont que suggérées et en le faisant d'une façon plus présentable pour l'académicien à qui je l'enverrai et moins difficile à reproduire par l'imprimeur.

Commencez au milieu de la page, espacez vos lignes, écrivez avec une encre qui marque, et une écriture plus lisible, espacez vos formules et caligraphiez les. Si vous étiez à Paris je vous aurais demandé de la faire dactylographier, mais là bas, c'est impossible sans doute.

Il me semble qu'il n'y a pas d'inconvénient à introduire un petit retard. D'autant qu'il me permettra de vous envoyer l'article de Feller qui s'est trouvé retardé; il vaut mieux que vous l'ayez revu avant l'envoi de votre note (128).

Pour les initiés vous pourriez dire l'équation fonctionnelle de Chapman et l'équation aux dérivées partielles de Kolmogoroff.

Je compte bien retrouver la 2ème note quand vous en aurez besoin , mais j'espère que vous en avez gardé copie (129).

J'ai transmis votre dossier (en pointillé) au Centre National de Recherche (130). Pour cette 1ère demande la notice sur vos travaux remplacera celle sur les travaux en cours.

XXI

Calcul des probabilités. Sur la solution de M.Hostinský de l'équation de Chapman. Note de M.Doblin présentée par M.   (131)

Nous adoptons ici les notations d'une Note récente (1).

Soit $U(x, y, s, t)$ une solution de l'équation de Chapman satisfaisant à toutes les conditions du Calcul des Probabilités et aux conditions suivantes:

$$(1) \qquad \frac{1}{t - s} \int_{|y-x|>\eta} d_y U(x, y, s, t) \to 0$$

uniformément si x et t sont bornés,

$$(2) \qquad \lim_{x \to \infty} U(x, y, s, t) = 0, \qquad \lim_{x \to -\infty} U(x, y, s, t) = 1.$$

et enfin supposons que $U(x, y, s, t)$ est continue par rapport à x. Sous ces hypothèses U définira un mouvement presque sûrement continu.

Soit maintenant $G(x, y, s)$ une fonction de 3 variables satisfaisant aux conditions énoncées dans notre Note précédente, en particulier à (3) et (4) (loc.cit). En utilisant soit le groupe de formules (I), soit le groupe (H) de notre précédente note, on peut calculer à partir de U et G une même fonction F, solution de l'équation de Chapman avec

$$\lim_{t \to s} \frac{F(x, y, s, t)}{t - s} = G(x, y, s) \qquad pour \quad y < x$$

$$\lim_{t \to s} \frac{1 - F(x, t, s, t)}{t - s} = G(x, \infty, s) - G(x, y, s) \qquad pour \quad y > x$$

si y est point de continuité de $G(x, y, s)$.

Cette solution correspond à un mouvement, les conditions locales duquel étant les mêmes que pour celui défini par U sauf que maintenant le point mobile a une probabilité $c(X(s), s)dt$ d'avoir un déplacement brusque pendant l'intervalle de temps $(s, s + ds)$.

Soit maintenant $F(x, y, s, t)$ une solution de l'équation de Chapman continue par rapport à x satisfaisant aux conditions (1) et (2) que nous avons imposées à U, pour laquelle les limites (3) et (4) existent. Les grandeurs $G(x, y, s)$ seront mesurables au sens de M. Borel par rapport à x et s.

Supposons encore que F satisfait à

$$\lim_{t \to s} \int_{|y - x| > \eta} d_y F(x, y, s, t) = 0$$

uniformément par rapport à x et s (si x et s sont bornés).

Il résulte alors des formules générales sur l'équation de Chapman que nous avons établies qu'on a encore pour F les formules I_1 et I_2 de notre précédente Note, les termes F_n ayant la même signification et étant continues par rapport à x. Définissons U par les formules II

(II$_1$) $U(x, y, s, t) = \sum_{i=0}^{\infty} W_i(x, y, s, t)$

(II$_2$) $W_0 = F_0$

(II$_3$) $W_i(x, y, s, t) = \int_s^t d\tau \int_z W_{i-1}(z, y, \tau, t) c(z, \tau) d_z W_0(x, z, s, \tau)$

$F_0(x, y, s, t)$ satisfaisant à l'équation de Chapman $(mais\ F_0(x, \infty, s, t) \leq 1!)$ il résulte des recherches de Hostinský que $U(x, y, s, t)$ est une solution de l'équation de Chapman et satisfait à toutes les conditions du Calcul des Probabilités. U définit un mouvement presque sûrement continu et on a, en tout point de continuité y de $G(x, y, s)$, si $y < x$,

$$U(x, y, s, t) = (t - s)G(x, y, s) - \int_s^t d\tau \int_z G(z, s, \tau) d_z U(x, z, s, \tau) + o(t - s)$$

et une formule analogue pour $y > x$. Dans tous les cas simples et en particulier si $G(x, y, s)$ ou $G'(x, h, s) = G(x, x + h, s)$ sont continues U satisfait à (1). De toute façon, de la fonction U ainsi calculée à partir de F_0 on tire de nouveau F_0 par les formules I_3, I_4 et les formules (H) sont applicables.

(1) Sur des mouvements mixtes.

XXII

Lettre de Doeblin à Fréchet, le 13 avril 1940.   (132)

Monsieur le Professeur

Je vous communique ci-joint encore quelques résultats complétant les 2 notes que je vous ai envoyées précédemment.

Soit $U(x, y, s, t)$ une solution de l'équation de Chapman

$$U(x, y, s, t) = \int U(z, y, u, t)\, d_z U(x, z, s, u)$$

monotone par rapport à y avec

$$U(x, -\infty, s, t) = 0, \qquad U(x, \infty, s, t) = 1$$

satisfaisant aux conditions

$$\int_{|y-x|>\eta} d_y U(x, y, s, t) \to 0$$

$$\frac{1}{t-s} \int_{|y-x|>a} d_y U(x, y, s, t) \to 0$$

uniformément par rapport à x et s. Faisons sur U des hypothèses entraînant que $\max_{s<u<t} |X(u)|$ est p.s. borné au sens du Calcul des Probabilités. Il résulte alors des résultats généraux sur l'équation de Chapman que $X(t)$ dépendant de U n'a presque sûrement pas de sauts $> a$ en valeur absolue. Supposons U continue par rapport à x.

Soit $G(x, y, s)$ une fonction arbitraire satisfaisant aux conditions suivantes:

G est mesurable au sens de Borel par rapport à x et s, monotone par rapport à y avec

$$0 = G(x, -\infty, s) \leq G(x, x - a \pm 0, s) = G(x, x + a + 0, s)$$
$$\leq G(x, \infty, s) = c(x, s)$$

Considèrons un mouvement régi par les mêmes conditions que le mouvement déterminé par U mais perturbé par la possibilité de sauts $\geq a$, la probabilité d'un saut a de x en y entre s et $s + ds$ étant $d_y G(x, y, s)ds$. Un tel mouvement est déterminé par

$$F(x, y, s, t) = U(x, y, s, t) + \sum_1^\infty U_i(x, y, s, t)$$

où U_i se déduit de U par les formules habituelles

(H) $$U_i(x, y, s, t) =$$

$$\int_s^t d\tau \int_z \Big|\int_{z'} \big[U_{i-1}(z', y, \tau, t) - U_{i-1}(z, y, \tau, t)\big] d_{z'} G(z, z', \tau) d_z U(x, z, s, \tau)$$

équivalentes aux formules d'Hostinský.

F satisfait à l'équation de Chapman, est continue par rapport à x et satisfait à toutes les conditions du calcul des probabilités.

Si G satisfait à la condition

(1) $$G(x, y, s) = \lim_{t \to s} \frac{1}{t-s} \int_s^t d\tau \int_z G(z, y, \tau) \, d_z U(x, z, s, \tau)$$

on a si $y < x - a$

(1) $$\lim_{t \to s} \frac{1}{t-s} F(x, y, s, t) = G(x, y, s)$$

$$\lim_{t \to s} \frac{1}{t-s} \big[1 - F(x, y, s, t)\big] = G(x, \infty, s) - G(x, y, s) \qquad pour \ \ y > x + a$$

On peut écrire aussi

(2) $$F(x, y, s, t) = F_0^{(a)}(x, y, s, t) + \sum_{i=1}^{\infty} F_i^{(a)}(x, y, s, t)$$

où $F_i^{(a)}(x, y, s, t)$ est la probabilité pour que la particule se trouve à gauche de y à l'instant t ayant eu exactement n sauts plus grands en valeur absolue que a. $F_i^{(a)}(x, y, s, t)$ se déduit de $F_0^{(a)}(x, y, s, t)$ par les formules de récurrence établies précédemment. $F_0^{(a)}(x, y, s, t)$ se déduit de $F(x, y, s, t)$ par les formules

(3) $$F_0^{(a)}(x, y, s, t) = F(x, y, s, t) + \sum (-1)^n \bar{H}_i(x, y, s, t)$$

$$\bar{H}_i(x, y, s, t) = \int_s^t d\tau \int_z \int_{z'} \bar{H}_{i-1}(z', y, \tau, t) \, d_z G(z, z', \tau) \, d_z F(x, z, s, \tau)$$

Inversement donnons nous une fonction F solution de l'équation de Chapman satisfaisant à toutes les conditions du Calcul des Probabilités avec

$$\int_{|y-x|>\eta} d_y F(x, y, s, t) \to 0$$

et satisfaisant aux formules (1), G ayant les propriétés indiquées.

En définissant $F_0^{(a)}(x, y, s, t)$ par la formule (3) et en calculant $F_i^{(a)}$, on obtient pour F la forme (2), les $F_i^{(a)}$ ayant la signification habituelle.

Définissons U par la série

$$U = W_0^{(a)} + \sum_{i=1}^{\infty} W_i^{(a)} \qquad W_0^{(a)} = F_i^{(a)}$$

$$W_i^{(a)} = \int_s^t d\tau \int_z W_{i-1}^{(a)}(z, y, \tau, t)\, c(z, \tau)\, d_z W_0^{(a)}(x, z, s, \tau)$$

on obtient pour F la formule (H), c'est à dire la forme de Hostinský généralisée et on peut calculer $F_0^{(a)}$ à partir de U par des formules indiquées dans ma note aux Comptes Rendus (133).

Dans les mouvements où il peut y avoir une infinité de sauts, les fonctions $G(x, y, s)$ définies comme ci-dessus existent très souvent et sont bornées, si $a \to 0$, les fonctions $G(x, \infty, s)$ ou plus exactement $G_a(x, \infty, s)$ tendent vers l'infini. Sous des hypothèses très générales on peut donc ramener l'étude de l'équation de Chapman à l'étude des solutions correspondant à des mouvements où il n'y a que des sauts $< \eta$, il se pose alors le problème du passage de η vers 0. J'ai obtenu des résultats dont je vous rendrai compte ultérieurement (134).

Je vous ai envoyé il y a quelques jours la "Note détaillée" sur mes travaux publiés (135).

Je vous prie, Monsieur le Professeur, de bien vouloir agréer l'expression de mes sentiments respectueux.

W. Doblin

Je m'excuse pour l'écriture, mais j'écris dans de fort mauvaises conditions matérielles.

XXIII

Monsieur le Professeur   (136)

J'ai reçu et regardé les tirages à part que vous m'avez envoyé. Je vous en remercie beaucoup. Comme je n'en ai plus besoin, je vous les renvoie. Je vous remercie beaucoup d'avoir transmis ma demande de bourse (ce n'est pas une allocation que je désire, la guerre peut se terminer avant la fin de 1941), je vous suis également très reconnaissant de m'avoir procuré cette allocation de secours universitaire, qui me permettra de travailler tranquillement quand on sera au repos. Quant à la première mensualité je vais la renvoyer ou la garder pour le mois prochain. Je suis en ligne, je ne pense guère travailler pour moi, il ne peut pas être question de louer une chambre. J'en avise aujourd'hui le Secours Universitaire. Le travail de Feller (137) contient une solution de l'équation de Chapman qui est absolument équivalente à la solution de Hostinský. Feller ne s'en est visiblement pas aperçu. En tout cas, le problème n'est pas tout à fait le même chez Hostinský et Feller. La note de Métadier ne me dit pas grand chose (138).

Je vous prie, Monsieur le Professeur, de bien vouloir agréer mes sentiments respectueux.

Doblin

2.2. LETTERS FROM W.DOEBLIN TO J.L.DOOB.

On the occasion of the Blaubeuren Conference, J.L. Doob kindly passed on to H. Cohn the following three extracts of letters addressed to him by W.D.

I

Paris le 6/10/38

Monsieur,

M.Fréchet vient de me signaler votre mémoire "Stoch. proc. with an integral valued param." (139). En le feuilletant j'ai trouvé la remarque p. 135: The latter authors omit...(140). Je regrette que vous n'avez ni cru utile de m'envoyer ce mémoire dans lequel vous nous accusez d'avoir commis une faute ni de nous écrire à ce sujet avant de publier cette remarque. Je le regrette d'autant plus que après avoir revu la note en question je ne vois absolument pas sur quoi vous vous basez pour affirmer que nous avons omis "mention of exceptional points, implying here the possible existence of R_3" (Je ne vois d'ailleurs pas pourquoi vous appelez ces points "exceptionnels"). Nous n'avons nullement affirmé que $W - \Sigma \mathcal{G}_\alpha$ est vide, il résulte même d'une proposition y donnée (141) ($P^{(n)}(X, W - \Sigma \mathcal{G}_\alpha) \to 0$) que cela n'a pas lieu dans le cas général. Quant à la décomposition de $\mathcal{G}_\alpha$ en sous ensembles cycliques, sans reste, cela est une propriété intervenant dans la définition de $\mathcal{G}_\alpha$. Les ensembles que j'ai appelés finals ne sont pas des ensembles quelconques, dans le cas d'un nombre fini d'états les matrices correspondant à ces ensembles sont nécessairement indécomposables. Une référence à un autre travail (142) a été donnée dans cette Note et si vous aviez été en doute sur un des énoncés vous auriez pu consulter cet autre travail.

Je viens de vous envoyer un autre de mes travaux sur les probabilités en chaîne.

En vous priant de bien vouloir m'envoyer ainsi qu'à M.Fortet (143) un exemplaire de votre travail, je vous prie, Monsieur, de croire à ma considération distinguée.

W.Doeblin   (144).

II

Carte postale (début 1939)   (145).

Cher Monsieur

Je n'ai reçu votre lettre qu'avec un long retard puisque l'on ne me l'avait fait suivre que le 11/10 et je n'ai pu y répondre avant n'ayant guère de temps libre. En ce qui concerne le point que vous m'avez signalé (146) concernant l'équation $A^{(t+s)} = A^{(t)}.A^{(s)}$ vous avez parfaitement raison, ma démonstration est fausse, puisque déjà l'équation $\lambda(t).\lambda(s) = \lambda(t + s)$ (t,s rationnels) a des solutions où $\arg \lambda(t)$ ne tend pas vers zéro. Toutefois si $A(t)$ est continu dans un intervalle quelconque (et un raisonnement direct permet de prouver que si $A(t)$ est mesurable, $A(t)$ est continu pour $t > 0$), le lemme I est correct. Des raisonnements directs permettent de même de démontrer le théorème II. En tout cas il faut changer assez radicalement les démonstrations. Je vous enverrai la démonstration complète dès que j'aurai quelques heures libres dans lesquelles je peux m'isoler et la rédiger. Si l'on considère $A(t + s)$ dans un intervalle $(-t, +t)$ B.v. Sz. Nagy m'a signalé qu'il a démontré le théorème I dans les Math. Ann. 1936 (147). J'espère que sa démonstration est plus complète que la mienne.

Avec mes remerciements

W. Doeblin

III

Fin d'une lettre datée du 16/4/39.

...je me proposais depuis octobre de rédiger le dimanche prochain la démonstration ci-dessus (148), mais pour des raisons multiples je l'ai toujours ajourné. Je ne m'occupe pour le moment qu'assez peu de mathématiques, le service militaire ne me laissant guère de loisirs. J'ai repris mes recherches sur l'ensemble de puissances d'une loi de probabilité que j'espère terminer avant octobre. Je vous enverrai quelques tirages à part ces jours-ci.

Bien cordialement

W. Doeblin

2.3. Correspondence between W. Doeblin and P. Lévy

The following letters were deposited by C. Doblin at the Schiller-National Museum / Deutschen Literaturarchivs, Marbach.

I

Draft of a letter from W. Doeblin to P. Lévy (September 1938). (149)

Monsieur le Professeur

Je vous remercie pour votre lettre. J'ai inscrit les deux questions (150) que vous m'avez signalées dans mon carnet de recherche ce qui me fera y penser de temps en temps. Elles me tentent effectivement parce qu'elles sont d'un genre tout à fait différent de mes recherches habituelles et j'aurais bien envie de vous suivre dans cette direction "last not least" parce que j'y rencontrerai Raikoff (151). Toutefois je ne crois pas que j'aurai l'occasion de les aborder assez longtemps, je serai incorporé le 19 ou 20 octobre, j'ai encore des choses à rédiger et surtout je suis engagé dans des recherches sur l'équation de Chapman que je voudrais d'abord finir provisoirement (elle m'occupera peut être toute une vie). Je ne sais d'ailleurs pas si j'aurai la possibilité de poursuivre mes recherches pendant mon service. C'est pour cela si vous tenez à ce que ces questions soient abordées rapidement il sera peut être préférable que je les refile à Loève qui est à la recherche de sujets ou à Raikoff avec lequel je désire de toute façon entrer en correspondance. Je vais aller demain lire votre note. La situation politique n'est évidemment pas très rassurante, je crois donc utile de publier les compléments à mon mémoire de Bull. Sci. Math. (152) qui paraîtra peut être avant 1940 mais certainement pas avant 6 mois dans le cas de plusieurs dimensions (153), je pourrai alors attendre tranquillement la publication du mémoire.

Par ailleurs vu le fait que, cette mesure de précaution étant prise, je ne gagne rien à m'énerver, je ne m'inquiète pas et me borne à restreindre ma consommation de journaux, informations radiophoniques - (les conversations ce n'est pas la peine, mes parents n'étant pas là, j'en ai eu une seule depuis un mois). Hostinský a trouvé préférable de remplacer la loi de Gauss enroulée par loi pliée dans son dernier mémoire. (154)

II

Letter from P. Lévy to W.D.   (155)

(Copie pour M. W. Doeblin d'une lettre addressée à Composito Mathematica).

Paris le 7 juin 1939

Monsieur

Je vous adresse ci-inclus un mémoire (156) de M. W. Doeblin sur les fractions continues.

M. Doeblin rattache à la théorie générale des probabilités en chaîne des résultats antérieurement connus de la théorie des fractions continues; il les complète en outre sur plusieurs points.

M. Doeblin m'a remis ce travail il y a un an; probablement à la fin de 1937. C'est par suite d'un malentendu, dont je suis sans doute le principal responsable, que je ne vous l'ai pas envoyé plus tôt.

Je vous serais très obligé de publier, en même temps que le travail de M. Doeblin, la phrase soulignée de la présente lettre, afin que la date au moins approchée de son travail soit connue.

P. Lévy

Comme vous le voyez, j'ai retrouvé votre mémoire, et le "posterai" tout à l'heure en même temps que cette lettre. Je cherche à m'expliquer. Sans doute m'avez vous demandé si je pouvais transmettre votre mémoire; j'en ai un vague souvenir. Mais vous n'avez dû me remettre votre travail qu'un ou deux mois après, et je n'aurais pas rapproché les deux choses. Je n'ai peut être pas beaucoup d'ordre, mais je n'aurais pas oublié une chose comme celle là si j'avais compris que j'avais entre les mains un mémoire dont vous comptiez que je le transmettrai à Composito Mathematica.

Je n'en suis pas moins confus. Je puis au moins vous certifier que je n'avais pas de mauvaise intention. Je puis au contraire vous raconter que mon mémoire sur la fonction de Green (Acta. Math. 1917) est un mémoire de 1914 qui n'aurait jamais vu le jour sans une réclamation que j'ai adressée au début de 1917 et on m'ôtera difficilement de l'idée que Fredholm et Mittag Lefler, qui se sont rejeté la faute l'un sur l'autre, n'aient pas - ou qu'au moins l'un d'eux - n'ait pas eu l'idée de servir la cause allemande en faisant disparaître mon mémoire. Si j'avais été tué, on n'en aurait rien su.

Bien cordialement et rappelez vous pour une autre fois qu'il faut prendre des précautions contre mes distractions.

P. Lévy

3. Notes

For W. Doeblin's works we adopt the reference conventions proposed by T. Lindvall in [92]: {i} refers to Doeblin's memoir i and {CRi} to Doeblin's "Note aux Comptes Rendus" i as reproduced at the end of the bibliography.

(1) For Wolfgang Doeblin's life in Berlin between 1915 and 1933, see [69], [72], [99], [104].

(2) W. Doeblin applied on the 19th of May, 1933, to the University of Zürich (n° 38241) for the following three courses: Practical Physics , Experimental Physics (Professor Meyer), Algebra (Professor Speiser). On the 19th of July,1933, he received from the University an "Abgangs-Zeugnis". W.D.'s notes on Speiser's Algebraic course are in Marbach (box 2). It is an elementary but rather intensive course from linear equations to Galois's theory.

Andreas Speiser (1885-1970) defended in 1909 a thesis on algebraic number theory in Göttingen University. He was professor at Zürich University from 1917 to 1944. He is well known for his works in history and philosophy of mathematics. Edgar Meyer (1879-1960) was professor of physics at Zürich University from 1916 to 1949.

W.D. does not seem to have been involved in algebraic number theory. Apparently he was on good terms with Marc Krasner, a Chevalley student at the Henri Poncaré Institute, who regularly attended with him the "Séminaire Hadamard" at the Collège de France, (e.g. see [100] p. 115).

(3) W.D. made strong efforts to improve his style. In his personal papers one can find quite a number of lecture notes, work time table planning and even copies of classical texts which he seems to have memorized. For example in preparing "mathématiques générales" he planned over one month to read [59] and work out all the exercises. [59] is the only book he seems to have possessed apart from a logarithmic table[11]. However, he had access to the Sorbonne's library and could borrow the classical treatises.

During his secondary studies in the "Königstädtisches Reform Realgymnasium" in Berlin, W.D. was generally top of his class, but was never regarded as exceptional, even in mathematics, where he received comments such as "good" or "satisfactory" but never "very good" (he was probably too good for that[12]).

(4) In Paris, W.D. followed a course on phonetics and French diction with Father Marcel Jousse, S.J.

(5) We can find in the library of the Institute of Statistics of the University of Paris, a copy of Doeblin's thesis which includes the following inscription:

"A Monsieur Georges Darmois

Avec mes remerciements sincères pour m'avoir donné des sujetsde

---

[11]Marbach box 1
[12]Marbach box 3

thèse trop difficiles me forçant à me familiariser avec les méthodes modernes du Calcul des Probabilités.

W. Doeblin"

The list of the books borrowed by W.D. at the Henri Poincaré Institute library gives very little information about his reading. He used to work at this library and did not bring home many books.

The register of the lending department of the IHP library indicates that between 1935 and 1938 W.D. borrowed the following:

July, 1935: Bliss, *Algebraic Functions*.

October, 1935: P. Lévy, *Calcul des Probabilités*.

March, 1936: H. Poincaré, *Dernières pensées*.

April, 1936: Bouasse, *Cristallographie*.

May, 1936: Borel, *Mécanique statistique*.

November, 1936: P. Lévy, *Analyse fonctionnelle*.

June, 1937: Eddington, *Nouveaux sentiers de la science*.

(6) National Archives AJ/16/5510. His marks were 8 and 16 for the written examination, 10 for the practical tests, 15 and 10 for the oral examination. So he succeeded without distinction (59 out of 100, the distinction "assez bien" required 60 out of 100). In a letter dated 30-7-1938, addressed to Dr Sigmund Pollag, a friend of his father in Zürich, W.D. wrote, "Je n'ai jamais été si fatigué depuis 3 ans, avant de venir en Suisse[13] j'ai passé en juin un examen de physique. Il m'a fallu le préparer en trois mois[14] et je ne savais absolument rien de physique avant, c'est le plus dur effort que j'ai dû fournir jusqu'ici".[15]

In order to undertake his thesis, W.D. needed only to pass Rational Mechanics, Differential and Integral Calculus and Statistics (licence de doctorat), but if he wanted to be elected professor in a lycée or at the University, then he had to pass General Physics (licence d'enseignement).

(7) One can find in Marbach[16] a copy of Drach's exam of June, 1935, with a few notes from Drach's course and a résumé of a Nöther's paper[17].

(8) See Part C of Appendices above.

(9) The first draft of the memoir was accepted by the Annales de l'ENS in March, 1938. W.D. was very anxious about his memoir, so he sent his draft to Sigmund Pollag in Zürich, just before he joined the French Army, with this note: "Dans le cas où je ne serais pas en mesure de réclamer ces démonstrations, prière de les donner à G. Pólya, professeur E.T.H." W.D. had met Pólya in Geneva in October, 1937[18].

(10) A first, very short, draft of this memoir, called "Sur l'équation de Chapman. Solution de Pospišil" and dated 23-1-1938, is in Marbach (box 1). The complete draft of {10} was sent by W.D. to Hostinský on the 27th of August, 1938, for publication in Časopis and sent back by Hostinský, according to W.D., with "une lettre insolente sans argument sérieux" on the 20th of September. Then W.D. sent his manuscript to Cramér and

---

[13]Where he was hiking, alone

[14]Just after he attended his thesis, the 26th of March 1938

[15]Marbach box 3

[16]Marbach box 1

[17]Math. Ann., 37

[18]Marbach box 1

Feller in Stockholm who accepted it for publication. The proofs of {10} were sent by Feller to W.D. in October, 1939, and returned quickly to Feller's address in Stockholm[19]. Curiously, Feller in [42] p. 492 note (10) wrote that he did not know the paper {10} in Febuary, 1940, when he presented his paper. Curiously also, Hostinský, after the war, in a letter to Fréchet dated the 4th of August, 1949, wrote, "j'admire les travaux de Doeblin, surtout le travail dans Skand. Akt. Tidsk..." In fact it seems that very few people had really understood or even read Doeblin's works on the Chapman equation before the war.

(11) W.D. gave a pension to his mother of $2,250^F$ between 15-1-37 and 31-5-37, and $2,700^F$ for 5 months and one month of holidays[20]:

In a letter to S. Pollag, undated but certainly of October 1938[21], with which he sent all his "brouillons" (rough work) just before he was enrolled in the Army, W.D. writes, "Mein Vater verdient etwa 12,000 französiche frs im Jahr, diese Summe wird eher abnehmen als zunehmen, er kann sich also kaum als einseher erhalten. Die Reserven werden, wie ich aus häuslichen Sienen weiss, bei Streckung noch höchstens zwei Jahre danern".

W.D. concludes that after his military service he will have to maintain his parents and his younger brother, Steve. He thinks he could expect $18,000^F$ from a research grant and $6,000^F$ for giving lectures. He explains he has to work very hard if he wants to have a career in a French University because it will be much more difficult for him, as a naturalized citizen. Thus money was a real problem for the Doeblin family.

(12) The first talks were about the axioms of the probability calculus. W.D. spoke about Kolmogorov's axioms and Tornier's axioms. Tornier's book [135] was praised by Lévy who considered Tornier's conceptions of the fundamental principles as "exactement les siennes". (See "Comptes Rendus" of Tornier's book by Lévy in Bull. Sci. Math., 1936, see also Feller [41].) In Spring 1938, W.D. lectured on Markov Chains and the Chapman equation (interesting Résumés of these talks are in box 1 of W.D.'s papers in Marbach).

(13) In the Marbach Archives, one can find at least two plans of unwritten papers, one about Lindeberg's method and one about Volterra's theory of mathematical biology.

(14) As early as 1935 Wolfgang Doeblin joined the "Ligue Française pour les Auberges de Jeunesse", founded in 1930 by Marc Sangnier (1873-1950) (see [33]). We can find in the Marbach Archives a draft of a letter intended for one of his friends, an "ajiste", which indicates fairly well his state of mind when he was at Givet.

W. Doeblin to an unknown person, the 22/2/1939.

"Mon cher ami

Aujourd'hui pour la première fois depuis mon arrivée au régiment j'ai un peu de temps libre avant 5 heures, j'en profite pour t'écrire.

J'espère que le Club réorganisé marchera mieux qu'avant, je n'y crois pas beaucoup, personnellement je suis de l'avis suivant: tant que le club

---

[19]Marbach box 2
[20]Marbach box 3
[21]Marbach box 3

s'occupe presqu'uniquement de sorties théatrales ou de sorties le samedi-dimanche agrémentées de palabres sur la vie du Club il est impossible de faire quoique ce soit de ce Club. La formule des Clubs de province (au moins des Clubs du Jura) me paraît plus heureuse: s'occuper d'auberges, améliorer les auberges existantes, en créer d'autres, faire des itinéraires pour les usagers des autres régions, voilà un véritable travail ajiste. J'ai vu trop d'<ajistes> dans les différents Clubs <unreadable> et qui font des vacances sur le tas ou même qui passent les vacances dans un hôtel. Il y a aussi le problème des campeurs dans les Clubs. Au fond tu connais bien mon opinion sur toutes ces questions, mais ne crois tu pas aussi que la meilleure façon d'éliminer les indésirables c'est de forcer tous les membres du Club à consacrer deux dimanches par an à l'amélioration ou à la création d'auberges, à peindre les murs, à balayer une A.J., à gratter les tables et à les cirer (comme on le fait ici le samedi)?

Je suis soldat (élève-caporal) dans l'infanterie, laquelle arme était évidemment bien indiquée pour un mathématicien, au 91ème R.I. 11ème Cie à Givet dans les Ardennes (il y a une A.J. à Givet, elle est fermée maintenant). Les environs de Givet sont très jolis ce qui est appréciable. Matériellement on n'est pas bien ici, même par rapport aux autres compagnies installées à Givet. Quant à la discipline, je ne peux pas comparer, en tout cas depuis qu'on est ici on n'a pas encore eu quartier libre, même à Noël, il paraît qu'on pourra compter les jours de quartier libre sur les cinq doigts de la main gauche.

Pour les permissions c'est la même chose: les anciens partent maintenant en permission ils n'y ont pas été depuis 10 mois. Le colonel avait offert dès la démobilisation d'envoyer des hommes en renfort à Givet pour que les anciens puissent partir en permission, le commandant a refusé, pendant quelques semaines en octobre le bataillon avait à fournir journellement 104 soldats, de tout point de vue, alors les supressions de permissions pleuvent. Les premiers mois je souffrais beaucoup, maintenant je suis arrivé à un degré de non-foutisme parfait. J'attends avec calme ma première punition. Il est certain que je serai bien entraîné pour la marche lorsque je sortirai d'ici, il paraît qu'on aura en juin une marche d'épreuve de 54 km avec chargement complet.

J'ai amené une lampe à alcool, alors je me prépare souvent des repas supplémentaires. Je finis cette lettre la nuit, à la lueur d'une lampe à pétrole dans un poste de garde. En été la garde sera plus agréable. Aujourd'hui je suis relativement tranquille, je suis fonctionnaire-caporal de poste. Je me borne à relever les sentinelles toutes les heures, évidemment je ne pense pas à dormir avec cela (de 9^h à 1^h on a repos quand même)".

(15) On an undated post card from W.D. to his parents, which is in Marbach, W.D. writes:

"Mes chers parents

J'ai reçu votre carte de St Germain. J'espère qu'il y a pas trop d'alertes et que cela se borne à une descente à la cave ... <unreadable>

291ème R.I.

Ordre de régiment n°20 du 19 mai 1940

Citation à l'ordre du régiment n°25.

Le soldat Doblin Wolfgang téléphoniste de la C.A. B3, soldat brave et dévoué, toujours volontaire pour occuper les postes les plus exposés, a assuré la liaison de jour et de nuit à l'intérieur du bataillon en réparant les lignes sous le feu de l'artillerie ennemie.

Cette citation comprend l'attribution de la croix de guerre.

La prochaine fois que je viendrai en permission j'aurai droit à 13 jours (la croix de guerre est la décoration la plus basse, au dessus il y a la croix de guerre avec palme, la médaille militaire, la légion d'honneur) en tout cas il n'y a pas beaucoup de gens ici qui ont la croix de guerre sans être blessés et je n'ai pas une égratignure.

A l'heure actuelle je suis encore assez fatigué. Ces damnées coliques y sont certainement pour quelque chose. En tout cas maintenant je me soigne. Je pourrai maintenant vous écrire régulièrement. J'aimerais avoir l'adresse de Pierre".

(16) W.D. showed considerable bravery in these fights, as is attested by the 1946 "Extrait de l'ordre général" below, which accompanied W.D.'s posthumous "Croix de Guerre avec Palme". It has not been possible to obtain more precise details about the facts reported below. "Beneng" may be Bening hamlet which is situated in the Commune of Viller, in Moselle, occupied by the Germans on the 16th of June [123]; it may also be the Bening farm just south of Bertring-Grostenquin, where the 291st I.R. was stationed between the 21st of May and the 15th of June.

"EXTRAIT DE L'ORDRE GENERAL N° 1912 C
Le Ministre des Armées Edmond Michelet cite
A l'ordre de l'Armée
DOBLIN Wolgang dit Vincent - soldat au 291ième R.I.

A combattu les 16, 17 et 18 juin avec un réel mépris du danger.

Le 16 juin à Beneng, il est resté seul, armé d'un fusil mitrailleur pour couvrir la retraite d'un groupe de soldats. Ne s'est replié qu'après avoir retardé le débouché des motocyclistes ennemis du village et permis au groupe de s'installer sur une nouvelle position.

A trouvé la mort le 21 juin 1940 à Housseras (Vosges)".

La présente citation comporte l'attribution de la croix de guerre avec palme.

A Paris le 21 novembre 1946".

This citation, probably written by captain Renard, was proposed by colonel Berck, staff-officer of the 291st I.R. in 1940[22].

W.D. was decorated with the "Médaille Militaire" by a decree dated the 11th of February, 1948.

(17) Father François Renard (1901-1985), captain of the C.A. 3, Doeblin's Company, had been, before the war, head of the minus-seminary St Léger in Soissons (a religious highschool for future priests). Marie-Antoinette Tonnelat tried to contact him in 1941 to get news of Wolfgang, in vain as F. Renard was a prisoner of war in Germany until 1945. As soon as he came back, he sent a letter[23] to her in which he wrote, "Je tiens à

---

[22]Marbach box 3
[23]Marbach box 3

vous témoigner toute l'admiration que j'ai eue pour sa conduite, dont j'ai été à plusieurs reprises le témoin, soit au cours de notre séjour aux avant-postes en mai-juin 40, soit pendant les 8 derniers jours de combat en retraite qui ont précédé notre captivité. Il a fait preuve d'un mépris de la mort, d'une conscience professionnelle, d'un dévouement et d'un sang froid incomparables, soit dans sa tâche obscure et ingrate de téléphoniste, soit en combattant avec acharnement jusqu'à la dernière heure. C'est un des soldats les plus courageux que j'ai eu sous mes ordres".

In November 1945, Father Renard wrote to Erna Doeblin, Wolfgang's mother, another letter in which he tried to explain W.D.'s disappearance: "Très surpris de cette disparition, mes hommes et moi (qui nous étions promis de ne pas nous séparer et qui faits prisonniers le 22, nous évadions tous ensemble le 23 au soir, malheureusement pour être trahis et repris dans le sud de l'Alsace le 30), nous avons alors supposé que Vincent, voulant à tout prix échapper à la captivité, avait tenté sa chance seul et avait essayé de franchir les lignes pour pouvoir combattre encore en France ou hors de France, ou bien qu'il était allé au devant de la mort, préférant disparaître qu'être fait prisonnier".

He speaks also of W.D.'s character during those days, "Je le connaissais peu avant notre montée en ligne. Là, c'est à dire au mois de mai, j'ai pu apprécier son courage, son dévouement infatigable, son sens du devoir, mais sans avoir avec lui de conversation personnelle et sérieuse. Du 15 au 20, il a combattu à côté de moi, avec un petit groupe d'hommes. Là encore, il a été un exemple continuel de bravoure et d'entrain. Mais pendant ces cinq jours, je n'ai jamais eu non plus l'occasion de le voir seul pour parler avec lui. Il ne semblait pas en avoir d'ailleurs le désir et je respectais trop la personnalité et l'indépendance des hommes sous mes ordres pour provoquer moi-même ces genres de conversations".

Father Renard was certainly a very good man. After the war, in poor health, he was vicar of a little village, Clastres, for almost forty years. For further details about him see, "Vie Diocésaine", Diocèse de Soissons, n°14, 15 juillet 1985.

(18) The identification or W.D.'s body in Housseras was made possible because of the efforts of his friend Marie-Antoinette Tonnelat. After writing to various official services in vain, she placed an advertisement in a parisian newspaper (Paris-Soir, 21 mars 1941, last page) asking for information about the third Battalion of the 291st R.I. and then, during several months, wrote to each address she obtained. The task was extremely difficult as all of the direct witnesses were either dead or prisoners in Germany.

To look for W.D. among the thousands of soldiers in Lorraine was a desperate and nearly impossible task, all the more because Marie-Antoinette Tonnelat could not reveal the real name of the man she was looking for, Wolfgang Doeblin, a German Jewish name, and Alfred Doeblin's son. So she said she was looking for her brother. When the Communications with Lorraine, then a German territory, were made possible, she managed to place another advertisement in the "Echo de Nancy" (1-6-1941) inquiring about unidentified bodies buried in the surrounding villages. Despite all the obstacles, she finally succeeded.

In a letter to Erna Döblin after the war, she wrote, "Quand j'ai commencé ces recherches je les ai faites parce que je ne pouvais plus rester dans cette incertitude et aussi dans l'espoir que, le cas échéant, il serait possible de faire quelque chose pour Wolf.

"Quelque fois il me semble aussi que la valeur que représentait Wolf, son courage, sa bonté, ne peuvent être complètement perdus et que tout cela est conservé et nous précède quelque part".

Let us note that Marie-Antoinette Tonnelat tried with Fréchet to protect Loève, during the Occupation, even when he was arrested in 1944, as a Jew, in Vichy and then in the Drancy camp near Paris (see Fréchet correspondence with Marie-Antoinette Tonnelat in the Archives of the Academy of Sciences).

(19) Box Fréchet 8, deposited at the Academy of Sciences Archives.

(20) This is a first draft of chapter 1 of W.D.'s thesis, a part of which is reproduced in the paper {8} published separately and summarized in the introduction of {5}, see {8} p. 2.

(21) {5} ch.I §6 p. 32-36 of {5}.

(22) It refers to {CR2} p.25.

(23) No doubt chapter 9 {5}.

(24) Box Fréchet 8.

(25) This also concerns {CR2} p.25. In Fréchet's Archives of the " Laboratoire de Probabilités" of the Paris 6 University, we can find a copy of that note corrected by W.D. himself. The errata of {CR2} have been published in the "Comptes Rendus de l'Académie des Sciences", 203, 1936, p.252 and 592.

(26) Box Fréchet 8, undated letter, probably written in September 1936. Fréchet did not generally return to Paris before October, and this can explain why most of W.D.'s letters between 1936 and 1938 are dated as summer or winter holidays (August, September or December). Fréchet had at his disposal an office at the IHP (Institut Henri Poincaré) and W.D. could meet him there throughout the University year, but only by appointment.

(27) This article [53] has been summarized in a "Note aux Comptes Rendus", vol.195, 1932, p. 590 and has been presented in full by Fréchet in his course at the IHP in 1933 on chain events.

(28) Chapter II of [53] which gives the proofs of the results announced by Hadamard at the Bologne Congress in 1928 [62] without demonstration and which W.D. proves again in his thesis Ch.II §8 p. 85-86 under less restrictive hypotheses than those given by Fréchet and Hadamard.

(29) This concerns §5 of {8} p.14 which was written in June-July 1936 {8} p.2. The $l(\alpha)$ are the cyclic subgroups of final groups. See also {5} p. 5 and {8} p.23.

(30) Box 8, the date appears on the letter.

(31) {5} p.64 has been announced in {CR3} p. 1211 in December 1936 and obtained "depuis longtemps" by Doeblin. About this condition see [31], [146]; see also {3} and [60], [61].

(32) Fortet [46] p.185 which refers back to Hadamard.

(33) Kolmogorov [83]. In box 17 of Fréchet's Archives can be found a Kolmogorov letter to Fréchet dated July the 30th (1937) in relation to this point:

Cher Monsieur Fréchet

"Il me semble que M. Doeblin doit publier des recherches sur les chaînes de Markoff indépendamment, comme il les a inventées. Ma note paraîtra dans le recueil Mathématique n°3 de l'année courante. Sur la réunion des probabilités à Moscou j'espère pouvoir parler définitivement pas plus tard que le 10 septembre, comme je pars après le 10 septembre au Caucase et en notre Asie centrale (Samarkand et dans les montagnes).

Mes hommages à votre épouse et à votre fille. Je regrette beaucoup de ne pouvoir venir à Léningrad pour vous voir".

Mr and Mrs Fréchet must have been in Leningrad where their daughter Hélène lived. The latter had married the biologist Edgar Lederer (1908-1988), head of a laboratory in the University of Leningrad from 1935 to 1937. On that point see the "Cahiers pour l'histoire du CNRS", 1989 vol 2. Doeblin's remarks on the countable case in {8} p.12-13 show better than does this letter how he recognized Kolmogorov's original contribution which, from his point of view, had "brillamment" solved the only "difficulté réelle" of the countable case.

Doeblin has reproduced the argument of this letter in his thesis {5} p.92-93.

(34) [53] see {5} p.64-65. The generalisation of the "cas de M. Fréchet" is one of the aims of paper {12}, which would only be written in 1938, cf {12} p.63-64. W.D. had already planned it in October 1936, and he would work on it again after the two notes [87] and [88] of Kryloff and Bogoliouboff and {CR4}.

(35) Box 8, the date appears on the letter.

(36) The Marquise Marie-Louise Arconati-Visconti (1840-1923), was the hostess of a quite brilliant "salon" in her parisian "hôtel". She had established the University of Paris as her sole legatee. One part of her fortune funded research fellowships. Her funds complemented Commercy's donation made for the same use in 1907, but which had been badly devalued after the franc's fall. Let us recall that the National Center of Scientific Research would only be created in 1939. Fortet and Ville were also holders of an Arconati-Visconti fellowship as well as Marie-Antoinette Baudot.

In box 8 of Fréchet's Archives can be found a note of December the 23rd, 1936, emanating from the Faculty of Sciences of Paris Secretariat which specifies that the first term grant would be at W.D.'s disposal on the 15th of January at the entrance gate n°1 of the Sorbonne's Faculty of Sciences Accountancy Department.

The "Alliance Israélite Universelle", whose center is in Paris, was founded in 1860 by J. Carvallo, N. Leven, Ch. Netter, I. Cahen, E. Manuel, and A. Astruc, in order to contribute to the "emancipation and the promotion of Jews living in countries where misery and discrimination prevail". Supported by legacies and donations, it opened institutions of French culture and Jewish tradition in all Mediterranean countries; it also gave scholarships and W.D. obtained one (see his letter V to Fréchet and [10]). In regard to Alfred Doeblin's relations with the Parisian Jewish circles in the thirties see [60]. When he arrived in Paris, A.D. joined for a short time the "territorialist" party whose purpose was to establish a Jewish Republic on a new territory, Australia or Abyssinia for example. A.D. left it in 1936 after having decided that this party had weak metaphysical and religious bases.

(37) Box 8; the date, August the 27th, appears on the letter which is obviously of 1937.

(38) This refers to the proofs of the manuscript of Fréchet's book [54]. As was the custom in Paris, Fréchet was helped by his students and "obligés" in the proofing of his book. Fréchet [54] was "advised" by Hostinský, Kolmogorov, von Mises, Onicescu and "helped" by Doeblin and Fortet.

(39) No doubt Fréchet [54] p.207-209.

(40) {5} p. 65 beginning of §2, the correction has been carried out.

(41) In box 24 of Fréchet's Archives, about the printing of Doeblin's thesis; the correspondence exchanged between O. Onicescu and Fréchet from February to April, 1937 is available. We are told that the "frais d'impression des exemplaires obligatoires pour la thèse seront diminués autant que possible" and that W.D.'s first manuscript had been sent for printing on the 14th of April 1937.

In a letter of March 36, Onicescu writes that Fréchet "traite le calcul des probabilités comme une branche de la physique. Cela ne me console pas du fait que vous n'acceptez pas du tout mon point de vue dans la question générale de la dépendance des éléments statistiques et des chaînes", [117].

(42) See next letter; these thoughts later became the paper {6}. The terminology "équation de Chapman" and "équation de Smoluchowski" in the stationary case was introduced in France as early as 1930 by Hostinský [63] who gives all the references.

(43) Box 8, Fréchet had written on the letter its date of arrival, 7th of September (1937).

(44) [87]. N.M. Krylov (1879-1955, Dictionary of Scientific Biography), and his student N.N. Bogolyubov, born in 1909, worked together in Kiev. Bogolyubov visited Paris at the beginning of the year 1936 and lectured on chains at the Société Mathématique de France.

(45) {5}, ch IV.

(46) This "observation" of Doeblin has been again taken into consideration in the book [54] p. 250-251.

(47) {5} p. 106-107.

(48) This result is proved in {6}. The paper {6} was written at the end of 1937; it was announced in Fréchet's book [54]. About the "groupes à liaisons instantannées", pointed out by W.D., see [104] and [13].

(49) See {6} p. 5.

(50) The countable case was treated by Doob [29] who, moreover, with the help of a general result of Auerbach, largely simplified W.D.'s proof.

(51) In {5} W.D. does not thank Onicescu but only his "maître M. Fréchet". O. Onicescu (1892-1983), Professor at Bucharest University, was the founder of the Rumanian school of probability, see [74].

(52) Box 8, the date appears on the letter. Doeblin's "fichier" (line 2) is in the Marbach Archives.

(53) This is definitely about Perron's memoir of 1905 which appears in Hostinský's Bibliography [63]; see [54] p. 258.

(54) Krylov and Bogolyubov [87].

(55) cite122, Theorem E p. 250. V.I. Romanovsky (1879-1954) had been A.A. Chuprov's (1874-1926) student in Saint Petersburg. He was Professor in Tachkent University in 1936. The printing of his memoir started on the 17th of October 1935.

In Fréchet's Archives of the "Laboratoire de Probabilités" of the Paris 6 University is a letter of G. Mihoc which indicates to Fréchet that his memoir [114] has been sent to the Bulletin de l'Université de Cernaüti in January 1936 and will be published before the 1st of July 1936.

(56) [114], G. Mihoc (1906-1983) was O. Onicescu's student in Bucharest. For details about him see Iosifescu [73]. The convergence theorem to the Gauss distribution had been proved in a particular case by Markoff as early as 1908; see Bernstein [1] p. 31. The limit distributions problem in the general case of simple chains was still an open problem when W.D. started his work in 1935. About that point see [54] p. 145, and also [118] and [128].

(57) [125], Günter Schulz (1903-1962) had been R. von Mises's student in Berlin before von Mises (1883-1953) left for Istanbul in 1933. In 1936 he was teaching in Berlin and contributed to the "Deutsche Mathematik" journal whose chief editor was L. Bieberbach (1886-1982). In this way Schulz managed to lift Markov chains to the dignity of "Mathématiques allemandes" although most of the founders of the aforementioned theory belonged to the biologically inferior type "S". See also [86] and [14].

(58) Bernstein [1] p. 32-40, Bernstein (1880-1968) was Professor in Kharkov. For details about him see Russian Math. Surveys 24, 3, 1969, p. 169-176.

(59) Thesis {5} ch.I §3 and 4. W.D. has dealt with the countable case in {7}; see [75].

(60) Schulz [124]

(61) [52] which Fréchet presented in his lecture at the IHP in 1931-1932; see also [120] and {5} p. 16.

(62) J. Ville born in 1910, ENS in 1929, Fréchet's student, holder of an Asconati-Visconti scholarship. He worked in Vienna with Karl Menger and defended his thesis [137] in 1939; in it, he particularly applied martingale theory to the study of continuous Markov processes. After the war he worked in private industry and was appointed Professor of Econometry at the Paris Faculty of Sciences in 1958.

(63) E. Feldheim, is born in Kassa (Hungary) in 1912. As he was not admitted to the University in Budapest he studied mathematics in Paris from 1931 to 1934 and defended a University thesis in February, 1937, under G. Darmois's supervision. Back in Budapest in October, 1934, he tried in vain to integrate University while still researching in the probability calculus. In a letter dated January the 21st, 1935, addressed to P. Lévy (Fréchet Archives, Laboratoire de Probabilités), Feldheim explained that he was studying the conjecture formulated by Lévy in November, 1934, on the Gaussian distribution [93] and that he had come to the conclusion that it could be untrue. He added a note to his letter, "sur la décompostion des lois de probabilités", where he explained his reasons (not very convincing we must say, but not without interest and which will be used in [37]) and a second note extending a result of Khintchine: "Une loi limite du

calcul des probabilités" which would be published in Szeged in 1936 [34].
P. Lévy seems to have been interested in E. Feldheim, to whom he gave a
part of the proofs of his book [96], and whom he quoted in [96], [98] and
[101] p. 26 where he remarks on the "joli travail" [37] of Feldheim about
[91]. E. Feldheim published a lot between 1936 and 1942, especially on
orthogonal functions, "which he handled with virtuosity" (see P. Turán,
Mat. Lapok 25, 3-4, (1974) 262-263, and e.g. [38]).

Let us recall that H. Cramér (1893-1985) is the one who early in 1936
proved Lévy's conjecture, [15], and this decided the latter to write his
book [96] which would be finished during the summer of 1936, see [102]
p. 111-112.

We can imagine that W.D. had also studied Lévy's conjecture but did
not succeed in obtaining a direct demonstration.

At the begining of the summer of 1942, Ervin Feldheim was sent as a
worker to the Russian front, to the Fastow camp near Kiev. He remained
there up to the end of 1943, and was then repatriated to Hungary, seem-
ingly in very bad health [139]. Deported in May 1944 in a camp near
Bor (Serbia), he was probably killed in the massacre of Červenka (in the
north of Voivodina) with 3,400 other prisoners coming from Bor in the
direction of Germany, [139].

(64) R. Fortet was born on the 1st of May 1912, ENS in 1931, holder of an
Asconati-Visconti and later on of a CNS scholarship, student of Fréchet,
he defended his thesis [39] in May 1939. He succeeded G. Darmois in the
Probability Calculus and Mathematical Physics Sorbonne Chair in 1960,
held by Borel from 1919 to 1940 and by Fréchet from 1940 to 1949.

(65) This is about {8} which will only appear in 1938 and which W.D. had
written as preliminary to his thesis in June-July 1936.

(66) The note {CR5} where W.D. announces his general theory of chains,
presented at the Academy in July 1937.

(67) Hostinský [63] who gives a very complete bibliography of the works on
Markoff chains before 1930.

(68) {5} Ch.I §3 where W.D. makes use of Bernstein's methods [1]. See the
letter of the 19th of September.

(69) The first note of W.D. on this subject was only published at the end of
1938 ({CR9}, {CR10}).

(70) Box 8, undated letter between the 10th and the 19th of September.

(71) [54] p. 156 where Fréchet rectifies his text and cites Doeblin {8}. See
next letter.

(72) Ville wrote down in 1937 Borel's lectures on games [5] and we know that
Borel's lectures were particularly difficult to write down. At the end of
[5] Ville gives the first elementary proof of von Neumann's minimax theorem which Borel had vainly searched for in the begining of the twenties;
see Fréchet [57] for the history. Fréchet wanted to establish Borel's prece-
dence over von Neumann on games theory. Indeed Borel had approached
the study of games of null sums as early as 1921 but admits himself in
a letter to Fréchet of the 8th of May, 1952[24], "Je tiens à vous signaler
que j'ai bien étudié la même question que von Neumann avant lui, mais

---

[24]Fréchet box 8

que je n'ai pas démontré, ni même soupçonné le théorème fondamental auquel il est arrivé. Je vous demande donc d'examiner les choses de très près avant de donner suite à votre projet. Si je ne me trompe c'est René de Possel qui a le premier exposé en français les résultats de Neumann".

(73) A Colloquium on Probability Calculus was organized in Geneva from the 11th to the 16th of October 1937 by the Faculty of Sciences of Geneva. It was presided over by Fréchet. The proceedings of this Colloquium have been published in Paris by Hermann in 1938, Industrial and Scientific Actualities, n° 734 to 740. Doeblin was present at that Colloquium and he often contributed to the discussions, B. de Finetti (1906-1985) reports on that in [45] p. 25, 27, 36, 42, 54, 58. It was W.D. who wrote Glivenko's, Romanovsky's and Slutsky's summarized conferences, as they were not present in Geneva. The invited lecturers on Chain Probabilities and the Ergodic Principle were E. Hopf, B. Hostinský, O. Onicescu and V. Romanovsky. H. Steinhaus, G. Pólya, W. Feller, A. Wald, H. Cramér were also present in Geneva and W.D. met them on that occasion; see [45]. It was at this Colloquium that S. Bernstein's work on Stochastic Differential Equations was presented, work which motivated considerable research from 1938 on. W.D. has corrected and edited Bernstein's text as we are informed in the letter XII.

(74) A. Wintner (1903-1958, see the Dictionary of Scientific Biography), author of many papers on analysis in the thirties and the forties, deserves better. W.D. probably reproached him for publishing results which he himself considered without interest. In Fréchet's Archives of the "Laboratoire de Probabilités", can be found a reprint of [143] with the following summary written by W.D. for Fréchet: "X_n taking its values -1 and +1 with probability 1/2, Wintner considers the series $\sum_{n=1}^{\infty} a_n X_n$ where $\sum_{n=1}^{\infty} a_n^2 < \infty$. Remarks which seem useless". We can also find a reprint of [144] with a short note of W.D., "the theorem is obvious and well known". This note is written on the back of a draft of the paper {12}. See also [102] p. 116.

(75) [54], supplement to the appendix A, p. 271-273.

(76) {6} I §4. This proof of Doeblin contains an error as Doob advised him; see {6} correction p. 35 and the correspondence between W.D. and J.L. Doob included in the Appendix 2-2.

(77) Lebesgue in 1907 and independently Fréchet in 1913; see {6} p. 25 note (1) where W.D. gives a proof; see also the following letter.

(78) W.D. in chapter 4 of {5} does not assume continuity but the "hypothèses I et II" see {5} p. 105.

(79) {5} p. 20.

(80) {5} p. 21-2.

(81) {5} p. 55.

(82) Bernstein [1] p. 21, "lemme fondamental".

(83) Box Fréchet 11.

(84) [54] p. 22.

(85) See note (76), Fréchet must have communicated to Doeblin the reference [50]. During his entire life Fréchet had been a militant for the international Esperanto language. Putting into practice his convictions he had published some of his works in Esperanto, in particular this very one. We

might note that Rudolf Carnap (1891-1970) and Fréchet corresponded in Esperanto[25].

M. Fréchet had been a French-English interpreter during the First World War but had a limited knowledge of the German and Russian languages.

(86) In the case of finite chains, this result appears in {6} p. 28; see also [29] p. 38 which gives a counter example in the countable case.

(87) [54] p. 159-160. This is about the a.s. convergence of the frequencies $f_{hk}^{(n)}$, of transition from state E_h to state E_k in n steps. Fréchet proved that, at least in the regular case, the $f_{hk}^{(n)}$ were convergent in the mean. Kolmogorov, to whom Fréchet had sent his proofs, had indicated to him that the a.s. convergence in the general case resulted from the "theorem of Birkhoff-Khintchine". W.D. gives a direct proof here. Fréchet became interested in Ergodic Theory in 1940-1941; see [55].

(88) {5} §11 p. 60-61.

(89) [54] p. 305-306, "principe ergodique" eventually was placed under e and under p.

(90) [54] p. 307-308.

(91) {5} p. 16 note 4.

(92) See note (82). {5} p. 20-21 where, as a precaution, W.D. has put the fundamental lemma in quotation marks.

(93) National Archives AJ/16/5551.

(94) The Congress was held in Zürich. In fact Bernstein [2] writes: "Grâce aux développements qu'ont reçus les idées de Markoff dans les travaux récents de MM. Romanovsky, Hostinský, Fréchet et von Mises, la théorie des chaînes discrètes de Markoff est devenue un des chapitres les plus parfaits du calcul des probabilités, susceptibles de nombreuses généralisations et de diverses applications".

(95) P. Lévy [95].

(96) This letter is deposited in the Marbach Archives, ref.: D. Döblin / C.D. Wolfgang Döblin.

(97) L. Freymann was head of the Hermann publishing house. Inventive, eclectic and enterprising, he had founded the "Actualités Scientifiques et Industrielles" (ASI) collection.

(98) R. Wavre, 1896-1949, Professor at the University of Geneva, president of the Geneva Colloquium Committee organisation. For his works see [141].

(99) M. Loève, after having specialized in Theoretical Physics and then in Actuarial Statistics, defended in 1941 a thesis in Probability under M. Fréchet. Doeblin met Loève at the IHP library as early as 1935.

(100) [103].

(101) Box Fréchet 11. Fréchet has added on the letter: "répondu le 1er novembre". With no news from W.D. since the declaration of war, Fréchet had written to his mother asking for his new posting. Erna Döblin communicated to him W.D.'s new adress: 291st I.R., 3rd Battalion staff headquarters Sector 29, and added[26], "Nous avions assez souvent de ses nouvelles, maintenant il y a 9 jours que je n'ai rien entendu et je

---

[25]Box Fréchet 5

[26]box Fréchet 8

m'inquiète un peu. Il y a 3 semaines que je lui envoyais des épreuves qui venaient d'Amsterdam, il les a corrigées, il m'écrivait qu'il y avait peine à expédier il doutait si elles arriveraient. Il y a une semaine que je lui faisais suivre les épreuves de Stockholm ... "

We may suppose that Fréchet had proposed that W.D. works on Defence contracts in which he had been involved at the IHP.

As soon as war was declared all 200 French scientific laboratories had been placed at the CNRSA's (Centre National de la Recherche Scientifique, section de la Recherche Appliquée) disposal to carry out all the research considered necessary by the Defence Ministry Offices. The laboratories of the Faculty of Sciences of Paris formed Group 3 of the CNRSA, headed by Ch. Maurain, dean of the Faculty. IHP was its Section 11, headed by E. Borel. On the 10th of October, Section 11 was divided into 5 laboratories:

Laboratory of Mathematics: Head E. Cartan, Secretary F. Vasilesco, members: Montel, Bouligand, Lagrange, Julia, Hadamard, P. Lévy, H. Cartan.

Laboratory of Calculus: Head H. Mineur, Secretary Guintini, members: Couderc, Miss Canavaggio, Chapelon, and 9 calculators.

Laboratory of Theoretical Physics: Head L. de Broglie, Secretary: Loubet, members: L. Cartan, Reulos, Roubaud-Valette Miss Février, Potier, Miss Chevreux, Marie-Antoinette Tonnelat, 1 calculator.

Laboratory of Ballistics: Head G. Valiron, Secretary: Lelong, members: Lichnerowicz, Hadamard, Esclangon.

Laboratory of Statistics, Head: M. Fréchet, Secretary: Fortet, members: Bouligand, Halphen, P. Lévy, Chapelon, Delaporte, and 3 calculators: Miss Baud, Petit, Wyckaert.

Moreover, in November 1939 the "Laboratory of Mechanical Calculus" was created, whose Head was L. Couffignal who had just defended the first French thesis on Automatic Calculus; also the "Laboratory of Calculus"whose Head was H. Mineur was transformed into the "Laboratory of Astrophysics".

M. Fréchet, head of the Statistics Laboratory, "approached the different services in order for them to put questions to him, but it was generally answered that they had no questions for him but that they would be fairly interested in the questions he would ask them" (secret report of the 31st of October, 1939, of the G.3 S.11 L.5).

From the 10th of October, 1939, to the 1st of May, 1940, the questions asked to the laboratory of statistics or which Fréchet asked himself were the following:

Question 27.A.5: Best use of a small number of shots for adjustment firing, proposed by E. Borel, studied by Fortet.

Question 28.B.5: Tests to decide between the first or the second Laplacian distribution, proposed by Fréchet, studied by Fortet.

Question 30.C.5: Dispersion of salvos, presented by Dupuis, President of the "Commission de Gâvre", studied by Fortet.

Question 52.D.5: Homogenization of patterns, proposed by Divisia (Public Works Ministry), studied by Halphen.

Question 53.E.5: Probability distribution of an apparent correlation, proposed by Dupuis, de Gâvre, studied by Fréchet with the collaboration of Fortet, Halphen, R.A. Fisher.

Question 54.F.5: Shortest distances on railways.

Question 66.G.5: Random motion of a particle in a fluid; proposed by the Technical Artillery Section, studied by Fortet.

Question 75.H.5: Blood groups, proposed by R.A. Fisher, studied by Halphen.

The GR3 S11 archives, saved by J.L. Destouches, secretary of the section, are deposited at the Library of Mathematics of the University of Paris. They contain some interesting details about the questions studied in the 11th section.

We might also point out that the Fréchet Archives of the "Laboratoire de Probabilités" and of the "Institut de Statistiques" of the Paris 6 University contain the correspondence between Fréchet and Fisher, Wilks, E. Pearson, Deming, E.B. Wilson, Steffensen, Gini, E.L. Dodd, when he was director of the Laboratory 5 of the GR3S11.

We should finally recall that during the First World War, E. Borel was the head of the "Direction des Inventions" which interested the National Defence, created in 1915 by P. Painlevé. 25 years later he found himself responsible for similar functions, which not only showed Borel's exceptional qualities and energy, but also the disquieting datedness of the French élites. During the First World War, France had officially, out of 8,410,000 men mobilized, 1,357,800 dead and 4,266,000 wounded; after one month of fighting in 1940 there were soon 120,000 dead, 250,000 wounded and 1,900,000 prisoners out of 5 million soldiers.

(102) Results announced in {CR6}, {CR7} and {CR9}.

(103) The first version of the paper {13}.

(104) In September 1939, in accordance with the articles of the German-Soviet pact, the Soviet troops occupied part of Poland including Lwów, the city where the Studia Mathematica journal was published. The Soviet political police slaughtered the Polish officers but left the Universities open. Banach and Steinhaus (1887-1972) went on editing the journal which appeared normally in 1940, a Ukrainian summary accompanying each article. So the paper {13}, which W.D. had sent to Steinhaus whom he knew, having met him in Geneva, was published in 1940 although Fréchet was not informed, and later published this same paper with complements in the Annales de l'ENS in 1946. S. Banach, dean of the Faculty of Sciences of Lwów from 1939 to 1941, who had a great reputation in Moscow and Kiev, was appointed in 1940 Corresponding Member of the Academy of Kiev. When the German army invaded the Ukraine in June 1941, the University of Lwów was closed, Banach had to work as a louse-feeder in Professor Weigl's Bacteriological Institute, [131], and many Polish mathematicians were murdered, among whom were A. Lomnicki, P.J. Schauder, J. Schreier, and others (see [136]).

(105) Garrison town in the French Ardennes where W.D. was doing his military service in the 91st I.R. from the 3rd of November, 1938, to the 2nd of September, 1939, the date of the declaration of war. When his regiment was dissolved, W.D. was enroled in the 291st I.R., a regiment which was

to be joined to the 52nd Infantry Division, in Lorraine.

(106) Peter Doeblin, born in 1912, was Alfred and Erna Doeblin's eldest son. After a stay in England he emigrated to the United States on 1935, where he married and practised the profession of typographic compositor.

(107) The manuscript arrived safely. It was brought back to Paris in 1945 by W.D.'s parents who deposited it in the Academy of Sciences Archives where it still is today (Doeblin file).

(108) Nobody to the present date has been able to elucidate this manuscript's formulae.

(109) This might be about Kiyosi Iseki whose address is on one of W.D.'s notebooks. M. Fréchet gave the parcel to W.D.'s mother who passed it on to him[27]. Other Japanese students were at the IHP between 1937 and 1939[28] in particular H. Ioi, Okamura and Y. Tanaka.

(110) These are the proofs of {10}.

(111) Box Fréchet 11. The date is on the letter, which is written on headed notepaper of Erna Döblin, and on which Fréchet has added "answered the 15th of November".

(112) {10} see previous letter and note (10) above.

(113) The note {CR9} completed by the note {CR10} probably written at the same time but publication of which must have been delayed, as W.D. had already deposited, in 1938, 4 notes at the "Comptes Rendus". About the importance of {CR9} see for example [98] p. 77, and of {CR10} see [43] p. 30.

(114) Box Fréchet 11, the date is on the letter.

(115) W.D. was a telephone operator at the headquarters of the third Battalion.

(116) Box Fréchet 11. This letter has no date. It surely dates from the beginning of January 1940, as W.D. wishes Fréchet a happy new year. Fréchet must have asked W.D. in November or December 1939 to think about the approximation of the Binomial B(n,p) distribution for different values of n and p.

(117) {CR12} read at the session of March the 4th, 1940. See [47] p. 180.

(118) {CR12}, p. 367, Theorem 4.

(119) Box Fréchet 11. This letter of Fréchet to Doeblin is a draft reply to the previous letter. It has no date but probably dates from January 1940, as Fréchet answered his letters in two days.

(120) Box Fréchet 11. The date is not on the letter.

(121) W.D.'s sealed letter on the Kolmogoroff equation was registered by the Academy of Sciences on the 26th of Febuary 1940 with the number 11,668. It has never been opened. The "brouillon" of this memoir is among the papers deposited at the Schiller Nazional Museum of Marbach, by Claude Doblin, W.D.'s younger brother, born in 1917, interior designer in Nice[29].

(122) Hostinský [64] and [140]. Hostinský's offprints loaned to W.D. by Fréchet are in the Fréchet Archives deposited at the "Laboratoire de Probabilités" of the Paris 6 University. See also [58].

(123) Box Fréchet 12. the date is on the letter as is W.D.'s address: SP 29.

(124) {CR13} presented by E. Borel at the session of the 29th of April.

---

[27]The box Fréchet n°8 contains a letter from Erna Döblin about this

[28]Box Fréchet 5

[29]Marbach box 1

(125) [40]

(126) The summary in French of [145] by W.D., requested by Fréchet, is in box 20 of the Fréchet Archives of the Academy of Sciences. In the Fréchet Archives of the "Laboratoire de Probabilités" we can also find long summaries written in French by W.D. of three important memoirs of Kolmogorov written in German [78], [79], [80]. Fréchet made use of W.D.'s summaries for his course at the IHP. After 1940, Fréchet made use of Loève's services, in particular for his course of the winter 1941-1942 on Ergodic Theory. On Fréchet's request, Loève translated into French Hopf's conference in Geneva[30] and made a summary for him of literature on that subject, see [55].

(127) Box Fréchet 12.

(128) Feller [40], which is in Fréchet's Archives of the "Laboratoire de Probabilités".

(129) It also can be found in box 12.

(130) The "Centre National de la Recherche Scientifique" was created in 1939. It grouped together all the public bodies in charge of Scientific Research, particularly the "Office National de la Recherche" which had succeeded the "Direction des Inventions" of 1915 and the CNRSA created in 1938. For these questions consult the "Cahiers pour l'histoire du CNRS", CNRS's editions, 1988, 1989, et seq.

(131) Box Fréchet 12, this note has never been published.

(132) Box Fréchet 11, the date is on the letter.

(133) {CR13} p. 692.

(134) There are no traces of these results in Fréchet's Archives.

(135) Note which has been transmitted to the CNRS by Fréchet for a scholarship request.

(136) Box Fréchet 8. Post card sent to Fréchet on the 21st of April, 1940. At that time W.D.'s address was 291ème RI, 3ème Bat. EM, SP 21.

(137) [40] see {CR13} p. 692.

(138) One of the notes [107], [111]; J. Métadier's book [112] which expands them was published on the 28th of February, 1940. Fréchet must have received it and must have asked W.D. his opinion about it. In that book Métadier presents a definitely unitarian theory of "tout système physique en évolution". "Le premier", Jacques Métadier rallied to Général de Gaulle in London where he took up his pen for "la France libre", see [113], and published the journal "Solidarity", a platform for all those who could help plan a better world. J. Métadier was professor of physics at "l'Ecole de Médecine et de Pharmacie" of Tours in 1939.

(139) [28].

(140) [28] p. 135 note, "The latter authors omit to mention exceptional points, implying here the possible existence of R_3". This is about the note {CR7}, CRAS; 204, 1699-1701: "Sur deux notes de MM. Kryloff et Bogoliouboff" by W. Doeblin and R. Fortet.

(141) {CR7} p. 1700 property b.

(142) {2}, which doesn't include any proof. In 1937, date of {CR7}, none of Doeblin's works on chains had yet been completely written up or pub-

---

[30]Box Fréchet 26

lished.

(143) An offprint of [28] is in Fortet's Archives of the "Laboratoire de Proba-bilités" of Paris 6 University. The note of page 135 has been meticulously crossed out in blue ink.

(144) J.L. Doob has added this remark on the letter, "of course by now I don't have the slightest idea who was right! But of course the complaint is justified, I should have written him".

(145) Post card sent by W. Doeblin, soldier of the 291st I.R., 11th Company, Givet, to "Monsieur Doob, Docteur ès Sciences, University of Illinois, Urbana, Etats-Unis d'Am." The postmark is unreadable.

(146) This is about the proof of theorem {6}, Bull. Sci. Math., (1940), 64, 35-37.

(147) [132].

(148) This proof was published in 1940, {6}.

(149) The draft is undated[31].

(150) The two questions were certainly about the possible extensions of Raï-kov's theorem [121], and Cramér's theorem [15], respectively, to the Pois-son distribution and the Normal distribution rolled on the circle. W. Doe-blin solved the first one immediately and sent it to Lévy; see [97]. The second one was solved by the great Polish analyst J. Marcinkiewicz (1910-1940), who was at the IHP in the Autumn of 1938, [106]. During his stay in Paris Marcinkiewicz asked Lévy a question about brownian motion which was the starting point for Lévy's renewed interest in brownian motion and the three Lévy notes of the end of 1938. Lévy sent W.D., then in Givet, those notes (Arc sinus law, maximum, etc ...) with nu-merous corrections and comments[32] on local time.

The W.D. "carnet de recherche" seems to be neither in Paris nor in Marbach.

(151) D.A. Raïkov, born in 1905, sent W.D. all of his papers from 1937 to 1939[33]. W. Doeblin had learned a little Russian in 1938, so he could read them, and summarized them in his "fichier". Apparently W.D., unlike Lévy, was very quick to read and understand other workers. He made detailed "résumés" of the more important papers, some of which are in the Paris and Marbach Archives.

(152) {9}.

(153) {CR8}, the corresponding memoir has never been published, nor even written down by W.D.

(154) The draft is unfinished. Probably W.D. solved, before he ended it, the question asked by Lévy and wrote another letter.

(155) Undated letter in Marbach[34], but evidently of June 1939.

(156) {11} published in 1940. It seems that this paper had been written by W.D. just after {3}.

Let us note that W.D. has kept this letter of Lévy (the only real one in Marbach. There is another very short letter from Lévy, but of no interest). W.D. was very anxious about the fate of his works, if

---

[31]Marbach box 2

[32]Marbach box 4

[33]Marbach box 4

[34]Box 2

he died before their publication. This explains why he sent a sealed letter to the Academy and why he sent his "brouillons" to S. Pollag in Switzerland, a neutral country, and to his brother Peter in the U.S.A. Wolfgang Doeblin even specified in a letter to S. Pollag that he made his "brouillons" completely unintelligible to anybody except himself (and in that he was successful because nobody could ever understand either the theorems or the proofs they contained).

He also wrote, "Ich gestehe dass ich für meine Person keine grosse Angst habe, Mangel am Vorstellungsvarmögen, sehr wohl aber für meine Wissenschaftlichen Arbeiten. Im Falle eines Krieges würden sellstver-ständlich meine im Laufe befindlichen oder noch nicht redigierten Ar-beiten verloren gehen, damit habe ich mich abgefunder".

For the unwritten works, it is probably too late, but perhaps the Doeblin Conference of Blaubeuren, will be the occasion to publish the last written and unpublished memoirs of Doeblin.

Acknowledgments. I should like to thank very much M. Claude Doblin who authorized the publication of the mathematical correspondence of his brother Wolfgang. I should also like to thank the Secrétaires perpétuels de l'Académie des Sciences for permitting the reproduction of the letters between W.D. and M. Fréchet deposited in their archives with the Fréchet personal papers. I am indebted to many people for hints and help and particularly C. Balloy, M. Bar-but, A. Blanc-Lapierre, N. Boillot, M.F. Bru, R. Bruge, P. Casey, G. Cho-quet, K.L. Chung, J. Coret, H. Cohn, Sœur Desjardin, J.L. Doob, C. Dufrasne, R. Fortet, C.C. Gillispie, L. Huguet, G. Kunetz, T. Lindvall, C. Lozé, E. Mourier, L. Ruppel, E. Seneta, L. Schwartz, G. Simon, U. Steike, A. Tortrat, I. Vincze; to all of them and my other helpers I address my warmest thanks.

REFERENCES

1. S. Bernstein, *Sur l'extension du théorème limite du calcul des probabilités aux sommes de quantités dépendantes*, Math. Ann. **97** (1927), 1–59.
2. S. Bernstein, *Sur les liaisons entre les grandeurs aléatoires*, Verhandlungen des Int. Math. Kong. I (1932), Zürich, 288–309.
3. S. Bernstein, *Equations différentielles stochastiques*, Actualités Sci. Ind. **738** (1938), 5–32.
4. J. Bezanek, *B. Hostinský*, Česk. časopis pro fys 1 (1951), 90–95.
5. E. Borel, *Traité du calcul des probabilités et de ses applications, Applications aux jeux de hasard*, rédigé par J.Ville, Gauthier-Villars, Paris, 1938.
6. R.Bruge, *Histoire de la ligne Maginot*, vol 1, Fayard, Paris, 1973.
7. R. Bruge, *Les combattants du 18 juin*, 5 vol, Fayard, Paris, 1982-1989.
8. R. Bruge (1990).
9. E. Čech, *Vědecké prace Bedřicha Pospíšila,*, Časopis Pěst Mat. **72** (1947), 1–9.
10. A. Chouraqui, *L'Alliance Israélite Universelle et la renaissance juive contempo-raine*, PUF, Paris, 1965.
11. K.L. Chung, *Contributions to the theory of Markov chains*, I, Journal of Reasearch of the National Bureau of Standards **50** (1953), 203–208; II, Trans. Amer. Math. Soc. **76, 3** (1954), 397–419.
12. K.L. Chung, *The General Theory of Markov Processes According to Doeblin*, Z. Wahrs-chein. **2,3**, (1964), 230–254.
13. K.L. Chung, *Markov Chains with stationary transition probabilities*, second ed., Springer-Verlag, Berlin, 1967.
14. L. Collatz, *Günter Schulz*, Z. Angew. Math. Mech. **43** (1963), 96.

15. H. Cramér, *Sur les propriétés de la loi de Gauss*, C.R. Acad. Sci. Paris **202** (1936), 615–616.
16. P. Crépel, *Histoire de la théorie des martingales*, Séminaire de probabilités (1984), Rennes I.
17. A. Datzeff, *Sur les problèmes des barrières de potentiel et la résolution de l'équation de Schrödinger*, Masson, Paris, 1938.
18. C. Doblin.
19. S. Doblin (1990).
20. W. Doeblin, *Sur les probabilités en chaîne*, Soc. Math. France, Comptes Rendus des Séances (1936), 28–29.
21. W. Doeblin, *Sur certaines propriétés asymptotiques de l'équation de Smoluchowski*, ibid. (1936), 29.
22. W. Doeblin, *Sur les noyaux stochastiques*, ibid. (1936), 31.
23. W. Doeblin, *Sur certains résultats de MM. Fréchet et Hadamard*, ibid. (1936), 31.
24. W. Doeblin, *Sur les chaînes de Markoff*, ibid. (1936), 34.
25. W. Doeblin, *Sur l'extension de quelques théorèmes de Frobenius et Jentzsch*, ibid. (1937), 31.
26. W. Doeblin, *Sur la loi de Gauss et les chaînes dénombrables*, ibid. (1937), 31.
27. W. Doeblin, *Remarques sur la théorie métrique des fractions continues*, ibid. (1938), 28–29.
28. J.L. Doob, *Stochastic processes with integral-valued parameter*, Trans. Amer. Math. Soc. **44, 1** (1938), 87–150.
29. J.L. Doob, *Topics in the theory of Markoff chains*, Trans. Amer. Math. Soc. **52** (1942), 37–64.
30. J.L. Doob, *Markoff chains - Denumerable case*, Trans. Amer. Math. Soc. **58, 3** (1945), 455–473.
31. J.L. Doob, *Asymptotic property of Markoff Transition Probabilities*, Trans. Amer. Math. Soc. **63, 3** (1948), 393–421.
32. J.L. Doob, *Stochastic Processes*, Wiley, New York, 1953.
33. C. Dufrasne, *Le mouvement Ajiste, (un mouvement original d'éducation populaire)*, Centre de Recherches, tour 34/33, 2 Place Jussieu, 75005 Paris (1991).
34. E. Feldheim, *Une loi limite du calcul des probabilités*, Acta. Litt. Sci. Szeged **8** (1936), 55–63.
35. E. Feldheim, *Sur les probabilités en chaîne*, Math. Ann. **112, 5** (1936), 775–780.
36. E. Feldheim, *Etude sur la stabilité des lois de probabilité*, Thèse Université de Paris, Szeged, 1937.
37. E. Feldheim, *Sur une équation intégrale singulière*, Bull. Sci. Math. **61** (1937), 10–18.
38. E. Feldheim, *Relations entre les polynômes de Jacobi, Laguerre et Hermite*, Acta. Math. **75** (1942), 117–138.
39. W. Feller, *Über der zentralen Grenzwertsatz des Wahrscheinlichkeitsrechnung*, Math. Zeitsch. **40** (1935), 521–559.
40. W. Feller, *Zur Theorie der stochastischen Prozesse (Existenz und Eindentigkeitssätze).*, Math. Ann. **113, 1** (1936), 113–160.
41. W. Feller, *Sur les axiomatiques du calcul des probabilités et leurs relations avec les expériences*, Les Fondements du calcul des probabilités, Actualités Sci. Indust. 735, Hermann, Paris, 1938, pp. 7–21.
42. W. Feller, *On the integro-differential equations of purely discontinuous Markoff processes*, Trans. Amer. Math. Soc. **48, 3** (1940), 488–515.
43. W. Feller, *Diffusion processes in one dimension*, Trans. Amer. Math. Soc. **77, 1** (1954), 1–31.
44. W. Feller, *An Introduction to Probability Theory and its Applications*, vol. 2, second ed., Wiley, New York, 1971.
45. B. de Finetti, *Compte rendu critique du colloque de Genève sur la théorie des probabilités*, Actualités Sci. Indust. 766, Hermann, Paris, 1939.
46. R. Fortet, *Sur les probabilités en chaîne*, C.R. Acad. Sci. Paris **201** (1935), 124–126.
47. R. Fortet, *Sur les probabilités en chaîne*, C.R Acad. Sci. Paris **202** (1936), 1362–1364.

48. R. Fortet, *Sur l'itération des substitutions linéaires algébriques d'une infinité de variables et ses applications au problème des probabilités*, Thèse Paris, Rev. Cienc., Lima, 1938.

49. R. Fortet, *Les fonctions aléatoires du type de Markoff associées à certaines équations linéaires aux dérivées partielles de type parabolique*, J. Math. Pures Appl. (1943), 177–243.

50. M. Fréchet, *Pri la funkcia ekvacio f(x+y) = f(x) + f(y)*, Enseign. Math. **15, 5** (1913), 390–392.

51. M. Fréchet, *Les probabilités continues "en chaîne"*, Comment. Math. Helv. **5** (1933), 175–245.

52. M. Fréchet, *Compléments à la théorie des probabilités discontinues "en chaîne"*, Ann. Scuola Norm. Sup. Pisa II, II, XI (1933), 131-162.

53. M. Fréchet, *Sur l'allure asymptotique des densités itérées dans le probléme des probabilités "en chaîne"*, Bull. Soc. Math. France **62** (1934), 68–83.

54. M. Fréchet, *Méthode des fonctions arbitraires. Théorie des événements en chaîne dans le cas d'un nombre fini d'états possibles*, 1938; 2ème ed., Gauthier Villars, Paris, 1952, 1952.

55. M. Fréchet, *Les fonctions asymptotiques presque périodiques et leur application au problème ergodique*, Revue Scientifique (1941), 341–354 and 407–417.

56. M. Fréchet, *Les contributions françaises récentes au calcul des probabilités et à la statistique mathématique, (A.F.A.S., $64^{è}$ session, Paris, Congrès de la Victoire, 20-26 octobre 1945)*, Intermédiaire des Recherches Mathématiques **9** (1947), 107–120.

57. M. Fréchet, *Le rôle d'Emile Borel dans la théorie des jeux*, Revue d'Economie politique (1959), 139–167.

58. R.D. Gill, S. Johansen, *A survey of product-integration with a view towards application*, Ann. Stat. **18, 4** (1990), 1501–1555.

59. W.A. Granville, P.F. Smith, *Eléments de calcul différentiel et intégral*, 2ème ed., Vuibert, Paris, 1926.

60. Y. Guivarc'h, A. Raugi, *Products of Random Matrices: Convergence Theorems*, Contemp. Math., AMS **50** (1986), 31–54.

61. Y. Guivarc'h, J. Hardy, *Théorèmes limites pour une classe de chaînes et applications aux difféomorphismes d'Anosov*, Ann. Inst. H. Poincaré **24, 1** (1988), 73–98.

62. J. Hadamard, *Sur le battage des cartes et ses relations avec la mécanique statistique*, Atti. del Congr. Inter. Mat. **5** (1928), Bologne, 133–139.

63. B. Hostinský, *Méthodes générales du calcul des probabilités*, Mémorial des sciences mathématiques, 52, Gauthier-Villars, Paris, 1931.

64. B. Hostinský, *Application du calcul des probabilités à la théorie du mouvement brownien*, Ann. Inst. H. Poincaré **3** (1932), 1–74.

65. B. Hostinský, *Sur une équation fonctionnelle de la théorie des probabilités*, Publi. Fac. Sci. Univ. Masaryk, 1st part **156** (1932); 2nd part **194** (1934); 3rd part **261** (1938).

66. B. Hostinský, *Sur les progrès récents de la théorie des probabilités*, Comptes Rendus du 2^{ème}congrès des mathématiciens des pays slaves, Prague, 1935, pp. 94–106.

67. B. Hostinský, *Sur les probabilités relatives aux variables aléatoires liées entre elles. Applications diverses*, Ann. Inst.H. Poincaré **VII** (1937), 69–119.

68. B. Hostinský, *Equations fonctionnelles relatives aux probabilités continues en chaîne*, Hermann, Paris, 1939.

69. L. Huguet, *Alfred Döblin, éléments de biographie et de bibliographie systématique*, Thèse de 3ème cycle, 3 vol., Paris-Nanterre, 1968.

70. L. Huguet, *La jeunesse d'A. Döblin. Héritage et élection*, Revue d'Allemagne **3** (1973), 728–745.

71. L. Huguet, *Pour un centenaire (1878-1978). Chronologie A. Döblin*, Collège de France, Paris, 1980.

72. L. Huguet, *Fils de grand écrivain et grand mathématicien: Wolfgang Doeblin*, Runa **2** (1984), 77–91.

73. M. Iosifescu, *In memoriam G. Mihoc (1906-1981)*, Rev. Roumaine Math. Pures Appl. **27** (1982), 1001–1002.

74. M.Iosifescu, *O. Onicescu (1892-1983)*, Internat. Statist. Rev. **54, 1** (1986), 97–108.

75. D.G. Kendall, *A note on Doeblin's Central Limit Theorem*, Proc. Amer. Math. Soc. **8, n°6** (1957), 1037–1039.

76. D.G. Kendall, *Obituary A.N. Kolmogorov (1903-1987)*, Bull. London Math. Soc. **22** (1990), 31–100.

77. Á. Kenyeres, *Magyar életrajzi lexikon*, vol. 1, Akadémiai kiadó, Budapest, 1967.

78. A.N. Kolmogorov, *Ueber das Gesetz des iterierten Logarithmus*, Math. Ann. **101** (1929), 126–135.

79. A.N. Kolmogorov, *Über die analytischen Methoden in der Wahrscheinlichkeitsrechnung*, Math. Ann. **104** (1931), 415–458.

80. A.N. Kolmogorov, *Zur Theorie der stetigen zufälligen Prozesse*, Math. Ann. **108** (1933), 149–160.

81. A.N. Kolmogorov, *Aufangsgründe der Theorie der Markoffschen Ketten mit unendlich vielen möglichen Zustanden*, Mat. Sb. **1** (1936), 607–610.

82. A.N. Kolmogorov, *Zur Theorie der Markoffschen Ketten*, Math. Ann. **112** (1936), 155160.

83. A.N. Kolmogorov, *Markov Chains with countably many possible states*, Bull. Univ. Moscou **1, n°3** (1937), 1–11.

84. A.N. Kolmogorov, *Deux théorèmes asymptotiques uniformes pour des sommes de variables aléatoires indépendantes*, Séminaire de calcul des probabilités, Inst. H. Poincaré (1958), 17 avril 1958..

85. A.N. Kolmogorov, *Sur les propriétés des fonctions de concentration de M.P. Lévy*, Ann. Inst. H. Poincaré **16** (1958), 27–34.

86. U. Krengel, *Wahrscheinlichkeitstheorie.*, Ein Jahrundert Mathematik 1890-1990 (1990), ed. G. Fischer, F. Hirzebruch, W. Scharlau und W. Töring, Braunschweig, Vieweg..

87. N. Kryloff, N. Bogoliouboff, *Sur les propriétés ergodiques de l'équation de Smoluchowski*, Bull. Soc. Math. France **64** (1936), 49–56.

88. N. Kryloff, N. Bogoliouboff, *Sur les probabilités en chaîne*, C.R. Acad. Sci. Paris **204** (1937), 1386–1388.

89. N. Kryloff, N. Bogoliouboff, *Les propriétés ergodiques des suites de probabilités en chaîne*, C.R. Acad. Sci. Paris **204** (1937), 1454–1456..

90. G. Kunetz, *Sur quelques propriétés des fonctions caractéristiques*, Thèse Université de Paris, Rodstein, Paris, 1937.

91. P. Lévy, *Sur une équation intégrale considérée par M. Picard*, Bull. Sci. Math. **52** (1928), 156–160.

92. P. Lévy, *Sur les intégrales dont les éléments sont des variables aléatoires indépendantes*, Ann. Scuola Norm. Sup. Pisa II **3** (1934), 337–366.

93. P. Lévy, *Sur la loi de Gauss*, Soc. Math. France, Comptes Rendus des Séances , 1934, séance du 28 nov. 1934, 1934, pp. 48–49.

94. P. Lévy, *Propriétés asymptotiques des sommes de variables aléatoires indépendantes ou enchaînées*, J. Math. Pure Appl. (VII) **14** (1935), 347–402..

95. P. Lévy, *Sur les solutions de l'équation de Chapman*, C.R. Acad. Sci. Paris **205** (1937), 1355-1357.

96. P. Lévy, *Théorie de l'addition des variables aléatoires*, 1937; 2ème ed. 1957, Gauthier Villars, Paris.

97. P. Lévy, *Sur l'arithmétique des lois de probabilité enroulées*, Soc. Math. France, Comptes Rendus des Séances, 1938, pp. 30-34.

98. P. Lévy, *Processus stochastiques et mouvement brownien*, Gauthier-Villars, Paris, 1948.

99. P. Lévy, *W. Doeblin (V. Doblin) (1915-1940)*, Rev. Histoire Sci. (1955), 107–115.

100. P. Lévy, *Le dernier manuscrit inédit de W. Doeblin*, Bull. Sci. Math. **80** (1956), 61–64.

101. P. Lévy, *Nouvelle notice sur les travaux scientifiques*, Paris, 1964.

102. P. Lévy, *Quelques aspects de la pensée d'un mathématicien*, Blanchard, Paris, 1970.

103. Ta Li, *Über die allgemeine lineare Differentialgleichung*, Comment. Math. Helv. **12** (1939), 1–19.

104. T. Lindvall, *W. Doeblin 1915-1940*, Ann. Prob. **19, 3** (1991), 929–934.
105. M. Loève, *Probability Theory*, 3rd ed., Van Nostrand, Princeton, 1963.
106. J. Marcinkiewicz, *Sur les variables aléatoires enroulées*, Soc. Math. France, Comptes Rendus des Séances, 1938, pp. 34–36.
107. J. Métadier, *Sur l'équation générale du mouvement brownien*, C.R. Acad. Sci. Paris **193** (1931), 1173.
108. J. Métadier, *Sur l'étude du mouvement brownien dans un champ de forces*, C.R. Acad Sci. Paris **195** (1932), 649.
109. J. Métadier, *Sur la théorie du mouvement brownien et la méthode opérationnelle*, C.R. Acad. Sci. Paris **197** (1933), 29.
110. J. Métadier, *Action du champ magnétique sur le mouvement brownien*, C.R. Acad. Sci. Paris **199** (1934), 1196.
111. J. Métadier, *Mouvement brownien dans l'espace de Hilbert, hyperquantification et superquantification*, C.R. Acad. Sci. Paris **200** (1935), 807.
112. J. Métadier, *La théorie de l'agitation chaotique, mouvement brownien, agitation moléculaire, dispersion, floculation, etc . . .*, Actualités Sci. Indust. 661, Hermann, Paris, 1938.
113. J. Métadier, *France*, Mac Donald, London, 1943.
114. G. Mihoc, *Sur les lois limites de probabilités liées en chaîne*, Bull. Fac. Sti. Cernaŭti **10** (1936), 1–26.
115. H. Neumann, *Alfred Döblin: Leben und Werk, Krankheit und Tod*, Kircheim, Mainz, 1987.
116. O. Onicescu, *Sur la notion de chaîne et l'idée de loi naturelle*, Mathematica **13** (1937), 66-71.
117. O. Onicescu, *Les applications de l'idée de chaîne statistique, XXIIIème session de l'Institut International de Statistique, Athènes*, Bull. Inst. Int. Stat. **29, 2** (1937), 297–303.
118. M. Petruszevich, *Chaînes de Markov discrètes dans le domaine linguistique*, Séminaire d'histoire des mathématiques au XXème siècle (1981), Rennes I.
119. B. Pospišil, *Sur un problème de MM. S. Bernstein et A. Kolmogoroff*, Časopis Pěs. Mat. **65, 2** (1936), 64–76.
120. J. Potoček, *Sur la dispersion dans la théorie des chaînes de Markoff*, Pub. Fac. Sci. Univ. Masaryk **154** (1932).
121. D.A. Raikov, *On the decomposition of Poisson laws*, Dok. Akad. Nauk. **14** (1937), 9–11.
122. V. Romanovsky, *Recherches sur les chaînes de Markoff*, Acta. Math. **66**, 147–251.
123. L. Ruppel, *Personal Communication*, 1992.
124. G. Schulz, *Über des summenproblem bei Markoffchen Ketten*, Actes. Congrès Int. Math. Zürich **2** (1932), 230–231.
125. G. Schulz, *Grenzwertsätze für die Wahrscheinlichkeiten verketteter Ereignisse*, Deutche Mathematik **1, 5** (1936), 665–699.
126. L. Schwartz, *Personal Communication*, 1991.
127. E. Seneta, *On the historical development of the theory of finite inhomogenous Markov chains*, Proc. Camb. Phil. Soc. **74** (1973), 507–513.
128. O.B. Sheynin, *A.A. Markov's Work on Probability*, Arch. Hist. Exact Sci. **39, 4** (1989), 337–377.
129. G. Simon, *Un soldat français inconnu*; unpublished, Housseras.
130. G. Simon, *Personal Communication*, 1992.
131. H. Steinhaus, *Stefan Banach, 1892-1945*, Scripta Math. **26, 2** (1961), 93–100.
132. B.V. Sz. Nagy, *Über messbare Darstellungen Liescher Gruppen*, Math. Ann. **112** (1938), 286–296.
133. M.A. Tonnelat, *Histoire du principe de la relativité*, Flammarion, Paris, 1971.
134. M.A. Tonnelat, *Retour à Pasargada*, Belfond, Paris, 1984.
135. E. Tornier, *Wahrscheinlichkeits Rechnung und Allgemeine Integrations Theorie*, Teubner, Leipzig, 1936.
136. S.A. Ulam, *Adventures of a Mathematician*, Scribner's, New York, 1976.
137. J. Ville, *Etude critique de la notion de collectif*, Gauthier-Villars, Paris, 1939.
138. J. Ville, *Conversation avec P. Crépel*, Langon, 1984.

139. I. Vincze, *Personal Communication*, 1992.

140. V. Volterra, B. Hostinský, *Opérations infinitésimales linéaires, applications aux équations différentielles et fonctionnelles*, Gauthier-Villars, Paris, 1938.

141. R. Wavre, *Sur le mouvement des astres fluides*, Ann. Inst. H. Poincaré **III, IV** (1933), 491–510.

142. W. Winkler, *Doeblin's and Harris's Theory of Markov Processes*, Z. Wahrschein. **31, 1** (1974), 79–88.

143. A. Wintner, *On analytic convolutions of Bernoulli distributions*, Amer. J. of Math. **54, 4** (1934), 659–663.

144. A. Wintner, *On the densities of infinite convolutions*, Amer. J. of Math. **59, 2** (1937), 376–378.

145. H. Wold, *A study in the analysis of stationary time series*, Almqwist, Uppsala, 1938.

146. K. Yosida, S. Kakutani, *Operator theoretical treatment of Markoff's Process and mean ergodic theorem*, Ann. of Math. **42, 1** (1941), 188–228.

THE DOEBLIN BIBLIOGRAPHY, [104] P.933-934

Memoirs

{1}. *Le cas discontinu des probabilités en chaîne*, Publ. Fac. Sci. Univ. Masaryc **236** (1937), 3–13, 592 (a correction).

{2}. *Sur le cas continu des probabilités en chaîne*, Rend. Accad. Lincei **25** (1937), 170–176.

{3}. *Sur les chaînes à liaisons complètes*, (R. Fortet co-author), Bull. Soc. Math. France **65** (1937), 132–148.

{4}. *L'équation de Smoluchowski*, Prakt. Akad. Athēnōn **12** (1937), 116–119.

{5}. *Sur les propriétés asymptotiques des mouvements régis par certains types de chaînes simples*, Bull. Soc. Math. Roumaine Sci. **39(1)**, 57–115; **39(2)** (1937), 3–61.

{6}. *Sur l'équation matricielle $A(t+s) = [A(t)A(s)]$ et ses applications aux probabilités en chaînes*, Bull. Sci. Math. **62** (1938), 21–32; **64** (1940), 35–37 (a correction).

{7}. *Sur deux problèmes de M. Kolmogoroff concernant les chaînes dénombrables*, Bull. Soc. Math. France **66** (1938), 210–220.

{8}. *Exposé de la Théorie des Chaînes simples constantes de Markoff à un nombre fini d'Etats*, Rev. Math. de l'Union Interbalkanique **2** (1938), 77–105.

{9}. *Sur les sommes d'un grand nombre de variables indépendantes*, Bull. Sci Math. **63** (1939), 23–32, 35–64.

{10}. *Sur certains mouvements aléatoires discontinus*, Skand. Aktuarie Tidskr. **22** (1939), 211-222.

{11}. *Remarques sur la théorie métrique des fractions continues*, Compositio Math. **7** (1940), 353–371.

{12}. *Elément d'une théorie générale des chaînes simples constantes de Markoff*, Ann. Sci. Ecole Norm. Sup. **57** (1940), 61–111.

{13}. *Sur l'ensemble de puissances d'une loi de probabilité*, Studia Math. **9** (1940), 71–96; Reprinted with a complement in, Ann. Sci. Ecole Norm. Sup. **63** (1947), 317–350.

Notes in Comptes Rendus de l'Académie des Sciences, Paris.

{CR1}. *Sur les sommes de variables aléatoires indépendantes à dispersions bornées inférieurement*, (P. Lévy co-author), CRAS **202** (1936), 2027–2029.

{CR2}. *Sur les chaînes discrètes de Markoff*, CRAS **203** (1936), 24–26.

{CR3}. *Sur les chaînes de Markoff*, CRAS **203** (1936), 1210–1211.

{CR4}. *Sur deux notes de MM. Kryloff et Bogoliouboff*, (R. Fortet co-author), CRAS **204** (1937), 1699–1701.

{CR5}. *Eléments d'une théorie générale des chaînes constantes simples de Markoff*, CRAS **205** (1937), 7–9.

{CR6}. *Premiers éléments d'une étude systématique de l'ensemble de puissances d'une loi de probabilité*, CRAS **206** (1938), 306–308.

{CR7}. *Etude de l'ensemble de puissances d'une loi de probabilité*, CRAS **206** (1938), 718–720.

{CR8}. *Sur les sommes d'un grand nombre de vecteurs aléatoires*, CRAS **207** (1938), 511–513.

{CR9}. *Sur l'équation de Kolmogoroff*, CRAS **207** (1938), 705–707.

{CR10}. *Sur certains mouvements aléatoires*, CRAS **208** (1939), 249–250.

{CR11}. *Sur un problème de calcul des probabilités*, CRAS **209** (1939), 742–743.

{CR12}. *Sur l'équation de Kolmogoroff*, CRAS **210** (1940), 365–367.

{CR13}. *Sur les mouvements mixtes*, CRAS **210** (1940), 690–692.

UNIVERSITE RENE DESCARTES, 45 RUE DES SAINTS PERES, 75006 PARIS.

E-mail address: bru@mathp7.jussieu.fr

Contemporary Mathematics
Volume **149**, 1993

REMINISCENCES OF ONE OF DOEBLIN'S PAPERS

KAI LAI CHUNG

In June of 1940, shortly after the fall of France in which Doeblin died, I received a package of reprints sent from France by Fréchet. It was probably among those reprints that I first saw Doeblin's paper [1] cited. I was then a student at Kunming, China and had read Fréchet's book [2]. Kolmogorov's paper [3] which laid the foundation of the theory of Markov chains with denumerable states was not accessible to me, although the journal containing it might have been under the caves of Yunnanfu to escape Japanese bombing, as I reminisced at the Paul Lévy memorial in 1987[1]. I read it later after learning Russian, and translated it in an exercise book which I still have. Kolmogorov mentioned Doeblin's independent work but did not state the two problems treated in [1]. I think Feller first showed that paper to me. Later Mrs. Doeblin sent me several of her son's reprints, among which was this one, with a cover but without the year of publication nor the number of the volume, testifying to the wartime conditions in France. What I remembered distinctly is that I asked a visitor, M.L., to report its content to a small audience, including Doob and Snell. Afterwards we had a good chuckle over it because the oral presentation did not sound any different from the rather obscure original script. The following year I was living in New York City and giving a course on limit theorems for sums of independent random variables at the Department of Mathematical Statistics of Columbia University, and the Korean war broke out. I must have tired of those sums and been looking for something else to think about. It was in this way that I returned to the first problem in [1]. In the standard notation of Markov chains the problem may be reduced to the existence of the following limit for two

---

AMS 1991 Classification: 60-03, 60-02, 60F05, 60J10.

This paper is in final form and no version of it will be submitted for publication elsewhere.

communicating states i and j

$$\lim_{N \to \infty} \frac{\sum_{n=0}^{N} p_{jj}^{(n)}}{\sum_{n=0}^{N} p_{ii}^{(n)}} \tag{1}$$

If the class is "positive recurrent" (Feller's terminology), then it is plain that the limit is equal to π_j/π_i where

$$\pi_j = \lim_{N \to \infty} \frac{1}{N+1} \sum_{n=0}^{N} p_{jj}^{(n)} \tag{2}$$

Kolmogorov had proved that the limit in (2) exists in fact for every j, and it is 0 unless j is a positive-recurrent state. Moreover, in a positive-recurrent class I all $\pi_j > 0$, and the set $\{\pi_j; j \in I\}$ is a stationary distribution. But in a null-recurrent or nonrecurrent class the limit in (1) is the indeterminate $0/0$; hence the problem: does it actually exist? Doeblin proved that it does, and is finite and strictly positive. The key idea in his proof is to consider the *rencontres* of the set of two states (i, j). As far as I know, there has been no subsequent exposition of his argument. My efforts brought me to uncoupling i and j, considering their rencontres not indiscriminately as suggested by Doeblin, but one after the other. Now for a linear, so-called "continuous" random walk, if one starts somewhere between i and j then the rencontre of (i, j) is the end of the gambler's game while the hitting of one before the other represents the ruin of Peter or Paul. The same scheme may be couched in terms of linear Brownian motion, which analogue was brought to my attention just at that time by T.E. Harris, a classmate in Princeton. One thing led to another, but for me the prepotent step was to write down the notation[2], for $n \geq 1$:

$$_k p_{ij}^{(n)} = P\{X_\nu \neq k \text{ for } 1 \leq \nu < n; X_n = j | X_0 = i\} \tag{3}$$

where $\{X_n, n \geq 0\}$ is the Markov chain associated with the transition matrix (p_{ij}) on I and i, j, k are three arbitrary elements of I. I call the quantity in (3) a "taboo probability", with the state k taboo. Once this is set down, one sees at once that when $k = j$ it reduces to the old first entrance probability often denoted by a new symbol such as $f_{ij}^{(n)}$ ("f" for first). The latter was already used by Pólya in his celebrated paper on random walks (1922). In fact he had, albeit implicitly via power series (generating functions), established the following "first entrance decomposition" formula which Doeblin wrote down explicitly in general, for $i \neq j$ and $n \geq 1$:

$$p_{ij}^{(n)} = \sum_{\nu=1}^{n} {}_j p_{ij}^{(\nu)} p_{jj}^{(n-\nu)}. \tag{4}$$

With this in mind and the superior notation exhibited in (3), it could have not taken long before I asked myself the question: what if $k = i$ in (3), given than one is to do equal justice to i and j? Thus it became the most natural thing to look for a mate for (4), here it is:

$$p_{ij}^{(n)} = \sum_{\nu=1}^{n} p_{ii}^{(n-\nu)} {}_i p_{ij}^{(\nu)} \tag{5}$$

The symmetry between the two mates is stunning. On hindsight, of course, "first entrance into j" dualizes into "last exit from i"; and "j is taboo before its first entrance" dualizes into "i is taboo after its last exit". Sounds easy ?[3] The rest is simple algebra and analysis. Just as (4) yields

$$\lim_{N \to \infty} \frac{\sum_{n=1}^{N} p_{ij}^{(n)}}{\sum_{n=0}^{N} p_{ii}^{(n)}} = \sum_{n=1}^{\infty} {}_j p_{ij}^{(n)} \tag{6}$$

which is well-known except for the notation; merely by changing j to i in the denominator of (6) we obtain the dual:

$$\lim_{N \to \infty} \frac{\sum_{n=1}^{N} p_{ij}^{(n)}}{\sum_{n=0}^{N} p_{jj}^{(n)}} = \sum_{n=1}^{\infty} {}_i p_{ij}^{(n)} \tag{7}$$

which was the new missing link. Therefore the limit in (1) sought by Kolmogorov and Doeblin is equal to

$$_j p_{ij}^* / {}_i p_{ij}^* \tag{8}$$

where I have used my old notation of denoting the sum of a series by an asterisk, learned from an old statistics book. It remains to show that the expression in (8) is not $0/0$, provided i and j communicate. Now if $\max_n p_{ij}^{(n)} > 0$, then it follows by algebra from (4) that $\max_n {}_i p_{ij}^{(n)} > 0$. For the first expression there is a facile verbal argument: if one gets from i to j, then there is a first time that one gets there. For the second assertion the verbal argument becomes: if one can go from i to j, then one can get there by avoiding going through i again. For example, since we can go from Blaubeuren to Ulm (Harry Cohn and I did that) then we can do so without making another stop in Blaubeuren. Isn't this trivial? But the assertion is false if the process is not Markovian! Next $_j p_{ij}^* \leq 1$ by its interpretation as a probability, but what about $_i p_{ij}^*$ which is an expectation, not a probability. Here we need the taboo probability below valid for any positive integers m and n, and immediately provable by its probability meaning

$$_i p_{ij}^{(m)} {}_i p_{ji}^{(n)} \leq_i p_{ii}^{(m+n)}$$

It follows from this that if $\max_n p_{ji}^{(n)} > 0$ so that $\max_n {}_i p_{ji}^{(n)} > 0$ as just proved , then $_ip_{ij}^* < \infty$ (the one line detail being left as an exercise). Altogether we have proved that if i and j communicate then both numerator and denominator in (8) are finite and strictly positive. This finishes the proof of Doeblin's theorem.

There are at least sixteen different ways to express the limit (1) by means of taboo probabilities; indeed, the first I obtained is the following[4]

$$(1 +_i p_{jj}^*)/(1 +_j p_{ii}^*)$$

which may look neater but is less meaningful then the one given in (8). To appreciate the latter, let us return to a recurrent class I. The denominator in (8) is equal to 1 for all i and j in I, and so the limit is just $_ip_{ij}^*$ which I have also denoted by e_{ij}. Now the important thing is that the set $\{e_{ij}, j \in I\}$, for any fixed $i \in I$, is a stationary measure for (p_{ij}), namely for any i and k in I we have

$$\sum_{j \in I} e_{ij}p_{jk} = e_{ik}. \tag{9}$$

This measure is finite when I is positive-recurrent as already noted above because then $e_{ij} = \pi_j/\pi_i$; and it is infinite when I is null-recurrent. In both cases, the total mass of this measure is equal to $m_{ii} = \sum_{n=1}^{\infty} n \, _ip_{ii}^{(n)}$, the expected recurrence time of j. In this way the old, so-called "steady-state" theory can be extended to the case of a "phase space of infinite mass".

I now turn to Doeblin's second result which concerns the asymptotic normality of evaluated sums of random variables in a Markov chain:

$$S_n = \sum_{\nu=1}^{n} f(X_\nu) \tag{10}$$

where f is an arbitrary numerical function on the state space I, which is a positive recurrent class. We seek conditions under which there exist constants M and $\sigma > 0$ such that the normed sum $(S_n - Mn)/\sigma\sqrt{n}$ tends in distribution to the unit normal as $n \to \infty$. Unlike the first problem which was posed by Kolmogorov[5] this arose from a letter from the latter to Fréchet, cited by Doeblin, in which the following condition was stated, in the notation adopted here:

$$\sum_{n=1}^{\infty} n^2 \, _ip_{ii}^{(n)} < \infty \; ; \sum_{i \in I} \pi_i f(i)^2 < \infty. \tag{11}$$

Doeblin said that he wanted to apply a natural extension of Kolmogorov's basic idea in [3] to obtain a somewhat different set of conditions. He said also that due to the difference between the two sets of conditions, he figured that Kolmogorov's proof must be quite different from his. So far as I know, Kolmogorov never published his proof. This is not surprising because the conditions in (11) turned out to be inadequate. Certainly, I did not suspect this when I began to study Doeblin's proof, and stumbled on a counterexample to Kolmogorov's announcement after making Doeblin's condition more explicit, so that it becomes computable in a special case. This example was given in [4] and later strengthened in [5] to allow for any higher moment (namely replacing the second power of $f(i)$ in (11) by any higher one), and norming of S_n other than the one indicated above. This is possible after the problem is reduced to one for sums of independent and identically distributed random variables, for which Lévy-Feller criterion can be applied.

The said reduction is Doeblin's fundamental idea[6]. Fix an i as an initial state. Since i is recurrent it will be re-entered an infinite number of times, almost surely. Denote the successive entrance times by $T_1 < T_2 < \cdots < T_n < \cdots$ and call the random time interval a "cycle". Then the (strong) Markov property implies that the sums of $f(X_\nu)$ over the successive cycles form a sequence of independent and identically distributed random variables, $\{Y_m, m \geq 1\}$. For a fixed (and large n) the sum S_n is the sum of a random number of the Y_m's plus a remainder due to an uncompleted cycle. The latter is by no means trivial, but Kolmogorov's famous maximum inequality handles it nicely. The same inequality is used, together with the strong law of large numbers, to overcome the randomness of the number of summands in $Y_1 + Y_2 + \cdots$ Thus it transpires that the classical sufficient condition for the central limit theorem for sums of independent and identically distributed random variables, namely the finiteness of the second moment, is also sufficient for our S_n. This is the condition given by Doeblin:

$$E\{(\sum_{\nu=1}^{T_1} f(X_\nu))^2\} < \infty. \tag{12}$$

Actually he added another condition $E(T_1)^2 < \infty$, which is the same as the first part of (11), and later shown to be dispensable by D.G. Kendall. Observe that if f is bounded, in particular when I is finite, then the latter implies (12).

Why does Kolmogorov's condition (11) fail? Because there the second moment of $f(\cdot)$ is evaluated in terms of the stationary distribution $\{\pi_j\}$, with $\pi_j = 1/m_{jj}$. In contrast, the second moment in (12) is evaluated in terms of the tabooed process associated with a cycle. It can be expressed in terms of the taboo probabilities in (3), or rather their sums $_k p_{ij}^*$; see p.88 of [5], and p. 67 for the particular case where $f \equiv 1$, i.e. $E(T_1^2) < \infty$. I still do not understand why the latter "works". But it is through such cumbersome formulas that the counterexample to Kolmogorov's announcement was verified.

It is clear that Doeblin's method applies to other cases than normality. In fact he sketched such an extension and referred to his paper on sums of a large number of independent variables. As far as I am aware of, no real progress has been made since, even for the special case of a stable law.

Postscript

As the title indicates, I have confined myself in this article to one paper by Doeblin. At the Blaubeuren conference however I reminisced also a little on some other papers. The most relevant may be those on denumerable Markov chains in continuous time, a subject which later attracted the attention of Doob, Kolmogorov and Lévy. It is curious to note that despite much work on the "equations" (see Bru's article), Doeblin did not attempt to establish the differentiability of $p_{ij}(t)$, nor did Doob except at $t = 0$. Under the assumption $p'_{ii}(0+) > -\infty$ its existence was not proved untill 1956, by D.G. Austin, see p. 139 of [5].

Notes

1. See [6]. My remarks were also published in the August, 4, 1987 issue of the *International Herald Tribune*. Doeblin being a war casualty, it is most appropriate that the reader be reminded that the European war was preceded by the wanton Japanese invasion of China on July 7, 1937. Some say it began earlier on September 18, 1930.

2. Notation can be incredibly important! It is recognized that Gauss's notation $a \equiv b \pmod{m}$ was a signal advance in number theory.

3. It is so simple that I put both formulas in my *Elementary Probability Theory* (Springer-Verlag), though I have often wondered how many teachers as well as students have appreciated them. Let me add that the extension of (5) to a continuous time Markov chain is very hard, whereas that of (4) is relatively easy. Last exit is a deeper concept than first entrance.

4. This was first printed in the mimeographed lecture notes prepared by the Columbia Graduate Mathematical Society, Spring 1951. The second part of these Notes contain my preliminary exposition of Doeblin's "general theory".

5. Though Doeblin said so in the paper, no trace of it was found in the documents of the archives reproduced in Bru's account published in this volume.

6. Lévy's theory of Brownian excursion (circa 1939) is the continuous time analogue of this idea, although it is much subtler because the set of zeros of a Brownian path is not well-ordered and non-denumerable. Indeed, one of Lévy's basic formulas is really a last exit decomposition. He also needed a taboo second moment for the excursion, analogue of (12). Both these observations are explained in [6]. It is a pity that in my conversations with Lévy theses connections never came up. Lévy's way of developing his ideas has largely been lost on the next generation.

REFERENCES

[1] W. Doeblin: Sur deux problèmes de M. Kolmogorov concernant les chaînes dénomerables. *Bull. Soc. Math. France* 66, 210-220, 1938.

[2] M. Fréchet: **Méthode des fonctions arbitraires, Théorie des événements en chaîne dans le cas d'un nombre fini d'états possibles.** Gauthier-Villard, Paris, 1938.

[3] A.N. Kolmogorov: Markov chains with countably many positive states *Bull. Univ. Moscou*, 1, 3, 1-11, 1937. This is in Russian. There is an earlier abstract in German: Anfangengründe der Theorie der Markoffschen Ketten mit unendlich vielen möglichen Zuständen. *Math. Sbornik*, 1, 607-610, 1936.

[4] K.L. Chung: Contributions to the theory of Markov chains. *Trans. Amer. Math. Soc.* 76, 397-419, 1954. There is an earlier paper with the same title in the *Journal of Research of the National Bureau of Standards* 50, 4, 203-298, 1953, where Doeblin's first problem was treated.

[5] K.L. Chung: **Markov chains with stationary transition probablities.** Springer-Verlag, 1967.

[6] K.L. Chung: Reminiscences of some of Paul Lévy's ideas in Brownian Motion and in Markov chains. *Soc. Math de France. Asterisque* 137-158, 1988. This is also in **Seminar on Stochastic Processes**, 99-108, 1988, Birkhäuser. The double appearance was occasioned by the added Postscript, which belongs to the annals of unbelievable academic folly.

STANFORD UNIVERSITY, DEPARTMENT OF MATHEMATICS, STANFORD, CALIFORNIA 94305, U.S.A.

Contemporary Mathematics
Volume **149**, 1993

The Coupling Technique
in
Interacting Particle Systems

THOMAS M. LIGGETT

ABSTRACT. One of Doeblin's most important contributions to probability is the coupling technique. One of the areas in which this technique has been most successful is interacting particle systems. This paper is devoted to explaining the use of this technique in this area.

A coupling is simply a joint construction of two or more stochastic processes. The coupling technique is the use of such a joint construction in order to obtain properties of the individual processes.

The coupling method was first used by Doeblin in the 1930's. (At the meeting in honor of Doeblin, Professor Chung stated that it had been used somewhat earlier by Kolmogorov, but he provided no specific reference.) The power of this technique was not recognized to any significant extent until the early 1970's, when it became one of the most useful tools in the then developing theory of interacting particle systems. Since then, it has become a standard part of every working probabilist's collection of techniques. Its use is now described in textbooks at both graduate and undergraduate levels ([**3**], [**9**], and [**15**]). A new book ([**13**]) is devoted entirely to it. Among the many other applications of the coupling technique are those related to Orey's theorem for Markov chains (see [**8**]), renewal theory (e.g., [**2**]), and diffusions (see [**6**]).

In this paper, we will try to explain via several examples how the coupling technique is used in the area of interacting particle systems. We begin in the first section with a brief description of Doeblin's original use of coupling to prove the basic convergence theorem for Markov chains. In Sections 2 and 3, we describe

1991 *Mathematics Subject Classification.* Primary 60K35; Secondary 60J25.

Preparation of this paper was supported in part by NSF Grant DMS 91-00725.

This paper is in final form and no version of it will be submitted for publication elsewhere.

its use in two particle system contexts: some nonergodic spin systems of current interest, and exclusion processes. One of the main points of these examples is to see how useful the technique is, even when the Markov process to which it is applied has a very large state space and more than one invariant measure.

1. Doeblin's application to Markov chains

In his 1938 paper [7], Doeblin considers a discrete time Markov chain with finitely many states and transition probabilities p_{ij}. (We use his notation as much as possible.) He defines two independent copies of this chain, starting from states i and k respectively. (This is the first coupling. It is the simplest imaginable choice. More sophisticated couplings use cleverly chosen interactions between the two chains.) Doeblin lets A_{ik} be the probability that the two chains are ever at the same state at the same time, and proves the following preliminary result:

THEOREM 1.1. *Either (a) $A_{ik} = 1$ for all i and k, or (b) $A_{ik} = 0$ for some i and k.*

Doeblin calls case (a) of the above result the regular case (in this case, modern terminology would call this coupling "successful"), and observes that if for some ρ, the ρ step transition probabilities are bounded below by $a > 0$, then the probability that the chains have not hit in $n\rho$ steps is at most $(1 - a)^n$, and hence the chain is regular. His main result is then

THEOREM 1.2. *In the regular case, the n-step transition probabilities converge:*

$$\lim_{n \to \infty} P_{ij}^{(n)} = P_j.$$

The key inequality which is used in the proof is the following:

$$|P_{ij}^{(n)} - P_{kj}^{(n)}| \leq \text{probability the two chains have not hit up to time } n.$$

In the regular case, the right side tends to zero as $n \to \infty$. While Doeblin does not say it quite this way, the above inequality is most easily seen by modifying the coupling in such a way that the chains remain together after they hit for the first time. If the chains are denoted by X_n and Y_n respectively, then the left side of the inequality is

$$|P(X_n = j) - P(Y_n = j)|,$$

and the right side is $P(X_n \neq Y_n)$.

2. The threshold voter model and threshold contact process.

The d-dimensional threshold voter model with range N is the continuous time Markov process η_t on $\{0,1\}^S$, where $S = Z^d$, in which the coordinate $\eta_t(x)$ flips at rate 1 if $\eta_t(x) \neq \eta_t(y)$ for some y with $||y - x|| \leq N$, and rate 0 otherwise. Here $|| \ ||$ denotes any norm on R^d such that $||x|| = 1$ for any vector x with one coordinate equal to one and the others equal to zero. If the process has a nontrivial invariant probability measure (i.e., one which is not a mixture of the pointmasses on $\eta \equiv 0$ and $\eta \equiv 1$), then the process is said to coexist, since both opinions "coexist" in equilibrium. If, on the other hand, $P\{\eta_t(x) \neq \eta_t(y)\} \to 0$ as $t \to \infty$ for every initial configuration, then the process is said to cluster. In a recent paper, Cox and Durrett [5] showed that the process clusters if $N = d = 1$, and conjectured that it coexists if $(N, d) \neq (1, 1)$. For each d, they showed coexistence for sufficiently large N (e.g., for $N \geq 4$ if $d = 1$). They did this via comparisons with certain contact processes. In each case, survival of some contact process implies coexistence for some voter model. (Their conjecture has recently been proved in [12].)

The d-dimensional threshold contact process with range N and parameter λ is the continuous time Markov process ζ_t on $\{0, 1\}^S$ in which the coordinate $\zeta_t(x)$ flips at rate 1 if $\zeta_t(x) = 1$, at rate λ if $\zeta_t(x) = 0$ and $\zeta_t(y) = 1$ for some y with $||y - x|| \leq N$, and at rate zero otherwise. The process is said to survive if it has a nontrivial invariant measure (i.e., one other than the pointmass on $\eta \equiv 0$). Otherwise, it is said to die out. In this section we will see how coupling techniques give a natural relation between a statement about the threshold contact process and the conclusion of the Cox-Durrett conjecture for the threshold voter model. This relation is an important part of our proof of the conjecture. The first coupling will be described in some detail. Similar couplings which appear later are described more briefly.

First, we note that there is a natural coupling between two copies of the threshold contact process with parameters λ_1 and λ_2 respectively which preserves the relation $\zeta_{t,1}(x) \leq \zeta_{t,2}(x)$ for all x, provided that $\lambda_1 \leq \lambda_2$. It is defined by letting $\zeta_{t,1}(x)$ and $\zeta_{t,2}(x)$ flip independently at the appropriate rates if they are different and together if they are both 1. If they are both 0, then at rate λ_1 they both flip and at rate $\lambda_2 - \lambda_1$ only the second coordinate flips if $\zeta_{t,1}(y) = 1$ for some y with $||y - x|| \leq N$. On the other hand, only the coordinate of the second process flips (at rate λ_2) if $\zeta_{t,2}(y) = 1$ for some y with $||y - x|| \leq N$, but $\zeta_{t,1}(y) = 0$ for all y with $||y - x|| \leq N$. The coupling can be defined formally as a Markov process $(\zeta_{t,1}, \zeta_{t,2})$ on

$$\{(\zeta_1, \zeta_2) \in \{0, 1\}^S \times \{0, 1\}^S : \zeta_1(x) \leq \zeta_2(x) \, \forall x\}$$

with the transition rates described above. As a consequence of the existence of this coupling, survival of $\zeta_{t,1}$ implies survival of $\zeta_{t,2}$. Therefore, there is a critical value $\lambda_c(N, d)$ so that the threshold contact process survives if $\lambda > \lambda_c(N, d)$ and dies out if $\lambda < \lambda_c(N, d)$.

A similar coupling argument gives the first of the following results. It is enough to couple together two copies of the threshold contact process with the same parameter λ but different values of (N, d). Note that there is no obvious analogous statement in the context of the threshold voter model. That is, one cannot compare directly threshold voter models with different values of N and/or d in such a way that coexistence in one implies coexistence in the other. This is one of the reasons for the usefulness of the threshold contact process in the analysis of the threshold voter model.

THEOREM 2.1. $\lambda_c(N, d)$ *is a nonincreasing function of* N *and* d.

THEOREM 2.2. *For any* N *and* d, *if* $\lambda_c(N, d) < 1$, *then the corresponding threshold voter model coexists.*

THEOREM 2.3. $\lambda_c(1, 2) \leq \lambda_c(2, 1)$.

Combining these theorems, we conclude that the Cox-Durrett conjecture would be a consequence of $\lambda_c(2, 1) < 1$. (In fact, it is enough to know that the threshold contact process survives if $N = 2, d = 1$ and $\lambda = 1$.) Numerical work provides the estimate $\lambda_c \leq .82$ (see [4]). As a further illustration of the use of coupling, we will sketch below the proofs of Theorems 2.2 and 2.3. Full proofs of these results will appear in [12]. Theorem 2.2 is due to Cox and Durrett. Our proof is somewhat different, and is more elementary in that it does not rely on hard results about the contact process (e.g., the complete convergence theorem).

SKETCH OF THE PROOF OF THEOREM 2.2. The proof is based on comparisons which follow from couplings between pairs of the following three processes: the threshold voter model with range N, the threshold contact process with range N and $\lambda = 1$, and the independent flip process in which each coordinate flips at rate 1, independently of the other coordinates. When the initial distribution is a measure μ, let the distribution at time t of these processes be denoted by $\mu S_v(t)$, $\mu S_c(t)$, and $\mu S_i(t)$ respectively. Let ν be the upper invariant measure of the threshold contact process, which is nontrivial by assumption, and let $\nu_{1/2}$ be the product measure with density $1/2$.

Note that the contact process and the voter model have the same rates for the transition $0 \to 1$, while the contact process has larger rates than the voter model for the transition $1 \to 0$. Similarly, the contact process and the independent flip process have the same rates for the transition $1 \to 0$, while the contact process has smaller rates than the independent flip process for the transition $0 \to 1$. Note also that in all three cases, the $0 \to 1$ rates are increasing functions of the configuration, while the $1 \to 0$ rates are decreasing functions of the configuration. (This property is known as attractiveness.) We use the notation $\mu_1 \leq \mu_2$ to denote stochastic comparisons (i.e., there is a coupling measure which concentrates on the appropriate side of the diagonal and has marginals μ_1 and μ_2 respectively). Then the natural couplings between these processes give the inequalities which follow, while the asserted convergence is trivial:

(a) $$\nu = \nu S_c(t) \leq \nu S_i(t) \to \nu_{1/2}.$$

(Here $\to$ denotes weak convergence.) Therefore,

(b) $$\nu \leq \nu_{1/2}.$$

By part (b),

(c) $$\nu = \nu S_c(t) \le \nu S_v(t) \le \nu_{1/2} S_v(t).$$

Therefore,

(d) $$\nu \le \nu_{1/2} S_v(t).$$

It follows from part (d) that any limit of a Cesaro average of $\nu_{1/2} S_v(t)$ is stochastically larger than ν, and hence concentrates on configurations with infinitely many ones. Since it is invariant under interchange of zeros and ones (both the intial distribution and the evolution mechanism have this invariance), it also concentrates on configurations with infinitely many zeros. Therefore, there is coexistence.

SKETCH OF THE PROOF OF THEOREM 2.3. Let ζ_t be the threshold contact process on Z^1 with parameter λ and range $N = 2$, and γ_t the threshold contact process on Z^2 with parameter λ and range $N = 1$. Consider these as random subsets of the appropriate lattice via the identification $\{x : \zeta_t(x) = 1\}$. We need to show that the survival of ζ_t implies that of γ_t. In order to compare the two processes via coupling, define a mapping $\pi : Z^2 \to Z^1$ by $\pi(m, n) = m + 2n$. This can be represented visually by attaching labels to the points in Z^2 as follows:

$$
\begin{array}{ccccc}
+2 & +3 & +4 & +5 & +6 \\
0 & +1 & +2 & +3 & +4 \\
-2 & -1 & 0 & +1 & +2 \\
-4 & -3 & -2 & -1 & 0 \\
-6 & -5 & -4 & -3 & -2
\end{array}
$$

The important property of this mapping is that the four (two dimensional) neighbors of a site with label k have labels which are the four (one dimensional) neighbors of k. This means that we can couple ζ_t and γ_t so that $\pi(\gamma_t) \supset \zeta_t$ at all times. To perform the coupling, if $\pi(\gamma_t) \supset \zeta_t$ at a given time, use some consistent rule to identify each point x in ζ_t with some point y in γ_t whose image under π is x. Use the same unit exponential times to determine when x and y will be deleted from the corresponding configuration. On the other hand, if $x \notin \zeta_t$ but some neighbor $x' \in \zeta_t$ (so there is rate λ of adding x to ζ_t), identify x' with some point y' in γ_t whose image under π is x', and couple the addition of x to ζ_t with the addition to γ_t of the neighbor y of y' which satisfies $\pi(y) = x$. (If y is already in γ_t, no addition is necessary.)

3. The exclusion process

We will consider here a special case of the exclusion process—the one dimensional asymmetric system with nearest neighbor jumps. It is a continuous time

Markov process η_t on $\{0,1\}^Z$ which describes the motion of particles on Z with at most one particle per site. A particle at $x \in Z$ waits a unit exponential time. At that time, it chooses to go to $x + 1$ with probability p and to $x - 1$ with probability q, where $p + q = 1$ and $p > 1/2$. If the selected site is vacant, the particle moves there, while if the selected site is occupied, the particle stays at x. A treatment of more general exclusion processes can be found in Chapter VIII of [**11**].

It has been known since Spitzer's original 1970 paper [**14**] that the product measures ν_ρ with density ρ, $0 \le \rho \le 1$, are invariant for this system. Our first illustration of coupling in the exclusion context is a sketch of the proof that these are the only extremal invariant measures which are translation invariant. (There are also invariant measures which are not translation invariant in this case.) Details can be found in Section 2 of Chapter VIII of [**11**]. By contrast with Doeblin's original application, it is interesting to note that coupling can be used effectively to characterize invariant measures even when there are many of them, and hence one cannot expect the coupling to be successful.

Two (or more) copies of the process can be coupled in a natural way: particles at the same site in different copies use the same exponential times to decide when to attempt a transition, and make the same decisions about the direction in which to jump. Of course, whether or not the jump actually occurs depends on the copy of the process one is looking at. This coupling has the very useful property that the (coordinatewise) relation $\eta_t \le \zeta_t$ is preserved. The bivariate coupled process is also a Markov process, but now with state space $\{0,1\}^Z \times \{0,1\}^Z$. Let $\mathcal{I}$ denote the invariant measures for the original process which are translation invariant, and $\mathcal{I}^*$ be the invariant measures for the coupled process which are translation invariant. The proof that

$$\mathcal{I}_e = \{\nu_\rho, 0 \le \rho \le 1\}$$

is based on the following outline (the subscript e denotes extreme points):

1. If $\nu \in \mathcal{I}^*$, then $\nu\{(\eta,\zeta) : \eta \le \zeta \text{ or } \eta \ge \zeta\} = 1$. Proof: Let $\nu(t)$ be the distribution of the process at time t. A computation which uses the translation invariance of ν (but not its invariance) gives

$$\frac{d}{dt}\nu(t)\begin{pmatrix} 1 \\ 0 \end{pmatrix} = -\nu(t)\begin{pmatrix} 0 & 1 \\ 1 & 0 \end{pmatrix} - \nu(t)\begin{pmatrix} 1 & 0 \\ 0 & 1 \end{pmatrix},$$

where we have used some shorthand notation. The probability which is being differentiated, for example, is the probability that $\eta(x) = 0$ and $\zeta(x) = 1$. (It does not matter what x is, since the measure is translation invariant.) The probabilities on the right side correspond to events based on two adjacent sites. For example, the first one is the probability that $\eta(x) = 1, \zeta(x) = 0, \eta(x+1) = 0$, and $\zeta(x + 1) = 1$. When the differentiation is performed, two types of terms appear on the right: those which correspond to the motion of a discrepancy, and those which correspond to two discrepancies of opposite types annihilating each other. Terms of the first type occur with both positive and negative signs, and cancel by translation invariance of the measure. Those of the second type remain. It is important in what follows that all the remaining terms have the same sign.

If ν is also invariant, then the derivative is zero, so the terms on the right side are zero as well. So, ν puts no mass on configurations with opposite discrepancies at adjacent sites. Using invariance again, it is easy to see that it puts no mass on configurations with opposite discrepancies at any two sites. The main idea behind this is most easily seen by considering the following elementary statement about Markov chains: if π is a stationary distribution for the chain and x and y are two states so that there is a positive rate of going from x to y, then $\pi(y) = 0$ implies $\pi(x) = 0$. The assertion in #1 is equivalent to the statement that discrepancies of opposite types cannot coexist in ν, so we are done.

2. An immediate consequence of #1 and the fact that the coupled process cannot leave the sets $\{(\eta, \zeta) : \eta \leq \zeta\}$ and $\{(\eta, \zeta) : \eta \geq \zeta\}$ is: If $\nu \in \mathcal{I}_e^*$, then either

$$\nu\{(\eta, \zeta) : \eta \leq \zeta\} = 1$$

or

$$\nu\{(\eta, \zeta) : \eta \geq \zeta\} = 1.$$

Proof: If not, then we can write

$$\nu = \alpha\nu_1 + (1 - \alpha)\nu_2,$$

where $\nu_1 = \nu(\ |\eta \leq \zeta)$ and $\alpha = \nu((\eta, \zeta) : \eta \leq \zeta)$, for example. But this violates the extremality of ν .

3. If μ_1 and μ_2 are in $\mathcal{I}_e$, then there is a $\nu \in \mathcal{I}_e^*$ with marginals μ_1 and μ_2 respectively. Proof: Start the coupled process with initial distribution $\mu_1 \times \mu_2$. Then every weak limit (along a subsequence of times) of Cesaro averages of the distribution of the coupled process at time t is in

$$\{\nu \in \mathcal{I}^* : \nu \text{ has marginals } \mu_1 \text{ and } \mu_2 \text{ respectively,}\}$$

and therefore this set is nonempty. Any extreme point of this set has the desired properties.

4. Combining #2 and #3, it follows that: If μ_1 and μ_2 are in $\mathcal{I}_e$, then either $\mu_1 \leq \mu_2$ or $\mu_1 \geq \mu_2$.

5. Now take $\mu \in \mathcal{I}_e$. Since $\nu_\rho \in \mathcal{I}_e$ also (it is extremal already in the class of translation invariant measures), one can apply #4 to $\mu_1 = \mu$ and $\mu_2 = \nu_\rho$. Therefore, for every $\rho \in [0, 1]$, either $\mu \leq \nu_\rho$ or $\mu \geq \nu_\rho$. It is then easy to see that $\mu = \nu_\rho$ for some ρ, thus completing the proof.

The above proof is a rather soft application of coupling. Our final example is a significantly harder application. It involves the limiting behavior of the system when the initial distribution is not translation invariant. Fix $\lambda, \rho \in (0,1)$, and let μ be the product measure on $\{0,1\}^{\mathbb{Z}}$ with marginals

$$\mu\{\eta(x) = 1\} = \begin{cases} \lambda & \text{if } x < 0 \\ \rho & \text{if } x \geq 0. \end{cases}$$

It has been known for fifteen years that with this initial distribution, the limiting distribution as $t \to \infty$ is the product measure with density given by the following picture, in all cases except $\lambda + \rho = 1, \rho > 1/2$:

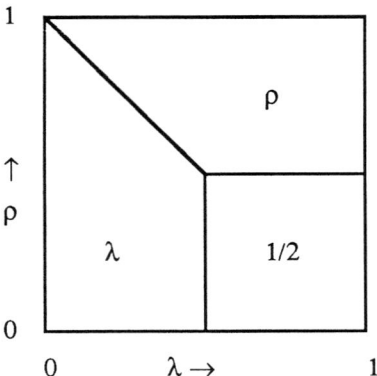

(See [10], for example. If $p = 1/2$, the situation is entirely different, and the limit is the product measure with density $(\lambda + \rho)/2$ for any λ, ρ.) The situation on the diagonal line above was unclear until recently, when Andjel, Bramson and Liggett ([1]) proved that the limit in this case is

$$\frac{1}{2}\nu_\lambda + \frac{1}{2}\nu_\rho.$$

One of the interesting features of this is that the limit is not an extremal invariant measure.

The proof of this result is rather long and involves a number of coupling arguments. An important part of this is the proof that any limit point of $\mu(t)$ as $t \to \infty$ is translation invariant. We will show now how coupling is used in the proof of this fact, along the lines of the following outline (the details of all of the steps can be found in the paper):

1. There is a coupling ν of μ with the translate of μ by one site with the property that μ concentrates on configurations (η, ζ) with finitely many discrepancies, and the same number of discrepancies of each type, i.e.,

$$\#\{x : \eta(x) = 0, \zeta(x) = 1\} = \#\{x : \eta(x) = 1, \zeta(x) = 0\}.$$

To see this, note that the marginals of μ and of its translate agree except at the origin. So, one can take $\eta(x) = \zeta(x)$ for $x < 0$, choose $\eta(0)$ and $\zeta(0)$ in the obvious way, $(\eta(x), \zeta(x)) = (0,0), (0,1), (1,0)$, and $(1,1)$ with probabilities

$\lambda^2, \lambda\rho, \lambda\rho$, and ρ^2 respectively for positive x's, until the numbers of discrepancies of the two types agree (this will happen in a finite number of steps because

$$\sum_{x=1}^{N} [\eta(x) - \zeta(x)]$$

is a symmetric random walk on the integers, and is therefore recurrent), and $\eta(x) = \zeta(x)$ thereafter. It is now enough to show that in the evolution of the coupled process, the discrepancies all disappear eventually.

2. To show that the discrepancies disappear eventually, we consider the simplest special case of an initial configuration. Suppose there is only one discrepancy of each type, and they are at the adjacent sites 0 and 1. This situation can be depicted visually in the following way:

$$\zeta : \qquad\qquad \cdot \quad \cdot \quad \cdot \quad 1 \quad 0 \quad \cdot \quad \cdot \quad \cdot$$

$$\eta : \qquad\qquad \cdot \quad \cdot \quad \cdot \quad 0 \quad 1 \quad \cdot \quad \cdot \quad \cdot$$
$$ X_0 \quad Y_0$$

The dots on the left represent Bernoulli random variables with parameter λ and those on the right Bernoulli random variables with parameter ρ ($\eta(x) = \zeta(x)$ for $x \neq 0, 1$). As the system evolves, there will always be at most one discrepancy of each type, and we let X_t and Y_t be the positions of the discrepancies, as long as they are present. The discrepancies disappear when for the first time, $X_t = Y_t$. After that time, we let X_t and Y_t evolve together as what is known as a "second class particle". This means that if another particle tries to move to the location of the second class particle, then the two particles exchange positions. (In fact, X_t and Y_t are second class particles even before they meet.)

More explicity, the following transition rates apply to transitions involving the second class particle, whose location is denoted by $\star$:

$$\star 0 \to 0 \star \qquad \text{at rate } p.$$

$$\star 1 \to 1 \star \qquad \text{at rate } q.$$
$$0 \star \to \star 0 \qquad \text{at rate } q.$$
$$1 \star \to \star 1 \qquad \text{at rate } p.$$

We see that X_t and Y_t are defined for all time, and we must show that

$$P(X_t = Y_t \text{ eventually}) = 1.$$

3. To show this, note that it is enough to show that the distribution of $Y_t - X_t$ is tight. The reason for this is that if $Y_t - X_t = k$, there is a probability $\geq \epsilon(k) > 0$ (independently of the rest of the configuration) that the two second class particles hit before they separate further. To show tightness, it is enough to show L_1 boundedness. Since $X_t \leq Y_t$ for all t, it suffices to show that

$$EY_t \leq EX_t + 1.$$

4. To obtain this inequality, shift the initial configuration of one of the processes, so that the second class particles are initially at the same site, say the origin. After the shift, the inequality we need to prove becomes

$$EY_t \leq EX_t.$$

The initial distributions of the two processes can now be described as follows:

$$\cdot \quad \cdot \quad \cdot \quad \lambda \quad \lambda \quad \lambda \quad \star \quad 0 \quad \rho \quad \rho \quad \cdot \quad \cdot \quad \cdot$$

for the process with second class particle at X_t, and

$$\cdot \quad \cdot \quad \cdot \quad \lambda \quad \lambda \quad 0 \quad \star \quad \rho \quad \rho \quad \rho \quad \cdot \quad \cdot \quad \cdot$$

for the process with second class particle at Y_t. The λ's, ρ's, and 0's indicate the parameters of the Bernoulli random variable at those sites, and the $\star$'s show the positions of the second class particles. Using this suggestive notation, letting Z_t be the position of the second class particle for a generic initial distribution, and using conditioning notation to indicate the initial distribution, we can now write

$$EX_t = \lambda E(Z_t| \cdots \lambda\,\lambda\,1 \star 0\,\rho\,\rho \cdots) + \rho E(Z_t| \cdots \lambda\,\lambda\,0 \star 0\,\rho\,\rho \cdots)$$

and

$$EY_t = \rho E(Z_t| \cdots \lambda\,\lambda\,0 \star 1\,\rho\,\rho \cdots) + \lambda E(Z_t| \cdots \lambda\,\lambda\,0 \star 0\,\rho\,\rho \cdots)$$

The first term on the right of each of these expressions is zero (it is the expected value of a random variable with a symmetric distribution) by a symmetry argument (the evolution of the process and the initial distribution are invariant under the transformation $\eta \to \eta'$, where η' is obtained from η by interchanging zeros and ones and reflecting the integers about the origin). Therefore,

$$EX_t - EY_t = (\rho - \lambda)E(Z_t| \cdots \lambda\,\lambda\,0 \star 0\,\rho\,\rho \cdots).$$

Since $\lambda < \rho$, we need to show that

$$E(Z_t| \cdots \lambda\,\lambda\,0 \star 0\,\rho\,\rho \cdots) \geq 0.$$

But

$$E(Z_t| \cdots \lambda\,\lambda\,1 \star 0\,\rho\,\rho \cdots) = 0$$

as observed earlier, so it suffices to carry out the comparison in step # 5:

5. $E(Z_t| \cdots \eta(-3)\,\eta(-2)\,\eta(-1) \star \eta(1)\,\eta(2)\,\eta(3) \cdots)$ is a decreasing function of the $\eta(k)$'s. To show this, another coupling argument is needed. We couple together two copies of the process with one second class particle, (η_t, X_t) and (ζ_t, Y_t), where again, X_t and Y_t are the locations of the second class particles. We want to do so in such a way that the following relation is maintained:

$$\eta_t \leq \zeta_t \quad \text{and} \quad X_t \geq Y_t.$$

The details of the coupling will be left to the reader, but we will consider one situation, so that it will become clear what is required. Suppose that at some time, the second class particles are at the same location, and the two configurations are:

$$\zeta_t : \qquad\qquad \cdot \quad \cdot \quad \cdot \quad \star \quad 1 \quad \cdot \quad \cdot \quad \cdot$$

$$\eta_t : \qquad\qquad \cdot \quad \cdot \quad \cdot \quad \star \quad 0 \quad \cdot \quad \cdot \quad \cdot$$

Then (recalling the transition rates for the second class particles in step #2 above) the second class particle in the top configuration moves to the right at rate q, while the second class particle in the bottom configuration moves to the right at rate p. Since $p > q$, the coupling can be defined so that both move to the right together at rate q, while only the bottom one moves to the right at rate $p - q$. This preserves the relation $X_t \geq Y_t$ as required. In the second case, for example, the configurations after the transition are as follows:

$$\zeta_t : \qquad\qquad \cdot \quad \cdot \quad \cdot \quad \star \quad 1 \quad \cdot \quad \cdot \quad \cdot$$

$$\eta_t : \qquad\qquad \cdot \quad \cdot \quad \cdot \quad 0 \quad \star \quad \cdot \quad \cdot \quad \cdot$$

REFERENCES

1. E. Andjel, M. Bramson and T. Liggett, *Shocks in the asymmetric exclusion process*, Prob. Th. Rel. Fields **78** (1988), 231–247.

2. K. Athreya, D. McDonald and P. Ney, *Coupling and the renewal theorem*, Amer. Math. Monthly **85** (1978), 809–814.

3. P. Billingsley, *Probability and Measure*, Wiley, New York, 1979.

4. L. Buttel, J. T. Cox and R. Durrett, *Estimating the critical values of stochastic growth models*.

5. J. T. Cox and R. Durrett, *Nonlinear voter models*, Festschrift in honor of Frank Spitzer, Birkhauser, 1991, pp. 189–201.

6. M. Cranston, *Gradient estimates on manifolds using coupling*, J. Funct. Anal. **99** (1991), 110–124.

7. W. Doeblin, *Exposé de la theorie des chaines simples constantes de Markoff a un nombre fini d'etats*, Rev. Math. Union Interbalkanique **2** (1938), 77–105.

8. D. Griffeath, *Coupling Methods for Markov Processes*, Studies in Probability and Ergodic Theory; Advances in Mathematics, Supplementary Studies, vol. II, Academic Press, New York, 1978, pp. 1–43.

9. P. Hoel, S. Port and C. Stone, *Introduction to Stochastic Processes*, Houghton Mifflin, Boston, 1972.

10. T. M. Liggett, *Ergodic theorems for the asymmetric exclusion process II*, Ann. Prob. **5** (1977), 795–801.

11. T. M. Liggett, *Interacting Particle Systems*, Springer, New York, 1985.

12. T. M. Liggett, *Coexistence in threshold voter models*.

13. T. Lindvall, *Lectures on the Coupling Method*, Wiley, New York, 1992.

14. F. Spitzer, *Interaction of Markov processes*, Adv. Math. **5** (1970), 246–290.

15. S. M. Ross, *Stochastic Processes*, Wiley, New York, 1983.

DEPARTMENT OF MATHEMATICS, UNIVERSITY OF CALIFORNIA, LOS ANGELES, CALIFORNIA 90024

E-mail address: tml@math.ucla.edu

Contemporary Mathematics
Volume **149**, 1993

Coupling and Shift-Coupling
Random Sequences

HERMANN THORISSON

1   Introduction

Doeblin is generally considered to be the inventor of the so called coupling method. In the 1938 paper *Exposé de la theorie des chaînes simple constantes de Markov à un nombre fini d'états* he proved the asymptotic stationarity of a regular finite state Markov chain X along the following lines. Let a differently started version X' jump independently of X until the two chains meet, at a time T say. From T onward let X and X' jump together. Regularity means that there is an m and an $\epsilon > 0$ such that $\mathbf{P}_i(X_m = j) \geq \epsilon$ for all initial states i and all j. This implies that $\mathbf{P}(T > km) \leq (1 - \epsilon)^k \rightarrow 0$ as $k \rightarrow \infty$. Thus the chains eventually coincide and we obtain

$$| \mathbf{P}(X_n = j) - \mathbf{P}(X'_n = j) | \leq \mathbf{P}(X_n \neq X'_n) \rightarrow 0 \quad \text{as } n \rightarrow \infty.$$

Add to this the observation that $\max_i \mathbf{P}_i(X_n = j)$ is non-increasing and $\min_i \mathbf{P}_i(X_n = j)$ is non-decreasing in n to deduce that $\mathbf{P}(X_n = j)$ has a limit. (Nowadays one usually takes X' stationary which makes the last sentence unnecessary.)

This idea of establishing limit results by pasting together the paths of two processes seems to have passed more or less unnoticed. During the last two decades, however, a revival of the method has lead to powerful new results and to simple proofs of known achievements in fields such as interacting particle systems, queuing, Markov theory, renewal theory and regeneration. For a survey the reader is referred to Lindvall's new book on the subject.

Here we shall focus on certain general aspects of coupling. From the end of the seventies it has been rather well-known that total variation convergence

1991 Mathematics Subject Classification. Primary 60H99, 60J05

This paper is in final form and no version of it will be submitted for publication elsewhere.

and the tail σ-algebra have to do with coupling. It has not been as well-known, however, that in the same way total variation convergence of *time-averages* and the *invariant* σ-algebra have to do with *shift*-coupling, a variation on coupling which means that the paths of the processes are pasted together modulo a random shift. Berbee (1979) established the time-average part and the invariant σ-algebra part is added in a recent paper by Aldous and the present author; see also Greven (1987).

We shall both review these results and present new material which fills a gap in the analogy between coupling and shift-coupling. The key new result is a *shift-coupling inequality* which plays a similar role for time-average asymptotics as the well-known coupling inequality for asymptotics. Extensions to continuous time are carried out in a forthcoming paper.

After establishing notation in Section 2, we collect the coupling results in Section 3, the shift-coupling analogs in Section 4 and new proofs in Section 5. For proofs of facts stated below and not proved here the reader is referred to Aldous and Thorisson (1993) unless otherwise indicated.

2   Notation

Let $(\Omega, \mathcal{F}, \mathbf{P})$ be the common probability space supporting all random elements in this paper. Let $Z = (Z_0, Z_1, \ldots)$ and $Z' = (Z'_0, Z'_1, \ldots)$ be discrete time stochastic processes on a Polish state space $(E, \mathcal{E})$. For $x = (x_0, x_1, \ldots) \in E^\infty$ define the shift-maps $\theta_n, 0 \le n < \infty$, by $\theta_n x = (x_n, x_{n+1}, \ldots)$; define θ_∞ by $\theta_\infty x = (\Delta, \Delta, \ldots)$ where Δ is a fixed state not in E. Put $\theta_t = \theta_{[t]}$ for $t \in [0, \infty)$.

Let $\mathcal{T}$ be the *tail σ-algebra*:

$$\mathcal{T} = \bigcap_{n=0}^{\infty} \mathcal{T}_n \quad \text{where} \quad \mathcal{T}_n = \theta_n^{-1} \mathcal{E}^\infty \quad \text{is the post-}n\ \sigma\text{-algebra.}$$

Let $\mathcal{I}$ be the the *invariant σ-algebra*:

$$\mathcal{I} = \{B \in \mathcal{E}^\infty : \theta_1^{-1} B = B\}. \quad (\text{note that}\, \mathcal{I} \subseteq \mathcal{T})$$

With $\mathcal{A}$ a sub-σ-algebra of $\mathcal{E}^\infty$ let $\mathbf{P}(Z \in \cdot)_{\mathcal{A}}$ denote the restriction of $\mathbf{P}(Z \in \cdot)$ from $\mathcal{E}^\infty$ to $\mathcal{A}$. Let $\| \cdot \|$ denote the *total variation norm* defined for bounded signed measures ν on $\mathcal{A}$ by

$$\| \nu \| = \sup_{A \in \mathcal{A}} \nu(A) - \inf_{A \in \mathcal{A}} \nu(A)$$
$$= \text{mass of}\, \nu^+ + \text{mass of}\, \nu^-.$$

Let $\to_{t.v.}$ denote total variation convergence of random elements in $\mathcal{E}^\infty$.

Throughout the paper let U be a random variable that is uniformly distributed on $[0, 1)$ and independent of Z and Z' and all other random elements introduced below.

3 Coupling

3.1. Definition. A pair of processes $\hat{Z}$ and $\hat{Z}'$ is a *coupling* of Z and Z' if

$$\hat{Z} \stackrel{D}{=} Z \quad \text{and} \quad \hat{Z}' \stackrel{D}{=} Z' \,.$$

A random time T is a *coupling epoch* if

$$\theta_T \hat{Z} = \theta_T \hat{Z}' \,.$$

Say that Z and Z' *admit coupling* if there exists a coupling of Z and Z' with coupling epoch T such that $\mathbf{P}(T < \infty) = 1$.

3.2. The coupling inequality. The following result is of basic importance for proving limit theorems: if T is a coupling epoch then

$$(3.1) \qquad \| \mathbf{P}(\theta_n Z \in \cdot) - \mathbf{P}(\theta_n Z' \in \cdot) \| \leq 2\mathbf{P}(T > n), \quad n \geq 0 \,.$$

3.3. Asymptotics. If T is a.s. finite the inequality yields

$$(3.2) \qquad \| \mathbf{P}(\theta_n Z \in \cdot) - \mathbf{P}(\theta_n Z' \in \cdot) \| \to 0 \quad \text{as} \quad n \to \infty$$

and in particular if Z' is stationary then (3.2) can be rewritten as

$$\theta_n Z \to_{t.v.} Z' \quad \text{as} \quad n \to \infty \,.$$

Further, if there is an $\alpha \in (0, \infty)$ such that $\mathbf{E}[T^\alpha] < \infty$ then $n^\alpha \mathbf{P}(T > n) \leq \mathbf{E}[T^\alpha 1_{\{T > n\}}] \to 0$ as $n \to \infty$ and thus

$$(3.3) \qquad n^\alpha \| \mathbf{P}(\theta_n Z \in \cdot) - \mathbf{P}(\theta_n Z' \in \cdot) \| \to 0 \quad \text{as } n \to \infty \,.$$

From $\mathbf{E}[T^\alpha] < \infty$ it is actually straightforward to deduce a stronger result:

$$\sum_{n=0}^{\infty} n^{\alpha-1} \| \mathbf{P}(\theta_n Z \in \cdot) - \mathbf{P}(\theta_n Z' \in \cdot) \| < \infty \,.$$

More general limit results follow by replacing n^α by an increasing non-negative function $\phi(n)$ such as $\phi(n) = e^{\alpha n}$ or $\phi(n) = n^\alpha \log^+ n$.

3.4. Maximal coupling. Total variation asymptotics are not only a consequence of coupling but actually equivalent to coupling since there exists a *maximal coupling*, i.e. a coupling with epoch T such that the coupling inequality is an identity:

$$(3.4) \qquad \| \mathbf{P}(\theta_n Z \in \cdot) - \mathbf{P}(\theta_n Z' \in \cdot) \| = 2\mathbf{P}(T > n), \quad n \geq 0 \,.$$

This result was established for countable state space Markov chains by Griffeath (1975) while Pitman (1976) gave a constructive picture. Griffeath (1978) went on to Polish state space Markov chains. Berbee (1979) noted that this actually covered the general non-Markovian Polish state space case since $(\theta_n Z : 0 \leq n < \infty)$ is Markovian with a Polish state space. Goldstein (1979) gave a direct proof in the general case. Thorisson (1986) gave a simplified proof using "distributional coupling". Greven (1987) and Harison and Smirnov (1990) constructed adapted maximal coupling in the Markovian case. The continuous time D-space case is treated by Svertchkov and Smirnov (1990). The last four papers use distributional coupling. It should be noted that the above facts and the upcoming results hold without the restriction to a Polish state space if we replace the coupling concepts by their distributional counterparts (cf. Section 5.4).

A straightforward reformulation of (3.4) is the following: for $n \geq 0$

$$(3.5) \qquad \mathbf{P}(\theta_n \hat{Z} \in \cdot, T > n) \quad \perp \quad \mathbf{P}(\theta_n \hat{Z}' \in \cdot, T > n)$$

where $\perp$ denotes mutual singularity of the two measures.

3.5. $\mathcal{T}$-maximality. With $\mathcal{A}$ a sub-σ-algebra of $\mathcal{E}^\infty$ call an event $C \in \mathcal{F}$ $\mathcal{A}$-coupling event if

$$\{\hat{Z} \in A\} \cap C = \{\hat{Z}' \in A\} \cap C, \qquad A \in \mathcal{A}.$$

It is readily checked that the following $\mathcal{A}$-coupling inequality holds:

$$\| \mathbf{P}(Z \in \cdot)_{\mathcal{A}} - \mathbf{P}(Z' \in \cdot)_{\mathcal{A}} \| \leq 2\mathbf{P}(C^c).$$

If T is a coupling epoch then $\{T \leq n\}$ is a $\mathcal{T}_n$-coupling event. Therefore (3.1) is a special case of the $\mathcal{T}_n$-coupling inequality and the maximal coupling is $\mathcal{T}_n$-maximal:

$$\| \mathbf{P}(Z \in \cdot)_{\mathcal{T}_n} - \mathbf{P}(Z' \in \cdot)_{\mathcal{T}_n} \| = 2\mathbf{P}(T > n).$$

Since $\mathcal{T}_n$ decreases to $\mathcal{T}$ one can from this deduce that $\{T < \infty\}$ is a $\mathcal{T}$-coupling event and that the maximal coupling is $\mathcal{T} - maximal$:

$$\| \mathbf{P}(Z \in \cdot)_{\mathcal{T}} - \mathbf{P}(Z' \in \cdot)_{\mathcal{T}} \| = 2\mathbf{P}(T = \infty).$$

We thus have the general limit result: as $n \to \infty$,

$$(3.6) \qquad \| \mathbf{P}(\theta_n Z \in \cdot) - \mathbf{P}(\theta_n Z' \in \cdot) \| \to \| \mathbf{P}(Z \in \cdot)_{\mathcal{T}} - \mathbf{P}(Z' \in \cdot)_{\mathcal{T}} \| .$$

3.6. Equivalences. Combining $\mathcal{T}$-maximality and (3.6) yields immediately that the following three statements are equivalent:

(i) Z and Z' admit coupling;

(ii) $\| \mathbf{P}(\theta_n Z \in \cdot) - \mathbf{P}(\theta_n Z' \in \cdot) \| \to 0$ as $n \to \infty$;

(iii) $\mathbf{P}(Z \in \cdot)_{\mathcal{T}} = \mathbf{P}(Z' \in \cdot)_{\mathcal{T}}$.

This result dates back to Goldstein (1979).

3.7. Markov chains. If X and X' are *homogeneous* Markov chains on a Polish state space with the same transition kernel P and initial distributions μ and μ', respectively, then the above equivalences easily yield that the following three statements are equivalent:

(a) X and X' admit coupling for all μ and μ';

(b) $\| \mu P^n - \mu' P^n \| \to 0$ as $n \to \infty$ for all μ and μ';

(c) $\mathbf{P}(X \in \cdot)_T = 0$ or 1 for all μ.

Griffeath (1978) added (a) while the equivalence of (b) and (c) had been known for some time and also that one can add the following two statements to the list:

(d) all space-time-harmonic functions are constant;

(e) X is mixing for all μ.

3.8. Inhomogeneous case. For *inhomogeneous* Markov chains the equivalence of (a), (b) and (c) still holds with a slight modification. Let P_n, $n \geq 0$, be transition kernels on a Polish state space $(E, \mathcal{E})$. For each $m \geq 0$ let $X^{(m)}$ and $X^{(m)\prime}$ be inhomogeneous Markov chains with the same transition kernels $P_{m+n}, n \geq 0$, and arbitrary initial distributions μ and μ', respectively. Then the following statements are equivalent:

(a$'$) $X^{(m)}$ and $X^{(m)\prime}$ admit coupling for all μ and μ' and all $m \geq 0$;

(b$'$) $\| (\mu - \mu') P_{m+1} \ldots P_{m+n} \| \to 0$ as $n \to \infty$ for all μ and μ' and all $m \geq 0$;

(c$'$) $\mathbf{P}(X^{(m)} \in \cdot)_T = 0$ or 1 for all μ and all $m \geq 0$.

4   Shift-coupling

4.1. Definition. Let $\hat{Z}$ and $\hat{Z}'$ be a coupling of Z and Z'. Two random times, T and T', are *shift-coupling epochs* if

$$\theta_T \hat{Z} = \theta_{T'} \hat{Z}' \,.$$

Since Δ is not in E this implies

$$\{T < \infty\} = \{T' < \infty\} \,.$$

When $T < \infty$ then $T - T'$ is the *shift*. There is no shift if $T \equiv T'$ and then T is an ordinary coupling epoch. Say that Z and Z' *admit shift-coupling* if there exists a coupling of Z and Z' with shift-coupling epochs T, T' such that $\mathbf{P}(T < \infty) = 1$.

4.2. The shift-coupling inequality. In Section 5.1 we prove the following result: if T and T' are shift-coupling epochs then

$$(4.1) \qquad \| \, \mathbf{P}(\theta_{Un} Z \in \cdot) - \mathbf{P}(\theta_{Un} Z' \in \cdot) \, \| \leq 2\mathbf{P}(T \vee T' > Un), \quad n \geq 0.$$

The analogy between the coupling inequality and this shift-coupling inequality is stressed in different ways by the following straightforward reformulations: for $n \geq 0$

$$\| \, \mathbf{P}(\theta_{Un} Z \in \cdot) - \mathbf{P}(\theta_{Un} Z' \in \cdot) \, \| \leq 2\mathbf{P}(\frac{T \vee T'}{U} > n),$$

$$\| \, \frac{1}{n} \sum_{k=0}^{n-1} \mathbf{P}(\theta_k Z \in \cdot) - \frac{1}{n} \sum_{k=0}^{n-1} \mathbf{P}(\theta_k Z' \in \cdot) \, \| \leq 2\frac{1}{n} \sum_{k=0}^{n-1} \mathbf{P}(T \vee T' > k),$$

$$\| \, \sum_{k=0}^{n-1} \mathbf{P}(\theta_k Z \in \cdot) - \sum_{k=0}^{n-1} \mathbf{P}(\theta_k Z' \in \cdot) \, \| \leq 2\mathbf{E}[(T \vee T') \wedge n].$$

4.3. Time-average asymptotics. If T is a.s. finite then the inequality yields

$$(4.2) \qquad \| \, \mathbf{P}(\theta_{Un} Z \in \cdot) - \mathbf{P}(\theta_{Un} Z' \in \cdot) \, \| \to 0 \quad \text{as} \quad n \to \infty$$

or equivalently

$$\| \, \frac{1}{n} \sum_{k=0}^{n-1} \mathbf{P}(\theta_k Z \in \cdot) - \frac{1}{n} \sum_{k=0}^{n-1} \mathbf{P}(\theta_k Z' \in \cdot) \, \| \to 0 \quad \text{as} \quad n \to \infty$$

and in particular if Z' is stationary then

$$\theta_{Un} Z \to_{t.v.} Z' \quad \text{as} \quad n \to \infty.$$

In Section 5.2 we prove the following rate result: if there is an $\alpha < 1$ such that $\mathbf{E}[T^\alpha] < \infty$ and $\mathbf{E}[T'^\alpha] < \infty$ then

$$(4.3) \qquad n^\alpha \| \, \mathbf{P}(\theta_{Un} Z \in \cdot) - \mathbf{P}(\theta_{Un} Z' \in \cdot) \, \| \to 0 \quad \text{as} \quad n \to \infty.$$

Note that this rate result differs from (3.3) by restricting α to be strictly less than one. In the case $\alpha = 1$ we only obtain the following boundedness result from the shift-coupling inequality: if $\mathbf{E}[T] < \infty$ and $\mathbf{E}[T'] < \infty$ then

$$\sup_{0 \leq n < \infty} \| \, \sum_{k=0}^{n-1} \mathbf{P}(\theta_k Z \in \cdot) - \sum_{k=0}^{n-1} \mathbf{P}(\theta_k Z' \in \cdot) \, \| < \infty.$$

4.4. On maximality. Since the left-hand side of (4.1) need not be non-increasing while the right-hand side is, we cannot in general obtain an identity in the shift-coupling inequality. In order to obtain the equivalence of time-average asymptotics and shift-coupling one might try to establish the conjecture that there exists a shift-coupling such that

$$(4.4) \qquad \lim_{n \to \infty} \frac{\| \mathbf{P}(\theta_{Un} Z \in \cdot) - \mathbf{P}(\theta_{Un} Z' \in \cdot) \|}{2\mathbf{P}(T \vee T' > Un)} \overset{?}{=} 1 \,.$$

Although there is no immediate analog of maximal coupling, Greven (1987) shows that the reformulation (3.5) has a shift-coupling analog: there exists a shift-coupling with epochs T and T' such that

$$(4.5) \qquad \sum_{n=0}^{\infty} \mathbf{P}(\theta_n \hat{Z} \in \cdot, T > n) \quad \perp \quad \sum_{n=0}^{\infty} \mathbf{P}(\theta_n \hat{Z}' \in \cdot, T > n) \,.$$

In the Markov case this shift-coupling is such that T and T' are randomized stopping times with respect to Z and Z', respectively.

For further discussion of the maximality problem (e.g. why (4.5) is not a full-fledged shift-coupling analog of maximal coupling) see Aldous and Thorisson (1993).

4.5. $\mathcal{I}$**-maximality.** It is easily seen that if T and T' are shift-coupling epochs then $\{T < \infty\}$ is an $\mathcal{I}$-coupling event. Moreover, one can establish (either by recursively maximal coupling Z and Z' shifted with respect to each other in all possible ways or by an application of (4.5)) that there exists an $\mathcal{I}$-*maximal* shift-coupling:

$$\| \mathbf{P}(Z \in \cdot)_{\mathcal{I}} - \mathbf{P}(Z' \in \cdot)_{\mathcal{I}} \| = 2\mathbf{P}(T = \infty) \,.$$

Based on this and the shift-coupling inequality we prove in Section 5.3 the following convergence result: as $n \to \infty$,

$$(4.6) \qquad \| \mathbf{P}(\theta_{Un} Z \in \cdot) - \mathbf{P}(\theta_{Un} Z' \in \cdot) \| \to \| \mathbf{P}(Z \in \cdot)_{\mathcal{I}} - \mathbf{P}(Z' \in \cdot)_{\mathcal{I}} \| \,.$$

As a corollary we have the following: if Z and Z' are stationary and

$$\mathbf{P}(Z \in \cdot)_{\mathcal{I}} = \mathbf{P}(Z' \in \cdot)_{\mathcal{I}}$$

then Z and Z' have the same distribution.

4.6. Equivalences. Combining $\mathcal{I}$-maximality and (4.6) immediately yields that the following three statements are equivalent:

  (i) Z and Z' admit shift-coupling;

  (ii) $\| \mathbf{P}(\theta_{Un} Z \in \cdot) - \mathbf{P}(\theta_{Un} Z' \in \cdot) \| \to 0 \quad$ as $n \to \infty$;

  (iii) $\mathbf{P}(Z \in \cdot)_{\mathcal{I}} = \mathbf{P}(Z' \in \cdot)_{\mathcal{I}}$

The equivalence of (i) and (ii) was established by Berbee (1979) while that of (i) and (iii) is from Aldous and Thorisson (1993). Note that the above way of adding (ii) to (i) and (iii) relies on the shift-coupling inequality.

4.7. Markov chains. If X and X' are *homogeneous* Markov chains on a Polish state space with the same transition kernel P and initial distributions μ

and μ', respectively, then the above equivalences easily yield that the following three statements are equivalent:

(a) X and X' admit shift-coupling for all μ and μ';

(b) $\| (\mu - \mu') \frac{1}{n} \sum_{k=0}^{n-1} P^n \| \to 0$ as $n \to \infty$ for all μ and μ';

(c) $\mathbf{P}(X \in \cdot)_{\mathcal{I}} = 0$ or 1 for all μ.

We remark that it is well-known that (c) is equivalent to

(d) all space-harmonic functions are constant.

4.8. Inhomogeneous case. For *inhomogeneous* Markov chains the equivalences (a), (b) and (c) still hold with the same modification as in the coupling case. That is, with the notation of Section 3.8, the following statements are equivalent:

(a') $X^{(m)}$ and $X^{(m)\prime}$ admit shift-coupling for all μ and μ' and all $m \geq 0$;

(b') $\| (\mu - \mu') \frac{1}{n} \sum_{k=0}^{n-1} P_{m+1} \ldots P_{m+n} \| \to 0$ as $n \to \infty$ for all μ, μ' and $m \geq 0$;

(c') $\mathbf{P}(X^{(m)} \in \cdot)_{\mathcal{I}} = 0$ or 1 for all μ and all $m \geq 0$.

5   Proofs

5.1. Proof of the shift-coupling inequality (4.1). Put

$$
\begin{aligned}
Y &= (n - T \vee T')^+ \\
C &= \{ T \leq Un < T + Y \}, \\
C' &= \{ T' \leq Un < T' + Y \}.
\end{aligned}
$$

(5.1)

Then

$$
\begin{aligned}
\mathbf{P}(\theta_{Un} \hat{Z} \in \cdot\,; C) &= \frac{1}{n} \sum_{k=0}^{n-1} \mathbf{P}(\theta_k \hat{Z} \in \cdot, T \leq k < T + Y) \\
&= \frac{1}{n} \sum_{k=0}^{n-1} \mathbf{P}(\theta_k \theta_T \hat{Z} \in \cdot, k < Y) \\
&= \frac{1}{n} \sum_{k=0}^{n-1} \mathbf{P}(\theta_k \theta_{T'} \hat{Z}' \in \cdot, k < Y) \\
&= \mathbf{P}(\theta_{Un} \hat{Z}' \in \cdot\,; C')
\end{aligned}
$$

i.e. C and C' are *distributional coupling events* for $\theta_{Un} Z$ and $\theta_{Un} Z'$:

$$
\mathbf{P}(\theta_{Un} \hat{Z} \in \cdot\,; C) = \mathbf{P}(\theta_{Un} \hat{Z}' \in \cdot\,; C').
$$

Thus, with C^c the complement of C,

$$\| \, \mathbf{P}(\theta_{Un} Z \in \cdot) - \mathbf{P}(\theta_{Un} Z' \in \cdot) \, \|$$
$$= \| \, \mathbf{P}(\theta_{Un} \hat{Z} \in \cdot, C^c) - \mathbf{P}(\theta_{Un} \hat{Z}' \in \cdot, C'^c) \, \|$$
$$\leq \mathbf{P}(C^c) + \mathbf{P}(C'^c).$$

Now

$$\begin{aligned} \mathbf{P}(C'^c) &= \mathbf{P}(C^c) \\ &= 1 - \mathbf{P}(T \leq Un < T + Y) = 1 - \mathbf{P}(Y > Un) \\ &= \mathbf{P}(Y < Un) = \mathbf{E}[n - Y]/n . \end{aligned}$$

Noting that $n - Y = (T \vee T') \wedge n$ yields

$$\| \, \mathbf{P}(\theta_{Un} Z \in \cdot) - \mathbf{P}(\theta_{Un} Z' \in \cdot) \, \| \leq 2\mathbf{E}[(T \vee T') \wedge n]/n$$

and completes the proof.

5.2. Proof of the rate result (4.3). Since $\mathbf{E}[T^\alpha]$ and $\mathbf{E}[T'^\alpha]$ are both finite we have

$$\mathbf{E}[(T \vee T')^\alpha] = \mathbf{E}[T^\alpha \vee T'^\alpha] \leq \mathbf{E}[T^\alpha + T'^\alpha] < \infty$$

and since $\alpha < 1$

$$\mathbf{E}[U^{-\alpha}] < \infty .$$

Thus

$$\mathbf{E}[(\frac{T \vee T'}{U})^\alpha] = \mathbf{E}[(T \vee T')^\alpha]\mathbf{E}[U^{-\alpha}] < \infty$$

which yields

$$n^\alpha \mathbf{P}(\frac{T \vee T'}{U} > n) \to 0 \quad \text{as} \quad n \to \infty .$$

Applying the shift-coupling inequality completes the proof.

5.3. Proof of the limit result (4.6). Note that

$$\mathbf{P}(Z \in \cdot)_{\mathcal{I}} = (\frac{1}{n} \sum_{k=0}^{n-1} \mathbf{P}(\theta_k Z \in \cdot))_{\mathcal{I}} = \mathbf{P}(\theta_{Un} Z \in \cdot)_{\mathcal{I}}$$

which yields the identity in

$$\begin{aligned} \| \, \mathbf{P}(Z \in \cdot)_{\mathcal{I}} - \mathbf{P}(Z' \in \cdot)_{\mathcal{I}} \, \| &= \| \, \mathbf{P}(\theta_{Un} Z \in \cdot)_{\mathcal{I}} - \mathbf{P}(\theta_{Un} Z' \in \cdot)_{\mathcal{I}} \, \| \\ &\leq \| \, \mathbf{P}(\theta_{Un} Z \in \cdot) - \mathbf{P}(\theta_{Un} Z' \in \cdot) \, \| . \end{aligned}$$

Thus

$$\| \, \mathbf{P}(Z \in \cdot)_{\mathcal{I}} - \mathbf{P}(Z' \in \cdot)_{\mathcal{I}} \, \|$$
$$\leq \liminf_{n \to \infty} \| \, \mathbf{P}(\theta_{Un} Z \in \cdot) - \mathbf{P}(\theta_{Un} Z' \in \cdot) \, \| .$$

Conversely, apply the shift-coupling inequality and the existence of an $\mathcal{I}$-maximal shift-coupling to obtain

$$\limsup_{n \to \infty} \| \mathbf{P}(\theta_{Un} Z \in \cdot) - \mathbf{P}(\theta_{Un} Z' \in \cdot) \|$$

$$\leq \| \mathbf{P}(Z \in \cdot)_{\mathcal{I}} - \mathbf{P}(Z' \in \cdot)_{\mathcal{I}} \|$$

and complete the proof.

5.4. On distributional shift-coupling. Two random times T and T' are *distributional coupling epochs* if

$$(\theta_T \hat{Z}, T) \overset{D}{=} (\theta_{T'} \hat{Z}', T')$$

and *distributional shift-coupling epochs* if

$$(5.2) \qquad\qquad \theta_T \hat{Z} \overset{D}{=} \theta_{T'} \hat{Z}' \,.$$

All results presented in this paper, except one, hold without the restriction to a Polish state space if we add 'distributional' in front of 'coupling' and 'shift-coupling'. The only exception is the shift-coupling inequality. The proof is not valid since the events C and C' in (5.1) are no longer distributional coupling events. However, the following *distributional shift-coupling inequality* holds: for $n \geq 0$,

$$(5.3) \quad \| \mathbf{P}(\theta_{Un} Z \in \cdot) - \mathbf{P}(\theta_{Un} Z' \in \cdot) \| \leq 2\mathbf{P}(T > Un) + 2\mathbf{P}(T' > Un) \,.$$

Thus the limit results are still valid.

In order to establish (5.3) note that due to (5.2) and the fact that U is independent of both $\theta_T \hat{Z}$ and $\theta_{T'} \hat{Z}'$ we have

$$\mathbf{P}(\theta_{T+Un} \hat{Z} \in \cdot) = \mathbf{P}(\theta_{T'+Un} \hat{Z}' \in \cdot)$$

and thus

$$\| \mathbf{P}(\theta_{Un} \hat{Z} \in \cdot) - \mathbf{P}(\theta_{Un} \hat{Z}' \in \cdot) \|$$
$$\leq \| \mathbf{P}(\theta_{Un} \hat{Z} \in \cdot) - \mathbf{P}(\theta_{T+Un} \hat{Z} \in \cdot) \|$$
$$+ \| \mathbf{P}(\theta_{Un} \hat{Z}' \in \cdot) - \mathbf{P}(\theta_{T'+Un} \hat{Z}' \in \cdot) \| \,.$$

Further, $\{T \leq Un\}$ and $\{Un \leq n - T\}$ are distributional coupling events for $\theta_{Un} \hat{Z}$ and $\theta_{T+Un} \hat{Z}$ and thus

$$\| \mathbf{P}(\theta_{Un} \hat{Z} \in \cdot) - \mathbf{P}(\theta_{T+Un} \hat{Z} \in \cdot) \| \leq 2\mathbf{P}(T > Un) \,.$$

Similarly

$$\| \mathbf{P}(\theta_{Un} \hat{Z}' \in \cdot) - \mathbf{P}(\theta_{T'+Un} \hat{Z}' \in \cdot) \| \leq 2\mathbf{P}(T' > Un) \,.$$

Combining these three inequalities yields (5.3).

REFERENCES

Aldous, D. and Thorisson, H., Shift-coupling, Stoch. Proc. Appl. (1993), to be published.

Berbee, H. C. P., *Random walks with stationary increments and renewal theory*, Math. Centre Tract **112**, Centre for Mathematics and Computer Science, Amsterdam, (1979).

Doeblin, W., Exposé de la theorie des chaînes simple constantes de Markov à un nombre fini d'états, Rev. Math. Union Interbalkan. **2** (1938), 77-105.

Goldstein, S., Maximal coupling, Z. Wahrscheinlichkeitsth. **46** (1979), 193-204.

Greven, A., Coupling of Markov chains and randomized stopping times, Part I, Probab. Th. Rel. Fields **75** (1987 a), 195-212.

Greven, A., Coupling of Markov chains and randomized stopping times, Part II, Probab. Th. Rel. Fields **75** (1987 b), 431-458.

Griffeath, D., A maximal coupling for Markov chains, Z. Wahrscheinlichkeitsth. **31** (1975), 95-106.

Griffeath, D., Coupling methods for Markov processes, Studies in Probability and Ergodic Theory. Adv. Math. Supplementary Studies **2** (1978).

Harison, V. and Smirnov, S. N., Jonction maximale en distribution dans le cas markovien, Probab. Th. Rel. Fields **84** (1990), 491-503.

Lindvall, T., *Lectures on the Coupling Method*, Wiley, New York, (1992).

Pitman, J. W., On coupling of Markov chains, Z. Wahrscheinlichkeitsth. **35** (1976), 315-322.

Svertchkov, M. Yu. and Smirnov, S. N., Maximal coupling of D-valued processes, Proceedings of Soviet Academy of Sciences, v. 311, n. 5 (1990).

Thorisson, H., On maximal and distributional coupling, Ann. Probab. **14** (1986), 873-876.

Thorisson, H., Shift-coupling in continuous time, forthcoming.

SCIENCE INSTITUTE, UNIVERSITY OF ICELAND,
DUNHAGA 3, 107 REYKJAVIK, ICELAND

E-mail address: thoris@rhi.hi.is

Contemporary Mathematics
Volume **149**, 1993

DOEBLIN AND THE METRIC THEORY OF CONTINUED FRACTIONS: A FUNCTIONAL THEORETIC SOLUTION TO GAUSS' 1812 PROBLEM

MARIUS IOSIFESCU

Abstract

This paper begins by emphasizing the importance of a paper by Doeblin (1940) on the metric theory of continued fractions. Surprisingly, Doeblin appears to have missed the opportunity of deriving a functional theoretic solution, fully at his hand, to Gauss' famous 1812 problem. An elementary solution to that problem is also provided.

1 INTRODUCTION

Let Ω denote the set of irrational numbers in the unit interval $I = [0, 1]$. Given $\omega \in \Omega$, let $a_1(\omega), a_2(\omega), \ldots$ be the sequence of the partial quotients of the continued fraction expansion of ω constructed as follows. Define $\tau : \Omega \to \Omega$ as $\tau(\omega) = $ fractional part of $1/\omega$, $\omega \in \Omega$. Then $a_{n+1}(\omega) = a_1(\tau^n(\omega))$, $n \in \mathbb{N}^* = \{1, 2, \ldots\}$, with $a_1(\omega)$ being the integer part of $1/\omega$, $\omega \in \Omega$. Let $\mathcal{B}_I$ denote the σ-algebra of the Borel subsets of I. Clearly, the a_n, $n \in \mathbb{N}^*$, are random variables on $(I, \mathcal{B}_I)$. They are defined almost surely with respect to any probability measure on $\mathcal{B}_I$ assigning probability 0 to the set $I - \Omega$ of rational numbers in I. For our purposes here, the most important probability measures of this kind are the Lebesgue measure λ, and Gauss' measure γ defined as

$$\gamma(A) = \frac{1}{\log 2} \int_A \frac{\mathrm{d}x}{x+1}, \ A \in \mathcal{B}_I.$$

AMS 1991 Subject Classifications: 11 K 50, 47 A 35, 47B 38, 60 K 99.

This paper is in final form and no version of it will be submitted for publication elsewhere.

We note that writing

$$\cfrac{1}{a_1+\cfrac{}{\ddots\;\cfrac{1}{a_n}}} = [a_1, + \ldots, a_n], \; n \in \mathbb{N}^*,$$

we have $\omega = \lim_{n\to\infty}[a_1(\omega), \ldots, a_n(\omega)]$, $\omega \in \Omega$. This is precisely the meaning of the equations

$$\omega = \cfrac{1}{a_1(\omega) + \cfrac{1}{a_2(\omega) + \ldots}} = [a_1(\omega), a_2(\omega), \ldots].$$

Next, let us put $r_n = a_n + [a_{n+1}, a_{n+2}, \ldots]$, $n \in \mathbb{N}^*$. Clearly, $r_n(\omega) = 1/\tau^{n-1}(\omega)$, $n \in \mathbb{N}^*$, with $\tau^0(\omega) = \omega$, $\omega \in \Omega$. Finally, for $n \in \mathbb{N}^*$, let us write

$$y_n = \begin{cases} a_1, & \text{if } n = 1 \\ a_n + [a_{n-1}, \ldots, a_1], & \text{if } n \geq 2 \end{cases}$$

$$s_n = \frac{1}{y_n} = [a_n, \ldots, a_1],$$

$$u_n = r_n + s_{n-1}, \text{ with } s_0 = 0.$$

The metric theory of continued fractions consists of the study of the random sequence $(a_n)_{n\in\mathbb{N}^*}$, and related sequences such as $(r_n)_{n\in\mathbb{N}^*}$, $(y_n)_{n\in\mathbb{N}^*}$, $(s_n)_{n\in\mathbb{N}^*}$, $(u_n)_{n\in\mathbb{N}^*}$.

Two facts are basic in this theory, namely:

I. The sequence $(a_n)_{n\in\mathbb{N}^*}$ is strictly stationary under γ. (Equivalently, τ preserves γ.)

II. The sequence $(a_n)_{n\in\mathbb{N}^*}$ is exponentially ψ-mixing under γ (and under many other probability measures including λ), that is,

$$|\gamma(A_1 \cap A_2) - \gamma(A_1)\gamma(A_2)| \leq a\theta^n \gamma(A_1)\gamma(A_2), \; n \in \mathbb{N}^*,$$

for any $A_1 \in \sigma(a_1, \ldots, a_k)$ (the σ-algebra generated by the random variables $a_1, \ldots, a_k$), $A_2 \in \sigma(a_{n+k}, a_{n+k+1}, \ldots)$, and $k \in \mathbb{N}^*$, with suitable positive constants $\theta < 1$ and a.

The second property is a deep result implied by the solution of a slight generalization of Gauss' famous 1812 problem. This was to estimate the error

$$e_n(x) = e_n(x, \lambda) = \lambda(r_n^{-1} < x) - \gamma([0, x]), \; x \in I.$$

This was the first problem in the metric theory of continued fractions. It was solved almost simultaneously, by different methods, by R.O. Kuzmin (in 1928) and Paul Lévy (in 1929). As $n \to \infty$, uniformly in $x \in I$, Kuzmin gave $e_n(x) = O(\theta^{\sqrt{n}})$ for some $0 < \theta < 1$, while Lévy obtained $e_n(x) = O(\theta^n)$, with $\theta < 0.68$. Actually, the ψ-mixing property **II** is a consequence of an error estimate

$$\frac{e_n(y, \mu) - e_n(x, \mu)}{\gamma([x, y])} = O(\theta^n), \ x, y \in I, \ x < y,$$

for all measures $\mu = \lambda(\cdot | a_k = i_k, \ 1 \leq k \leq m)$, $i_k \in \mathbb{N}^*$, $1 \leq k \leq m$, $m \in \mathbb{N}^*$, rather than λ. For details, precise references, and proofs, the reader is referred to Iosifescu and Grigorescu (1990, Section 5.2).

The solution of Gauss' problem provided a strong impetus to the development of the metric theory of continued fractions. Subsequently, important results were obtained in the 1930's by A. Ya. Khinchin and Paul Lévy (see Khintchin (1964) and Lévy (1954)). However, as we have argued in a recent paper (Iosifescu, 1990), the most important work of the 1940's on the metric theory of continued fractions is that of Wolfgang Doeblin (1940). While the name of Doeblin is well known for his many contributions to the theory of Markovian dependence, his 1940 paper testifies to other important achievements of which few people today are aware. As a matter of fact, Doeblin's 1940 paper could almost be viewed as a concise picture of the state of the metric theory of continued fractions today: Doeblin took up and made important contributions to all the main topics which were studied in the following fifty years. Thus, his 1940 paper could also be considered as suggesting a working program in this field (even if it was largely ignored by subsequent authors). Our paper (Iosifescu, 1990) is about how this program has been carried out.

Among the most important achievements of Doeblin (1940) are: (i) a solution to the (generalized) Gauss problem; (ii) central limit theorems and laws of the iterated logarithm for certain types of weakly dependent random variables including ψ-mixing ones. Note that this was achieved many years before mixing had become fashionable, and that Doeblin's predecessors did not go beyond the law of large numbers; (iii) results on the sequence $(u_n)_{n \in \mathbb{N}^*}$, which were ignored by subsequent authors and became the object of conjectures. These results show that Doeblin was able to develop ideas and concepts intuitively, which have only been made precise quite recently; (iv) the use of the Birkhoff-Khinchin ergodic theorem in the metric theory of continued fractions, an idea for which subsequent authors were credited with; (v) results concerning convergence to stable laws.

Doeblin died at the age of 25; this is a premature age to die for one destined to be a creative mathematician. Yet even a single piece of work like Doeblin's 1940 paper is enough to ensure first rank status among mathematicians to its author.

In this note, we comment on an opportunity surprisingly missed by Doeblin, namely a functional theoretic solution, fully at his hand, to the generalization alluded to of Gauss' 1812 problem. We shall give an entirely elementary solution to this problem.

2 DOEBLIN'S SOLUTION TO GAUSS' PROBLEM

As we have already noted, one of the main results of Doeblin (1940), to which its first section is devoted, is a derivation of the (generalized) Gauss formula by using elements of the theory of dependence with complete connections, then in its inception. In our notation, the formula in question (see page 357 of Doeblin's paper) is

$$\lambda(s < r_{m+n} < t | a_1, \ldots, a_m) = \frac{1}{\log 2}(\log \frac{2t}{t+1} - \log \frac{2s}{s+1})(1 + a\theta^{\sqrt{n}}) \quad (1)$$

for all $m, n \in \mathbb{N}^*$ and $t > s \geq 1$, with $|a| < c$, where $\theta < 1$ and c are positive constants independent of m, n, s, t and the values of $a_1, \ldots, a_m$. Doeblin's reasoning is not entirely clear, but what he intended to say seems to be the following (cf. Iosifescu, 1974, and Iosifescu and Grigorescu, 1990, Subsection 5.2.2). Let us start from the so-called Brodén-Borel-Lévy formula

$$\lambda(r_{m+1} > s | a_1, \ldots, a_m) = \frac{y_m + 1}{sy_m + 1}, \quad m \in \mathbb{N}^*, \ s \geq 1, \quad (2)$$

which implies that

$$\lambda(a_{m+1} = i | a_1, \ldots, a_m) = \lambda(i < r_{m+1} < i + 1 | a_1, \ldots, a_m)$$

$$= \lambda(r_{m+1} > i | a_1, \ldots, a_m) - \lambda(r_{m+1} > i + 1 | a_1, \ldots, a_m)$$

$$= \frac{y_m + 1}{iy_m + 1} - \frac{y_m + 1}{(i+1)y_m + 1} = \frac{y_m(y_m + 1)}{(iy_m + 1)((i+1)y_m + 1)}, \quad m, i \in \mathbb{N}^*.$$

This shows that $(y_m)_{m \in \mathbb{N}^*}$ is a homogeneous Markov chain on $(I, \mathcal{B}_I, \lambda)$, with the following transition mechanism: from state $y \geq 1$ the only possible one-step transitions are those to states $y + i$, $i \in \mathbb{N}^*$, with corresponding probabilities $y(y + 1)/(iy + 1)((i + 1)y + 1)$, $i \in \mathbb{N}^*$. Let U be the transition operator associated with $(y_m)_{m \in \mathbb{N}^*}$, defined as

$$Uf(y) = E_\lambda(f(y_2)|y_1 = y) = \sum_{i \in \mathbb{N}^*} \frac{y(y+1)f(y+i)}{(iy+1)((i+1)y+1)}, \ y \ge 1,$$

for any bounded measurable complex-valued function f on $[1, \infty)$. We have

$$\lambda(r_{m+n} > t | a_1, \ldots, a_m) = U^{n-1} f_t(y_m), \ m, n \in \mathbb{N}^*, \ t \ge 1,$$

with U^0 denoting the identity operator, and

$$f_t(u) = \frac{u+1}{tu+1}, \ u \ge 1.$$

This can easily be checked by induction with respect to n, as for $n = 1$ the equation above reduces to the Brodén-Borel-Lévy formula (2). Doeblin's computation, based on a typical dependence-with-complete-connections argument, shows that

$$U^{n-1} f_s(y_m) - U^{n-1} f_t(y_m) = \frac{1}{\log 2} (\log \frac{2t}{t+1} - \log \frac{2s}{s+1})(1 + a\theta^{\sqrt{n}})$$

for all $m, n \in \mathbb{N}^*$ and $t > s \ge 1$, so that (1) holds.

Clearly, (1) is equivalent to

$$\lambda(x < r_{m+n}^{-1} < y | a_1, \ldots, a_m) = \gamma([x, y])(1 + a\theta^{\sqrt{n}}) \tag{1'}$$

for all $m, n \in \mathbb{N}^*$ and $x < y$, $x, y \in I$, with $|a| \le c$, where $\theta < 1$ and c are positive constants independent of m, n, x, y and the values of $a_1, \ldots, a_m$.

It is rather curious that Doeblin did not use the ergodic theorem derived by him and R. Fortet in their joint 1937 paper (cf. Iosifescu, 1992a) to obtain θ^n instead of $\theta^{\sqrt{n}}$ in (1'), as in Iosifescu (1974). The only plausible explanation is that Doeblin overlooked the possibility of using that theorem in the special case considered. This oversight may have been due to his Lévy-type approach to Gauss' problem. A Kuzmin-type approach to the problem, which we describe below, would have allowed him to derive the optimal results by using the theorem we have alluded to.

3 The Proof that Doeblin Overlooked

Let μ be an arbitrary non-atomic probability on $\mathcal{B}_I$ and define

$$F_n(x) = F_n(x, \mu) = \mu(r_{n+1}^{-1} < x),\ n \in \mathbb{N} = \{0\} \cup \mathbb{N}^*,\ x \in I.$$

Clearly, $F_0(x) = \mu([0, x))$. Since $0 < r_{n+2}^{-1} < x$ iff $(x + a_{n+1})^{-1} < r_{n+1}^{-1} < a_{n+1}^{-1}$, we can write Gauss' equation as

$$F_{n+1}(x) = \sum_{i \in \mathbb{N}^*} (F_n(\frac{1}{i}) - F_n(\frac{1}{x+i})),\ n \in \mathbb{N},\ x \in I.$$

Assuming that for some $m \in \mathbb{N}$ the derivative F_m' exists everywhere in I and is bounded, it is easy to see by induction that F_{m+n}' exists and is bounded for all $n \in \mathbb{N}^*$, and we have

$$F_{n+1}'(x) = \sum_{i \in \mathbb{N}^*} \frac{1}{(x+i)^2} F_n'(\frac{1}{x+i}),\ n \geq m,\ x \in I.$$

Further, write $f_n(x) = (x + 1)F_n'(x)$, $x \in I$, to get $f_{n+1} = U f_n$, $n \geq m$, with U being the linear operator defined as

$$U f(x) = \sum_{i \in \mathbb{N}^*} \frac{(x+1)}{(x+i)(x+i+1)} f(\frac{1}{x+i}),\ f \in B(I),\ x \in I,$$

where $B(I)$ is the Banach space of bounded measurable complex-valued functions f on I under the supremum norm

$$|f| = \sup_{x \in I} |f(x)|.$$

Hence

$$F_{m+n}(x) = \int_0^x \frac{U^n f_m(u)}{u + 1} du,\ n \in \mathbb{N},\ x \in I, \tag{3}$$

with $f_m(u) = (u + 1)F_m'(u)$, $u \in I$.

Note that here U is the transition operator of the homogeneous Markov chain $(s_n)_{n \in \mathbb{N}^*}$ on $(I, \mathcal{B}_I, \lambda)$, with the following transition mechanism: from state $s \in I$ the only possible one-step transitions are those to states $1/(s + i)$, $i \in \mathbb{N}^*$, with corresponding probabilities $(s + 1)/(s + i)(s + i + 1)$, $i \in \mathbb{N}^*$. (Cf. the Markov chain $(y_n)_{n \in \mathbb{N}^*}$ in the preceding section and remember that $s_n = 1/y_n$, $n \in \mathbb{N}^*$.) Also, it appears that γ is a stationary probability measure for $(s_n)_{n \in \mathbb{N}^*}$, which implies that

$$\int_I U^n f(x) \gamma(\mathrm{d}x) = \int_I f(x) \gamma(\mathrm{d}x), \ n \in \mathbb{N}. \tag{4}$$

Now putting

$$p_i(x) = \frac{x+1}{(x+i)(x+i+1)}, \quad i \in \mathbb{N}^*, \ x \in I,$$

for any $f \in B(I)$ and $x', x'' \in I$, $x' \neq x''$, we can write

$$\frac{Uf(x') - Uf(x'')}{x' - x''} = \sum_{i \in \mathbb{N}^*} p_i(x') \frac{f(\frac{1}{x'+i}) - f(\frac{1}{x''+i})}{\frac{1}{x'+i} - \frac{1}{x''+i}} \frac{\frac{1}{x'+i} - \frac{1}{x''+i}}{x' - x''}$$

$$+ \sum_{i \in \mathbb{N}^*} \frac{p_i(x') - p_i(x'')}{x' - x''} f(\frac{1}{x''+i}).$$

It follows that if $f \in L(I)$ (for the notation see Iosifescu, 1992a, where W is replaced by I), then we have

$$\mathrm{s}(Uf) \leq \Big(\sum_{i \in \mathbb{N}^*} \frac{1}{i^2} \sup_{x \in I} p_i(x) \Big) \mathrm{s}(f) + \Big(\sum_{i \in \mathbb{N}^*} \sup_{x \in I} |p_i'(x)| \Big) |f|.$$

Noting that

$$\sum_{i \in \mathbb{N}^*} \frac{1}{i^2} \sup_{x \in I} p_i(x) = \frac{1}{2} + \frac{3 - 2\sqrt{2}}{4} + 2 \sum_{i \geq 3} \frac{1}{i^2(i+1)(i+2)}$$

$$\leq \frac{1}{2} + \frac{3 - 2\sqrt{2}}{4} + \frac{2}{9} \cdot \frac{1}{4} = q < 1,$$

$$\sum_{i \in \mathbb{N}^*} \sup_{i \in I} |p_i'(x)| < \sum_{i \in \mathbb{N}^*} \frac{1}{i^2} = Q,$$

we deduce that

$$\|Uf\| = |Uf| + \mathrm{s}(Uf) \leq q\|f\| + (Q + 1 - q)|f|$$

for all $f \in L(I)$.

Therefore the Doeblin-Fortet ergodic theorem (see Theorem 1 in Iosifescu, 1992a) applies to U. Moreover, it appears that 1 is the only eigenvalue of U of modulus 1. This follows from the uniform convergence of $U^n f$ as $n \to \infty$ to a constant function $U^\infty f$ depending on f for any $f \in L(I)$.

The latter fact is implicit in Doeblin (1940, p. 356). For an explicit treatment see Iosifescu (1974). Also, equation (4) in conjunction with that convergence implies that

$$U^\infty f = \int_I f(x)\gamma(\mathrm{d}x) \quad (= U_1 f).$$

So, the conclusion is that there exist positive constants $\theta < 1$ and c such that

$$\|U^n f - U^\infty f\| \le c\theta^n \|f\| \tag{5}$$

for all $f \in L(I)$ and $n \in \mathbb{N}^*$.

To derive (1'), with θ^n replacing $\theta^{\sqrt{n}}$, for any arbitrarily given $m \in \mathbb{N}^*$ take

$$\mu = \lambda(\cdot | a_1, \ldots, a_m),$$

that is,

$$F_{m+n}(x) = \lambda(r_{m+n+1}^{-1} < x | a_1, \ldots, a_m), \ n \in \mathbb{N}, \ x \in I.$$

In particular, by (2),

$$F_m(x) = \lambda(r_{m+1}^{-1} < x | a_1, \ldots, a_m) = \frac{(s_m + 1)x}{s_m x + 1}, \ x \in I. \tag{2'}$$

Note that the case $\mu = \lambda$ is obtained by formally putting $m = 0$ and $s_m = s_0 = 0$ in (2'). Then

$$f_m(x) = (x+1)F_m'(x) = \frac{(s_m + 1)(x + 1)}{(s_m x + 1)^2}, \ x \in I,$$

$$U^\infty f_m = \frac{s_m + 1}{\log 2} \int_0^1 \frac{\mathrm{d}x}{(s_m x + 1)^2} = \frac{1}{\log 2} \tag{6}$$

and

$$|f_m| < 2, \ \mathrm{s}(f_m) < 2, \ \|f_m\| = |f_m| + \mathrm{s}(f_m) < 4 \tag{7}$$

for all possible values of s_m. The computation of the bounds above, which are exact, though elementary, is tedious. Now, (1') with θ^n instead of $\theta^{\sqrt{n}}$ follows from (3), (5), (6) and (7).

Remarks.

1 The optimal value of θ in (1') and (5) has been obtained by E. Wirsing in 1971 (see Wirsing, 1974). It equals 0.30366300289873265860...

2 It is interesting to note that the solution of the corresponding Gauss

problem for the nearest integer continued fraction expansion requires the use of the Ionescu Tulcea-Marinescu ergodic theorem (see Theorem 2 in Iosifescu, 1992), which is an extensive generalization of the Doeblin-Fortet ergodic theorem. For details see Iosifescu and Kalpazidou (1992).

4 A SIMPLE SOLUTION TO GAUSS' 1812 PROBLEM

We have recently realized that an estimate more precise than (5) can be obtained using some simple algebra and calculus. The main idea is to study the action of the operator U on functions of bounded variation on I. Remember that the variation $\mathrm{var}_A f$ over $A \subset I$ of a function $f \in B(I)$ is defined as $\sup \Sigma_{i=1}^k |f(t_i) - f(t_{i-1})|$, the supremum being taken over all points $t_1 < \ldots < t_k$ in A for $k \geq 2$. We write simply $\mathrm{var} f$ for $\mathrm{var}_I f$ and, if $\mathrm{var} f < \infty$, then f is called a function of bounded variation. The collection $BV(I)$ of all complex-valued functions f of bounded variation on I is a Banach space under the norm

$$\|f\|_{\mathrm{V}} = \mathrm{var} f + |f|.$$

Clearly, $L(I) \subset BV(I)$. It is well known that a real-valued $f \in BV(I)$ can be written as $f = g - h$, where both g and h are non-decreasing (or non-increasing) functions on I. For example, we can take

$$\text{or} \quad \begin{array}{l} g(x) = \mathrm{var}_{[0,x]} f, \quad h(x) = \mathrm{var}_{[0,x]} f - f(x), \quad x \in I, \\ g(x) = \mathrm{var}_{[x,1]} f, \quad h(x) = \mathrm{var}_{[x,1]} f - f(x), \quad x \in I. \end{array} \quad (8)$$

Note that

$$\begin{array}{l} \mathrm{var} g = g(1) - g(0) = \mathrm{var} f, \quad \mathrm{var} h = h(1) - h(0) = \mathrm{var} f + f(0) - f(1) \\ \text{or} \quad \mathrm{var} g = g(0) - g(1) = \mathrm{var} f, \quad \mathrm{var} h = h(0) - h(1) \\ = \mathrm{var} f - f(0) + f(1) \end{array}$$

$$(9)$$

Proposition 1 *If $f \in B(I)$ is non-decreasing (non-increasing), then Uf is non-increasing (non-decreasing).*

Proof. To make a choice, assume f is non-decreasing. Let $y > x$, $x, y \in I$.

We have $Uf(y) - Uf(x) = S_1 + S_2$, where

$$S_1 = \sum_{i \in \mathbf{N}^*} p_i(y)(f(\frac{1}{y+i}) - f(\frac{1}{x+i})), \quad S_2 = \sum_{i \in \mathbf{N}^*} (p_i(y) - p_i(x))f(\frac{1}{x+i}).$$

Clearly, $S_1 \leq 0$. We shall prove that $S_2 \leq 0$, too. Since $\Sigma_{i \in \mathbf{N}^*} p_i(u) = 1$ for any $u \in I$, we can write

$$S_2 = -\sum_{i \in \mathbf{N}^*} (f(\frac{1}{x+1}) - f(\frac{1}{x+i}))(p_i(y) - p_i(x)).$$

It is easy to see that the function p_1 is decreasing while the functions p_i, $i \geq 3$, are all increasing. Therefore

$$-S_2 \geq (f(\frac{1}{x+1}) - f(\frac{1}{x+2})) \sum_{i \geq 2}(p_i(y) - p_i(x))$$

$$= (f(\frac{1}{x+1}) - f(\frac{1}{x+2}))(p_1(x) - p_1(y)) \geq 0.$$

that is, $S_2 \leq 0$, as claimed. Thus $Uf(y) - Uf(x) \leq 0$, and the proof is complete.

Remark. Proposition 1 is also proved in Szüsz (1961, p. 450), under the unnecessary assumption that f is differentiable.

Proposition 2 *If* $f \in B(I)$ *is monotone, then* $varUf \leq \frac{1}{2}varf$. *The constant* $\frac{1}{2}$ *cannot be lowered.*

Proof. Assume again that f is non-decreasing. Then by Proposition 1 we have

$$\begin{aligned} \mathrm{var}\, Uf \;=\; & Uf(0) - Uf(1) = \sum_{i \in \mathbf{N}^*} (p_i(0)f(\frac{1}{i}) - p_i(1)f(\frac{1}{1+i})) \qquad (10) \\ =\; & \frac{1}{2}f(1) - \frac{1}{3}f(\frac{1}{2}) + \frac{1}{6}(f(\frac{1}{2}) - f(\frac{1}{3})) + \sum_{i \geq 3}(p_i(0) - p_i(1))f(\frac{1}{i}) \\ & + \sum_{i \geq 3} p_i(1)(f(\frac{1}{i}) - f(\frac{1}{1+i})). \end{aligned}$$

But, by an earlier remark,

$$\sum_{i\geq 3}(p_i(0) - p_i(1))(f(0) - f(\frac{1}{i})) \geq 0,$$

whence

$$\sum_{i\geq 3}(p_i(0) - p_i(1))f(\frac{1}{i}) \leq f(0)\sum_{i\geq 3}(p_i(0) - p_i(1))$$

$$= f(0)(p_1(1) - p_1(0) + p_2(1) - p_2(0)) = -\frac{1}{6}f(0).$$

Next, as $p_i(1) \leq 1/10$ for all $i \geq 3$, we have

$$\sum_{i\geq 3}p_i(1)(f(\frac{1}{i}) - f(\frac{1}{1+i})) \leq \frac{1}{10}(f(\frac{1}{3}) - f(0)).$$

Therefore

$$\text{var}\,Uf \leq \frac{1}{3}(f(1) - f(\frac{1}{2})) + \frac{1}{6}(f(\frac{1}{2}) - f(\frac{1}{3})) + \frac{1}{10}(f(\frac{1}{3}) - f(0))$$

$$+\frac{1}{6}(f(1) - f(0)) \leq (\frac{1}{3} + \frac{1}{6})(f(1) - f(0)) = \frac{1}{2}\text{var}f.$$

If f is non-increasing, then $-f$ is non-decreasing and $\text{var}U(-f) = \text{var}Uf$, $\text{var}(-f) = \text{var}f$. Finally, for any non-decreasing f such that $f(1) > 0$ and $f(x) = 0$, $0 \leq x \leq 1/2$, by (10) we have $\text{var}Uf = \frac{1}{2}f(1) = \frac{1}{2}\,\text{var}f$.

Proposition 3 *For any real-valued $f \in BV(I)$ we have*

$$|U^nf - U^\infty f| \leq 2^{-n}(2\,\text{var}f - |f(0) - f(1)|), \ n \in \mathbb{N}.$$

Proof. Clearly, (4) amounts to

$$U^\infty U^n f = U^\infty f, \ f \in B(I), \ n \in \mathbb{N}. \tag{11}$$

Next, note that for any $f \in BV(I)$ and $u \in I$ we have

$$|f(u)| - |\int_I f(x)\gamma(dx)| \leq |f(u) - \int_I f(x)\gamma(dx)|$$

$$= |\int_I (f(u) - f(x))\gamma(dx)| \leq \text{var}f,$$

whence

$$|f| \leq |\int_I f(x)\gamma(dx)| + \text{var}f, \ f \in BV(I). \tag{12}$$

Finally, (11) and (12) imply that

$$|U^n f - U^\infty f| \leq \text{var}(U^n f - U^\infty f) = \text{var} U^n f, \ n \in \mathbb{N}, \ f \in BV(I). \quad (13)$$

Now, let $f \in BV(I)$ be real-valued. Then $f = g - h$, with monotone g and h defined as in (8). By Proposition 2 we have

$$\text{var} U^n g \leq 2^{-n} \, \text{var} g, \, \text{var} U^n h \leq 2^{-n} \, \text{var} h. \quad (14)$$

Then (13) and (14) imply that

$$|U^n f - U^\infty f| \leq |U^n g - U^\infty g| + |U^n h - U^\infty h|$$

$$\leq \text{var} U^n g + \text{var} U^n h \leq 2^{-n}(\text{var} g + \text{var} h).$$

Hence by (9)

$$|U^n f - U^\infty f| \leq 2^{-n} \min((2 \, \text{var} f + f(0) - f(1)), (2 \, \text{var} f - f(0) + f(1)))$$

$$= 2^{-n}(2\text{var} f - |f(0) - f(1)|),$$

as claimed.

Corollary 4 *For all $x < y$, $x, y \in I$, all $m, n \in \mathbb{N}$, and all values of $a_1, \ldots, a_m$ we have (when $m = 0$, conditioning vanishes)*

$$|\lambda(x < r_{m+n+1}^{-1} < y | a_1, \ldots, a_m) - \gamma([x, y])| \leq (\log 2) 2^{-n} \gamma([x, y]). \quad (15)$$

Proof. By (3) and (6), (15) follows from Proposition 3 since

$$2 \, \text{var} \, f_m - |f_m(0) - f_m(1)| \leq 1,$$

as is easy to see.

5 A FEW FINAL REMARKS

It is clear that (15) is a much more precise result than (1′) (with θ^n in place of $\theta^{\sqrt{n}}$). Moreover, there is good reason to believe that 2^{-n} in (15) (but not in Proposition 3!) can be replaced by $(c_n c_{n+1})^{-1}$, where

$$c_n = \frac{(1 + \sqrt{5})^{n+1} - (1 - \sqrt{5})^{n+1}}{2^{n+1}\sqrt{5}}, \ n \in \mathbb{N},$$

that is, the Fibonacci numbers defined by $c_0 = c_1 = 1$, $c_n = c_{n-1} + c_{n-2}$, $n \geq 2$. In this respect, see Iosifescu (1992b). Note that Proposition 3 implies that the spectral radius of $U - U^\infty$ in $BV(I)$ defined as

$$\lim_{n \to \infty} \left(\sup_{\|f\|_V = 1} \|U^n f - U^\infty f\|_V \right)^{1/n}$$

does not exceed $1/2$. It is an open question whether it equals $1/2$. We believe that the answer is 'no'.

It remains to try to understand how such a simple solution to Gauss' problem has been overlooked for 180 years. Perhaps, the answer is that Lipschitz functions appeared to be the most convenient and suitable tools for solving the problem. This may have prevented workers from studying the action of the operator U on other classes of functions.

REFERENCES

DOEBLIN, W. (1940) Remarques sur la théorie métrique des fractions continues. *Compositio Math.* 7, 353-371.

IOSIFESCU, M. (1974) On the applications of random systems with complete connections to the theory of f-expansions. In: J. Gani et al, Eds. *Progress in Statistics* (European Meeting of Statisticians, Budapest 1972), Colloq. Math. Soc. János Bolyai, Vol. 9, 335-365, North-Holland, Amsterdam.

IOSIFESCU, M. (1990) A survey of the metric theory of continued fractions, fifty years after Doeblin's 1940 paper. In: B. Grigelionis et al, Eds. *Probability Theory and Mathematical Statistics* (Proc. V Internat. Conf. Vilnius, Lithuania, 1989), Vol. 1, 550-572, VSP, Zeist, The Netherlands.

IOSIFESCU, M. (1992a) A basic tool in mathematical chaos theory: the 1937 ergodic theorem of Doeblin and Fortet and its generalization of Ionescu Tulcea and Marinescu. In: these Proceedings.

IOSIFESCU, M. (1992b) A very simple proof of a generalization of the Gauss-Kusmin-Lévy theorem on continued fractions and related questions. *Rev. Roumaine Math. Pures Appl.* 37, no. 10. To appear.

IOSIFESCU, M. & GRIGORESCU. S. (1990) *Dependence with Complete Connections and Its Applications.* , Cambridge Univ. Press, Cambridge.

IOSIFESCU, M. & KALPAZIDOU S. (1992) The nearest integer continued fraction. expansion: an approach in the spirit of Doeblin. In: these Proceedings.

KHINCHIN, A.YA. (1964) Continued Fractions. Univ. of Chicago Press, Chicago. (first Russian edition published in 1937.)

KUZMIN, R.O. (1928) On a problem of Gauss. *Dokl. Akad. Nauk SSSR Ser. A* 375-380 (Russian; French version in: *Atti Congr. Internaz. Mat. (Bologna, 1928)*, Vol. VI, 83-89, Zanichelli, Bologna, 1932.)

LÉVY, P. (1929) Sur les lois de probabilité dont dépendent les quotients complets et incomplets d'une fraction continue. *Bull. Soc. Math. France* 57, 178-194.

LÉVY, P. Théorie de l'addition des variables aléatoires, second edition, (1954) *Gauthier-Villars, Paris.* Second edition (First edition published in 1937.)

SZÜSZ, P. (1961) Über einen Kusminschen Satz. *Acta Math. Acad. Sci. Hungar.* 12, 447-453.

WIRSING, E. (1974) On the theorem of Gauss-Kusmin-Lévy and a Frobenius type theorem for function spaces. *Acta Arith.* 24, 507-528.

ROMANIAN ACADEMY, CENTRE OF MATHEMATICAL STATISTICS, BD. MAGHERU 22, RO-70158 BUCHAREST, ROMANIA.

Contemporary Mathematics
Volume **149**, 1993

A BASIC TOOL IN MATHEMATICAL CHAOS THEORY: DOEBLIN AND FORTET'S ERGODIC THEOREM AND IONESCU TULCEA AND MARINESCU'S GENERALIZATION

MARIUS IOSIFESCU

Abstract

An ergodic theorem of Doeblin and Fortet (1937), and a generalization of it due to Ionescu Tulcea and Marinescu (1951), are discussed. Both are quite useful in what is usually called 'mathematical chaos theory', a fact which has been overlooked by writers in that field.

0 INTRODUCTION

Let (W, d) be a metric space. For any function $f \epsilon C(W) =$ the collection of all bounded continuous complex-valued functions on W, set

$$|f| = \sup_{w \epsilon W} |f(w)|, \ \mathrm{s}(f) = \sup_{w_1 \neq w_2} \frac{|f(w_1) - f(w_2)|}{d(w_1, w_2)},$$

$$\|f\| = |f| + \mathrm{s}(f).$$

Let $L(W) = \{f : f \epsilon C(W), \|f\| < \infty\}$ be the collection of all bounded Lipschitz complex-valued functions on W. As is well known, $C(W)$ and $L(W)$ are Banach spaces under $|\cdot|$ and $\|\cdot\|$, respectively. Let $C_c(W)$ be the collection of all continuous complex-valued functions on W with compact support; we recall that the support of $f \epsilon C(W)$ is the closure of the set

---

AMS 1991 Subject Classifications: 47 A 35, 47 B 38, 60 K 99, 70 K 50.

This paper is in final form and no version of it will be submitted for publication elsewhere.

$\{w \epsilon W : f(w) \neq 0\}$. In general, $C_c(W) \subseteq C(W)$, with the equality holding when (W, d) is compact. Note that, in general, $C_c(W)$ is a subalgebra of $C(W)$, which is not closed in the norm $|\cdot|$. Finally, if the metric space (W, d) is locally compact, let $C_0(W)$ be the collection of all continuous complex-valued functions on W that vanish at infinity. More precisely, $f \epsilon C_0(W)$ means that for any $\epsilon > 0$ there exists a compact subset K_ϵ of W such that $|f(w)| < \epsilon$ for all $w \epsilon W - K_\epsilon$. Clearly, $C_c(W) \subset C_0(W)$, and as is well known, $C_0(W)$ is a Banach subalgebra of $C(W)$, which is just the closure of $C_c(W)$ in $C(W)$.

Let $p_i, 1 \leq i \leq r$, be non-negative elements of $C(W)$ such that $\sum_{i=1}^{r} p_i(w) = 1$ for any $w \epsilon W$, and let $\Phi = \{\phi_i\}_{1 \leq i \leq r}$ be a collection of continuous mappings from W into itself. Consider the Markov process $\zeta = (\zeta_n)_{n \geq 0}$ with $\zeta_0 = w$ (arbitrarily given in W) and the following transition mechanism: given ζ_n, we have $\zeta_{n+1} = \phi_i(\zeta_n)$ with probability $p_i(\zeta_n)$, $n \geq 0$, $1 \leq i \leq r$. Therefore ξ is a process of random composition of the mappings of Φ. This is precisely the basic object of study in what is usually called 'mathematical chaos theory'.

It should be noted that the framework just described, in which ζ has been defined, is a special instance of what is called 'dependence with complete connections'. This topic, whose study was initiated and mainly developed by Romanian probabilists, allows a unifying treatment of subjects as varied as stochastic models for learning, the metric theory of continued fractions, and piecewise monotonic transformations. For details we refer the reader to Iosifescu and Grigorescu (1990). Mention should also be made of the fact that the writers on mathematical chaos theory have not so far noticed the relevance of dependence with complete connections to their work.

The transition operator U of ζ, defined as

$$Uf(w) = E(f(\zeta_1)|\zeta_0 = w) = \sum_{i=1}^{r} p_i(w)f(\phi_i(w)), \quad w \epsilon W,$$

for $f =$ any measurable complex-valued function on W, takes boundedly $C(W)$ into itself. (In other words, ζ is a Feller process.) Moreover, we have $|Uf| \leq |f|$ for any $f \epsilon C(W)$. It is possible to give simple conditions for the convergence in distribution of ζ to a limit probability μ on $\mathcal{B}_W$, the σ-algebra of Borel subsets of W, regardless of the initial distribution ν of ζ. In terms of the operator U, such eqa condition is

$$\lim_{n\to\infty} \int_W U^n f d\nu = \int_W f d\mu, \tag{1}$$

for all $f \epsilon C(W)$ and $\nu \epsilon \Phi r(W) = $ the collection of all probabilities on $\mathcal{B}_W$, where U^n is the nth power of U and is given by the equation

$$U^n f(w) = E(f(\zeta_n)|\zeta_0 = w), \ w \epsilon W, \ n \geq 1.$$

Proposition 0.

  (i) Assume that the metric space (W, d) is separable and complete and (1) holds. Then

   (i1) for some $w \epsilon W$, whatever $\epsilon > 0$, there is a compact set $K_\epsilon = K \subset W$ such that

$$|U^n f(w)| \leq |f|_K + \epsilon|f|$$

   for all $n \geq 1$ and $f \epsilon C(W)$, where $|f|_K = \sup_{w \epsilon K} |f(w)|$;

   (i2) for any f belonging to a subset of $C(W)$, which is everywhere dense in $C_0(W)$, $U^n f$ converges pointwise as $n \to \infty$ to a constant (depending on f).

  (ii) Assume that the metric space (W, d) is locally compact and (i1) and (i2) hold. Then (1) holds.

The proof can be found in Elton and Yan (1989, p.71). Note that Proposition 0 holds for any Feller Markov process on (W, d) not just for our ζ. But, on the other hand, any Markov process on (W, d) can be viewed as a process of random composition of some mappings from W into itself. See, e.g. Kifer (1986, Ch.1).

Several recent papers, see, e.g. Barnsley et al. (1988) and Barnsley and Elton (1988), give conditions on (W, d), as well as the p_i and ϕ_i, $1 \leq i \leq r$, which imply (i1) and (i2) in Proposition 0, thus ensuring the existence of a stationary distribution μ for ζ for which (1) holds.

The aim of the present paper is to show that the ergodic theorem of Doeblin and Fortet (1937), and a generalization of it due to Ionescu Tulcea and Marinescu (1951), can be used to derive deeper properties of μ, in particular, the rate of convergence to μ of the distribution of ζ_n as $n \to \infty$, as well as classical and functional limit theorems for ζ.

The paper is divided into six sections. In Sections 2 and 3 the Doeblin-Fortet and Ionescu Tulcea-Marinescu ergodic theorems are presented. In Section 4 the Doeblin-Fortet theorem is applied to the study of the Markov process ζ in the case where the metric space (W, d) is compact. Even if the results presented are not new, their statement for the process ζ would have a certain interest. In Section 5 the case of a locally compact metric space (W, d) is considered, a context in which the Ionescu Tulcea-Marinescu theorem can be used. Note that the results in this section are new. Finally, in Section 6 we offer some concluding remarks.

1 THE DOEBLIN-FORTET ERGODIC THEOREM

The ergodic theorem stated below was motivated by the study of certain processes with complete connections, to which Doeblin and Fortet's 1937 paper is devoted. This theorem, as well as its generalization discussed in the next section, essentially shows that the operators considered are quasi-compact under the assumptions made. Note that in Sections 2 and 3 the letter U will be used to denote operators which include the transition operator of ξ as a special case.

Assume that the metric space (W, d) is compact and consider a linear operator U on $L(W)$ such that

$$|Uf| \leq |f|, \ f\epsilon L(W),$$

and there exist two positive constants $q < 1$ and Q such that

$$\|Uf\| \leq q\|f\| + Q|f|, \ f\epsilon L(W) \tag{2}$$

Recall that the norms $|\cdot|$ and $\|\cdot\|$ have been previously defined. Since $|f| \leq \|f\|$, the operator U takes $L(W)$ boundedly into itself.

Theorem 1, Doeblin-Fortet. Under the above assumptions on (W, d) and U, the following properties hold.

   (i) The set E of eigenvalues of U of modulus 1 is finite and each of them is of finite multiplicity.

(ii) There are compact operators U_σ on $L(W)$, $\sigma \epsilon E$, such that $U_\sigma^2 = U_\sigma$, $UU_\sigma = U_\sigma U = \sigma U_\sigma$, $U_\sigma U_{\sigma'} = 0$, $\sigma \neq \sigma'$, σ, $\sigma' \epsilon E$.

(iii) The operator $T = U - \Sigma_{\sigma \epsilon E} \sigma U_\sigma$ is quasi-compact and has spectral radius

$$r(T) \doteq \lim_{n \to \infty} \|T^n\|^{1/n} < 1.$$

Further, $U_\sigma T = T U_\sigma = 0$ for any $\sigma \epsilon E$.

(iv) $U^n = \Sigma_{\sigma \epsilon E} \sigma^n U_\sigma + T^n$ for all $n \geq 1$.

Theorem 1 is essentially a precise formulation of the theorem concluding the section entitled Note sur une équation fonctionnelle in Doeblin and Fortet (1937, pp. 142-148). The proof given there, though correct in principle, misses some details. Moreover, the original theorem as well as Theorem 1 are valid assuming, more generally, that U takes $L(W)$ boundedly into itself and that there exist two positive constants $q < 1$ and Q, and a natural integer $k \geq 1$ such that

$$\|U^k f\| \leq q\|f\| + Q|f|, \ f \in L(W). \tag{2'}$$

2 THE IONESCU TULCEA-MARINESCU ERGODIC THEOREM

The Ionescu Tulcea-Marinescu ergodic theorem is an abstract version of Theorem 1 above, in which the Banach spaces $C(W)$ and $L(W)$ are replaced by arbitrary (complex) Banach spaces $(X, |\cdot|)$ and $(Y, \|\cdot\|)$, such that Y is a linear subspace of X and Condition (ITM$_1$) holds.

Condition (ITM$_1$). If $y_n \in Y$, $n \geq 1$, $\sup_{n \geq 1} \|y_n\| = c < \infty$, and $\lim_{n \to \infty} |y_n - y| = 0$ for some $y \in X$, then $y \in Y$ and $\|y\| \leq c$.

It is easy to see that (ITM$_1$) holds in the Doeblin-Fortet special case.

Denote by $L_k(Y, X)$, $k \geq 1$, the collection of all linear operators U on Y which are bounded with respect to both $\|\cdot\|$ and $|\cdot|_Y$, where the latter is the restriction of $|\cdot|$ to Y, and which in addition, satisfy the following conditions.

Condition (ITM$_2$).

$$H = \sup_{n \geq 0} |U^n| Y < \infty.$$

Condition (ITM$_3$). There exist two positive constants $q < 1$ and Q such that

$$\|U^k y\| \leq q\|y\| + Q|y|, \quad y \in Y.$$

Condition (ITM$_4$). If A is a bounded subset of $(Y, \|\cdot\|)$, then $U^k A$ has compact closure in $(X, |\cdot|)$.

It can be shown that Conditions (ITM$_2$) and (ITM$_3$) imply that

$$J = \sup_{n \geq 0} \|U^n\| < \infty.$$

Also, it is easy to see that both (ITM$_2$) and (ITM$_4$) hold, while (ITM$_3$) is explicitly stated, with $k = 1$, in the Doeblin-Fortet special case. The reader should, nevertheless, refer to the remarks following Theorem 1.

Theorem 2 (Ionescu Tulcea-Marinescu). Let $U \in L_k(Y, X)$ for some $k \geq 1$. Then the following properties hold.

(i) The set E of eigenvalues of U of mudolus 1 is finite and each of them is of finite multiciplicity.

(ii) There are compact operators U_σ on Y, $\sigma \in E$, such that $|U_\sigma|_Y \leq H$, $\|U_\sigma\| \leq J$, $U_\sigma^2 = U_\sigma$, $UU_\sigma = U_\sigma U = \sigma U_\sigma$, $U_\sigma U_{\sigma'} = 0$, $\sigma \neq \sigma'$, σ, $\sigma' \in E$.

(iii) The operator $T = U - \Sigma_{\sigma \in E} \sigma U_\sigma$ belongs to $L_k(Y, X)$, and has spectral radius

$$r(T) = \lim_{n \to \infty} \|T^n\|^{1/n} < 1.$$

Further, $U_\sigma T = T U_\sigma = 0$ for any $\sigma \in E$.

(iv) $U^n = \Sigma_{\sigma \in E} \sigma^n U_\sigma + T^n$ for all $n \geq 1$.

3 THE COMPACT CASE

If the metric space (W, d) is assumed to be compact, then Doeblin and Fortet's Theorem 1 can be immediately applied to the study of the Markov process ζ.

Theorem 1'. Assume that

  (i) the transition operator U of ζ satisfies the assumptions of Theorem 1 (see also the remarks following Theorem 1);

 (ii) for any $f \in L(W)$, $U^n f$ converges pointwise as $n \to \infty$ to a constant (depending on f).

Under these assumptions there exist a probability measure μ on $\mathcal{B}_W$ and positive constants $\theta < 1$ and c such that

$$\|U^n f - \int_W f \mathrm{d}\mu\| \leq c\theta^n \|f\| \tag{3}$$

for any $n \geq 1$ and $f \in L(W)$.

Proof. Assumption (i) implies that Theorem 1 holds for U. Assumption (ii) implies that the only eigenvalue of U of modulus 1 is $\sigma = 1$ and this eigenvalue is simple (the corresponding eigenfunction is $f_1 \equiv 1$). Thus, we can write

$$U^n = U_1^n + T^n, \tag{4}$$

where U_1 and T are defined in Theorem 1. Note that the projection U_1 can be uniquely extended by continuity to the closure of $L(W)$ in $C(W)$, that is, to $C(W)$ itself. Since U_1 is a positive linear functional and $U_1 1 = 1$, by Riesz's representation theorem [see, e.g. Mukherjea and Pothoven, 1984, Theorem 5.20, we have $U_1 f = \int_W f \mathrm{d}\mu$ for all $f \in C(W)$, where μ is a probability measure on $\mathcal{B}_W$. Now, it is clear that (3) follows from (4) as $r(T) < 1$.

Remarks.

  1. Assumption (ii) implies assumption (i2) in Proposition 0. Since assumption (i1) there is automatically satisfied in our case (with

$K_\epsilon = W$), it follows that (ii) alone guarantees the existence of a (unique) probability measure μ on $\mathcal{B}_W$ such that (1) holds, whence in particular

$$\lim_{n\to\infty} U^n f(w) = \int_W f \, \mathrm{d}\mu \tag{5}$$

for all $w \in W$ and $f \in C(W)$. (Take $\nu = \delta_w$, the probability measure concentrated at $w \in W$.) Under the additional assumption (i), Theorem 1 specifies the rate of convergence in (5) and also in (1) since (3) implies

$$|U^n f - \int_W f \, \mathrm{d}\mu| \le c\theta^n \|f\|$$

for any $n \ge 1$ and $f \in L(W)$, whence

$$|\int_W U^n f \, \mathrm{d}\nu - \int_W f \, \mathrm{d}\mu| \le c\theta^n \|f\|$$

for any $n \ge 1$, $f \in L(W)$, and $\nu \in \mathrm{pr}(W)$. Actually, (3) says even more. Recall [for details see Iosifescu and Grigorescu (1990, p.83) that $\triangle$ defined as

$$\triangle(\nu, \mu) = \sup |\int_W f \, \mathrm{d}\nu - \int_W f \, \mathrm{d}\mu|, \qquad \nu, \mu \in \mathrm{pr}(W),$$

where the supremum is taken over all real-valued $f \in L(W)$ such that $\|f\| = 1$, is a metric on $\mathrm{pr}(W)$. Also, the topology of the weak convergence on $\mathrm{pr}(W)$ and the topology induced on $\mathrm{pr}(W)$ by the metric $\triangle$ are equivalent. Noting that

$$U^n f(w) = \int_W P^n(w, \mathrm{d}w') f(w'), \quad f \in C(W),$$

it follows that

$$\int_W U^n f \, \mathrm{d}\nu = \int_W \nu(\mathrm{d}w) \int_W P^n(w, \mathrm{d}w') f(w') = \int_W f \, \mathrm{d}(\nu P^n),$$

where P^n is the n-step transition probability of ζ, defined as

$$P^n(w, A) = \mathrm{prob}(\zeta_n \in A | \zeta 0 = w), \ w \in W, \ A \in \mathcal{B}_W.$$

Hence we see that $\triangle(\nu P^n, \mu) \le c\theta^n$ for any $\nu \in \mathrm{pr}(W)$ and $n \ge 1$.

2. A set of simple conditions which imply (2), that is assumption (i), is that

$$\sum_{i=1}^{r} p_i(w_1)\frac{d(\phi_i(w_1),\phi_i(w_2))}{d(w_1,w_2)}) \leq a < 1, \ w_1 \neq w_2 \in W, \quad (6)$$

and $p_i \in L(W)$, $1 \leq i \leq r$. The reader may refer to Iosifescu and Grigorescu (1990, Theorem 3.1.16), where a more general setting is considered.

A set of simple conditions which imply assumption (ii) is that (6) above holds, there exists $\delta > 0$ such that

$$\sum_{\{i:d(\phi_i(w_1),\phi_i(w_2))\leq ad(w_1,w_2)\}} p_i(w_1)p_i(w_2) \geq \delta^2, \quad w_1, w_2 \in W,$$

and, for some $\alpha > 0$,

$$\int_0^\alpha \frac{c_i(t)\mathrm{d}t}{t} < \infty, \ 1 \leq i \leq r, \quad (7)$$

where $c_i(t) = \sup |p_i(w_1) - p_i(w_2)|$, $t \geq 0$, the supremum being taken over all $w_1, w2 \in W$ for which $d(w_1, w_2) \leq t$. Clearly, if $p_i \in L(W)$, $1 \leq i \leq r$, then (7) holds. See Barnsley et al. (1988).

To be aware of the usefulness of Theorem 1, we now state some limit theorems for ζ. For the proofs see Iosifescu & Grigorescu (1990, Theorems 4.2.7 and 4.2.8), under a more general setting.

Let h be a fixed real-valued function in $L(W)$. Put $f_n = h(\zeta_n) - \int_W h\mathrm{d}\mu$, $S_n = \sum_{j=1}^n f_j$, $n \geq 1$. Let $P_\mu(P_w)$ denote the distribution (probability measure) of ζ when the initial distribution of ζ_0 is $\mu(\delta_w)$. Then the mean value operator with respect to $P_\mu(P_w)$ will be denoted $E_\mu(E_w)$. Put $\sigma^2 = E_\mu(f_1^2) + 2\sum_{n\geq 1}^\mu E_\mu(f_1 f_{n+1})$. It can be shown that the series defining σ^2 is absolutely convergent and $E_\psi(S_n^2) = n\sigma^2 + 0(1)$ as $n \to \infty$, where ψ stands for either μ or w (arbitrary in W). Thus $\sigma^2 \geq 0$.

Theorem 3. Assume that the assumptions of Theorem 1′ hold and $\sigma > 0$. Then there exists a constant $c > 0$ such that for any $w \in W$, $x \in \mathrm{R}$, and $n \geq 1$ we have

$$|P_w(\frac{S_n}{\sigma\sqrt{n}} \le x) - \frac{1}{\sqrt{2\pi}} \int_{-\infty}^{x} \exp(-\frac{u^2}{2})du| \le \frac{c}{\sqrt{n}}.$$

Next, assuming always $\sigma > 0$, define the random processes

$$\eta_n^C : \eta_n^C(t) = \frac{1}{\sigma\sqrt{n}}(S_{[nt]} + (nt - [nt])h_{[nt]+1} \quad n \ge 1,$$

$$\eta_n^D : \eta_n^D(t) = \frac{1}{\sigma\sqrt{n}}S_{[nt]} \quad , t \in [0, 1], \quad n \ge 1,$$

where $[x]$ denotes the integer part of $x \ge 0$. As usual, C is the metric space of real-valued continuous functions on $[0, 1]$ with the uniform metric and $C \subset D$ is the metric space of real-valued functions on $[0, 1]$, which are right-continuous and have left limits with the Skorohod metric.

Theorem 4. Assume that the assumptions of Theorem 1 hold and $\sigma > 0$. Then, under P_ψ, both η_n^C and η_n^D converge weakly to the standard Brownian motion process; further the sequence $(\eta_n^C/\sqrt{(2 \log \log n)})_{n \ge 3}$, viewed as a subset of C, is a relatively compact set, whose derived set coincides P_μ-almost surely with the collection of all absolutely continuous functions $x \in C$ for which $x(0) = 0$ and $\int_0^1 [(x'(t)]^2 dt \le 1$.

4 THE LOCALLY COMPACT CASE

If the metric space (W, d) is assumed to be locally compact rather than compact, then Theorem 1 can no longer be used. We shall show that Theorem 2 with an appropriate choice of the Banach spaces X and Y can be used instead.

Consider a non-zero function $\alpha \in L(W) \cap C_0(W)$ such that

$$\inf_{w_1 \ne w_2 \in K} \frac{|\alpha(w_1) - \alpha(w_2)|}{d(w_1, w_2)} > 0$$

whatever the compact subset K of W. For any $f \in C(W)$ put

$$s_\alpha(f) = \sup_{w_1 \neq w_2} |\frac{f(w_1) - f(w_2)}{\alpha(w_1) - \alpha(w_2)}|, \; \|f\|_\alpha = |f| + s_\alpha(f)$$

and let $L_\alpha(W) = \{f : f \in C(W), \|f\|_\alpha < \infty\}$. It is easy to see that $L(W) \cap C_c(W) \subset L_\alpha(W)$ for any function α with the stated properties. Also, it is not difficult to show that $L_\alpha(W)$ is a Banach space under $\|\cdot\|_\alpha$, and that the Banach spaces $(C(W), |\cdot|)$ and $(L_\alpha(W), \|\cdot\|_\alpha)$ satisfy Condition (ITM$_1$) (see Section 3). It is useful to note that

$$s(f) \leq s(\alpha)s_\alpha(f) \tag{8}$$

for any $f \in L_\alpha(W)$, which shows that $L_\alpha(W) \subset L(W)$ for any function α with the properties stated.

Theorem 2$'$. Assume that

(i) the transition operator U of ζ takes boundedly $L_\alpha(W)$ into itself and satisfies the property corresponding to (2$'$) in the compact case, i.e. there exists two positive constants $q < 1$ and Q, and a natural integer $k \geq 1$ such that

$$\|U^k f\|_\alpha \leq q\|f\|_\alpha + Q|f|, \; f \in L_\alpha(W);$$

(ii) for any $f \in L_\alpha(W)$, $U^n f$ converges pointwise as $n \to \infty$ to a constant (depending on f).

Under these assumptions there exist a probability measure μ on $\mathcal{B}_W$ and positive constants $\theta < 1$ and c such that

$$\|U^n f - \int_W f d\mu\|_\alpha \leq c\theta^n \|f\|_\alpha \tag{9}$$

for any $n \geq 1$ and $f \in L_\alpha(W)$.

Proof. Condition (ITM$_2$) obviously holds for U as an operator on $X = C(W)$. Assumption (i) implies that Condition (ITM$_3$) holds for U, with $X = C(W)$ and $Y = L_\alpha(W)$, two Banach spaces which satisfy Condition (ITM$_1$), as we have already noted. To verify Condition (ITM$_4$) let $L' = \{f : f \in L_\alpha(W), \|f\|_\alpha \leq M\}$ for an arbitrarily fixed $M > 0$. Clearly, L' is bounded in $L_\alpha(W)$ as well as the set $U^k L'$, which is contained in an L' corresponding to another $M > 0$; also, any L' is a closed bounded set in

$C(W)$. Notice that any L' is a collection of continuous functions which are equally bounded and equally uniformly continuous on any compact subset of W [by (8)] and at infinity, too (by the vanishing of α at infinity). It then follows from Theorem IV.6.5 in Dunford and Schwartz (1958, p.266) that any L' is compact in $C(W)$. Thus any bounded set in $L_\alpha(W)$ is contained in a compact subset of $C(W)$, hence its closure should be compact in $C(W)$, too. To conclude, Theorem 2 holds for U, with $X = C(W)$ and $Y = L_\alpha(W)$.

Next, assumption (ii) implies that the only eigenvalue of U of modulus 1 is $\sigma = 1$ and this eigenvalue is simple (the corresponding eigenfunction is $f_1 \equiv 1$). Thus, we can write

$$U^n = U_1^n + T^n, \tag{10}$$

where U_1 and T are defined in Theorem 2. Note that the projection U_1 can be uniquely extended by continuity to the closure of $L_\alpha(W)$ in $C(W)$. This set contains $C_c(W)$ and coincides with $C_\ell(W) = $ the collection of all functions $f \in C(W)$ with limit at infinity. More precisely, $f \in C_\ell(W)$ means that there exists a complex number $f(\infty)$ such that $f - f(\infty) \in C_0(W)$. Since U_1 is a positive linear functional and $U_1 1 = 1$, by Riesz's representation theorem we have $U_1 f = \int_W f \, d\mu$ for all $f \in C_0(W)$, hence for all $f \in C_\ell(W)$, where μ is a probability measure on $\mathcal{B}_W$. Now, it is clear that (9) follows from (10) as $r(T) < 1$.

Remarks.

1. Theorem $2'$ is not void. It applies, for example, when $W = \mathrm{R}_+$ under the Euclidean metric, $p_i(w) = p_i$ (constant), $1 \leq i \leq r$, $\alpha(w) = e^{-w}$, $w \in \mathrm{R}_+$, $\sum_{i=1}^{r} p_i s_\alpha(\alpha \circ \phi_i) < 1$. (One may take, e.g., $\phi_i(w) = \max(w + a_i)$, $a_i \in \mathrm{R}$, $1 \leq i \leq r$, $w \in \mathrm{R}_+$, with $\sum_{i=1}^{r} p_i e^{-a_i} < 1$.)

2. Remark 1 following Theorem $1'$ applies *mutatis mutandis* to the present case. This is also true of Theorems 3 and 4.

3. Theorem 2 contains Theorem $1'$. Indeed, if the metric space (W, d) is compact, then for any function α with the properties stated, the norms $\| \cdot \|$ and $\| \cdot \|_\alpha$ are equivalent.

4. It seems that the scope of Theorem $2'$ in the locally compact case is narrower than that of Theorem $1'$ in the compact case. Actually, in Theorem $2'$ the one-point compactification of (W, d) plays an important part, even if this does not appear explicitly.

5 CONCLUDING REMARKS

The more general context alluded to in Section 4, in which the theorems presented for the Markov process ζ are valid, is precisely that of the Markov process associated with a general random system with complete connections. For such a system, the finite collection Φ is replaced by an arbitrary collection of mappings from W into itself.

It is quite remarkable that a result obtained some 50 years ago turns out to be useful today in a fashionable context. This shows once again the depth of Doeblin's genius. It is typical of Doeblin to deal *avant la lettre* with concepts explicitly defined later on, and to solve problems restated as new ones after decades. In these respects our papers Iosifescu (1992a, 1992b) are relevant. Bearing this in mind, were Doeblin alive nowadays, could we speculate about his possible achievements? Without attempting a precise answer, we may suggest that with Doeblin active for a number of years after 1940, several chapters of probability theory and stochastic processes would nowadays have been well in advance of our present knowledge.

Acknowledgements. The ideas and results in this paper have been presented under various forms in Warsaw and Salonika (in 1990) and in Melbourne and Paris (in 1991). The author is grateful to Prof. S. Kwapień, Dr. Sofia Kalpazidou, Prof. Harry Cohn, Prof. Bui Trong Lieu and Prof. M. Schreiber for their kind invitations.

REFERENCES

Barnsley, M.F. and J.H. Elton, J.H. (1988) A new class of Markov processes for image encoding. *Adv. in Appl. Probab.* 20, 14-32.

M.F. Barnsley, M.F., Demke, S.G., Elton, J.H. & Geronimo, J.S. (1988) Invariant measures for Markov processes arising from iterated function systems with place-dependent probabilities. *Ann. Inst. H. Poincaré* Sec. B (N.S.)24, 367-394. Erratum, ibid. 25(1989), 509-590.

Doeblin, W. & Fortet, R. (1937). Sur des chaines à liaisons complètes. *Bull. Soc. Math. France* 65, 132-148.

Dunford, N. & Schwartz, J.T.(1958) *Linear Operators, Part I: General Theory*. Interscience, New York.

Elton, J.H. & Yan, Z. (1989) Approximation of measures by Markov processes and homogeneous affine iterated function systems. *Constructive Approximation* 5, 69-87.

Ionescu Tulcea, C.T. and Marinescu, G. (1951) Théorie ergodique pour des classes d'opérations non complètement continues. *Ann. of Math.* (2), 52, 140-147.

Iosifescu, M. (1992) A coupling method in the theory of dependence with complete connections according to Doelbin. *Rev. Roumaine Math. Pures Appl.* 37, 59-66.

Iosifescu, M. (1992b) Doeblin and the metric theory of continued fractions: a functional theoretic solution to Gauss' 1812 problem. In this volume.

Iosifescu, M & Grigorescu, S. (1990) Dependence with Complete Connections and Its Applications. *Cambridge Tracts in Mathematics* 96. Cambridge Uni. Press, Cambridge.

Kifer, Y. (1986) Ergodic Theory of Random Transformations. Birkhauser, Boston.

Mukherjea, A. and Pothoven, K. (1984) Real and Functional Analysis. Part A: Real Analysis. Second Edition. *Mathematical Concepts and Methods in Science and Engineering* 27. Plenum Press, New York.

ROMANIAN ACADEMY, CENTRE FOR MATHEMATICAL STATISTICS,
22 MAGHERU BLVD. RO-70158, BUCHAREST, ROMANIA.

Contemporary Mathematics
Volume **149**, 1993

THE NEAREST INTEGER CONTINUED FRACTION EXPANSION: AN APPROACH IN THE SPIRIT OF DOEBLIN

MARIUS IOSIFESCU AND SOFIA KALPAZIDOU

Abstract

We show the relevance of the Ionescu Tulcea-Marinescu ergodic theorem in the context of a Doeblin-type approach to the metric theory of the nearest integer continued fraction expansion.

1 INTRODUCTION

Doeblin (1940) used dependence with complete connections in the study of metric properties of the usual continued fraction expansion. For a detailed discussion of Doeblin's 1940 paper see Iosifescu (1990). Our purpose here is to develop Doeblin's ideas in the context of the nearest integer continued fraction (NICF) expansion. The main point is the appropriate use of the Ionescu Tulcea-Marinescu ergodic theorem (see Iosifescu, 1992). We think that our treatment makes the probabilistic aspects more transparent than do, e.g., Rieger (1979) and Rockett (1980). Notice that even in these two papers, the Ionescu Tulcea-Marinescu theorem could have been profitably used, as shown in Kalpazidou (1983, 1985, 1986a).

For everything concerning dependence with complete connections and, in particular, the concept of a random system with complete connections (RSCC), we refer the reader to Iosifescu and Grigorescu (1990). Especially, Subsections 5.5.1 and 5.5.2 there contain a treatment of the continued fraction expansion, which clarifies Doeblin's original ideas following Iosifescu (1974).

AMS 1991 Subject Classifications: 11 K 55, 47 A 35, 47 B 38, 60 K 99, 28 D 05.

This paper is in final form and no version of it will be submitted for publication elsewhere.

2 THE NEAREST INTEGER CONTINUED FRACTION EXPANSION

Let $I = [-\frac{1}{2}, \frac{1}{2}]$. Any irrational number $y \in I$ can be written in a unique way as

$$y = \frac{\epsilon_1(y)}{b_1(y) + \frac{\epsilon_2(y)}{b_2(y)+\cdots}} = \left[\begin{array}{l} \epsilon_1(y), \epsilon_2(y), \ldots \\ b_1(y), b_2(y), \ldots \end{array} \right], \tag{1}$$

where $\epsilon_n(y) \in \{-1, 1\}$ and $2 \leq b_n(y) \in \mathbb{N}^* = \{1, 2, \ldots\}$, $b_n(y) + \epsilon_{n+1}(y) \geq 2$, $n \in \mathbb{N}^*$. The integers $\epsilon_n(y)$ and $b_n(y)$ are determined as follows:

$$\epsilon_1(y) = \text{ sgn } y, \, b_1(y) = [|1/y| + 1/2],$$

and

$$\epsilon_{n+1}(y) = \epsilon_1(\tau^n(y)), \, b_{n+1}(y) = b_1(\tau^n(y)), \, n \in \mathbb{N}^*,$$

with

$$\tau(y) = |1/y| - [|1/y| + 1/2].$$

(Here $[\cdot]$ denotes integer part.) Clearly, $\tau(y) = \epsilon_1(y)/y - b_1(y) \in I$, so that $b_1(y)$ is the nearest integer to $\epsilon_1(y)/y$. This explains the name NICF which was given to expansion (1). Note that the b_n and $\epsilon_n, n \in \mathbb{N}^*$, can be viewed as random variables on $(I, \mathcal{B}_I)$, where $\mathcal{B}_I$ is the collection of Borel subsets of I. They are defined almost surely with respect to any probability on $\mathcal{B}_I$ which assigns probability 0 to the set of rational numbers in I (thus, in particular, with respect to Lebesgue measure λ).

The basic facts about the NICF expansion can be found in the classical monograph of Perron (1977), and we now reproduce those we shall need. Let us define $\mathbb{N}^*$-valued functions $p_n(y)$ and $q_n(y)$, $n \in \mathbb{N}^*$, by the recursions

$$p_n(y) = b_n(y)p_{n-1}(y) + \epsilon_n(y)p_{n-2}(y),$$

$$q_n(y) = b_n(y)q_{n-1}(y) + \epsilon_n(y)q_{n-2}(y), \, n \in \mathbb{N}^*,$$

with $p_{-1}(y) = q_0(y) = 1$, $q_{-1}(y) = p_0(y) = 0$ for any irrational number $y \in I$. Then

$$|p_n(y)q_{n-1}(y) - p_{n-1}(y)q_n(y)| = 1, n \in \mathbb{N} = \{0\} \cup \mathbb{N}^*,$$

and putting $r_n(y) = b_n(y) + \tau^n(y) = b_n(y) + \left[\begin{smallmatrix} \epsilon_{n+1}(y), \epsilon_{n+2}(y), \ldots \\ b_{n+1}(y), b_{n+2}(y), \ldots \end{smallmatrix} \right], \, n \in \mathbb{N}^*$, we have

$$y = \frac{r_{n+1}(y)p_n(y) + \epsilon_{n+1}(y)p_{n-1}(y)}{r_{n+1}(y)q_n(y) + \epsilon_{n+1}(y)q_{n-1}(y)}, \, n \in \mathbb{N},$$

for any irrational number $y \in I$. It follows that the set $E_{i_1 \cdots i_n}^{j_1 \cdots j_n}$ of irrational numbers $y = \left[\begin{smallmatrix} \epsilon_1(y), \epsilon_2(y), \ldots \\ b_1(y), b_2(y), \ldots \end{smallmatrix} \right] \in I$ for which $b_\ell(y) = i_\ell$ and $\epsilon_\ell(y) = j_\ell$,

$1 \leq \ell \leq n$, with $i_\ell \geq 2$, $j_\ell = \pm 1$, $1 \leq \ell \leq n$, $i_\ell + j_{\ell+1} \geq 2$, $1 \leq \ell < n$, $n \in \mathbb{N}^*$, is either the set of irrational numbers in the interval with end points

$$\frac{p_n - p_{n-1}/2}{q_n - q_{n-1}/2} \quad \text{and} \quad \frac{p_n + p_{n-1}/2}{q_n + q_{n-1}/2}$$

or the set of irrational numbers in the interval with end points

$$\frac{p_n}{q_n} \quad \text{and} \quad \frac{p_n + p_{n-1}/2}{q_n + q_{n-1}/2}$$

according as $i_n \geq 3$ or $i_n = 2$. Clearly, the values assumed here by p_{n-1}, q_{n-1}, p_n, and q_n are those corresponding to the values $i_1, \ldots, i_n$ of $b_1, \ldots b_n$ and the values $j_1, \ldots, j_n$ of $\epsilon_1, \ldots, \epsilon_n$. Putting $s_n = q_{n-1}/q_n$, $n \in \mathbb{N}^*$, the above results allow us to compute the conditional probability

$$\lambda\left(r_{n+1} > t, \epsilon_{n+1} = \epsilon \Big| E^{j_1 \ldots j_n}_{i_1 \ldots i_n}\right) = \frac{\lambda\left(\{rn + 1 > t, \epsilon_{n+1} = \epsilon\} \bigcap E^{j_1 \ldots j_n}_{i_1 \ldots i_n}\right)}{\lambda\left(E^{j_1 \ldots j_n}_{i_1 \ldots i_n}\right)}$$

$$= \begin{cases} \dfrac{\left|\frac{p_n t + p_{n-1}}{q_n t + q_{n-1}} - \frac{p_n}{q_n}\right|}{\left|\frac{p_n}{q_n} - \frac{p_n + p_{n-1}/2}{q_n + q_{n-1}/2}\right|} = \dfrac{2(q_n + q_{n-1}/2)}{q_n t + q_{n-1}} = \dfrac{s_n + 2}{s_n + t}, & \text{if } i_n = 2 \text{ and } \epsilon = 1 \\[3mm] 0, & \text{if } i_n = 2 \text{ and } \epsilon = -1 \\[3mm] \dfrac{\left|\frac{p_n t + \epsilon p_{n-1}}{q_n t + \epsilon q_{n-1}} - \frac{p_n}{q_n}\right|}{\left|\frac{p_n - p_{n-1}/2}{q_n - q_{n-1}/2} - \frac{p_n + p_{n-1}/2}{q_n + q_{n-1}/2}\right|} = \dfrac{q_n^2 - q_{n-1}^2/4}{q_n(q_n t + \epsilon q_{n-1})} = \dfrac{4 - s_n^2}{4(\epsilon s_n + t)}, & \text{if } i_n \geq 3 \text{ and } \epsilon = \pm 1, \end{cases}$$

where $t \geq 2$. Also, since $y = \epsilon_1(y)/r_1(y)$, we have

$$\lambda(r_1 > t, \epsilon_1 = \epsilon) = \frac{1}{t}, \ t \geq 2, \ \epsilon = \pm 1.$$

Consequently, since

$$\{b_n = i\} = \{i - \frac{1 - \delta(i, 2)}{2} < r_n < i + \frac{1}{2}\}, \ n \in \mathbb{N}^*, \ i \geq 2,$$

where δ is Kronecker's, we obtain

$$\lambda\left(b_{n+1} = i, \epsilon_{n+1} = \epsilon \Big| E^{j_1 \cdots j_n}_{i_1 \cdots i_n}\right)$$

$$= \begin{cases} \dfrac{(s_n + 2)(1 - \delta(i,2)/2)\delta(\epsilon,1)}{(s_n + i - (1 - \delta(i,2))/2)(s_n + i + 1/2)}, & \text{if } i_n = 2 \\[3mm] \dfrac{(4 - s_n^2)(1 - \delta(i,2)/2)}{4(\epsilon s_n + i - (1 - \delta(i,2))/2)(\epsilon s_n + i + 1/2)}, & \text{if } i_n \geq 3 \end{cases} \qquad (2)$$

and

$$\lambda(b_1 = i, \epsilon_1 = \epsilon) = (1 - \delta(i,2)/2)/(i - (1 - \delta(i,2))/2)(i + 1/2)$$

for all $n \in \mathbb{N}^*$, $i \geq 2$, and $\epsilon = \pm 1$.

Note that the equation $q_n = b_n q_{n-1} + \epsilon_n q_{n-2}$ implies that $1/s_n = b_n + \epsilon_n s_{n-1}$, whence

$$s_n = \frac{1}{b_n + \epsilon_n s_{n-1}}, \, n \in \mathbb{N}^*, \tag{3}$$

with $s_0 = 0$, that is

$$s_n = \begin{bmatrix} 1, \epsilon_n, \dots, \epsilon_2 \\ b_n, b_{n-1}, \dots, b_1 \end{bmatrix}, n \in \mathbb{N}^*.$$

Put $G = (\sqrt{5} + 1)/2 = 1.6180\dots$, $g = (\sqrt{5} - 1)/2 = 0.6180\dots$. Without further mentioning we shall frequently use identities like $g + 1 = G$, $g^2 = 1 - g$, $G^2 = G + 1$, $gG = 1$, $g + 2 = G^2$. It is not difficult to see that, whatever $n \in \mathbb{N}^*$, we have $0 < s_n < g^2$ iff (= if and only if) $b_n \geq 3$, and $g^2 < s_n < g$ iff $b_n = 2$. For the proof see Perron (1977, Satz 5.18(B) and Satz 5.20). This fact and equations (2) and (3) lead us to the RSCC

$$((W, \mathcal{W}), (X, \mathcal{X}), u, P), \tag{4}$$

where

$$W = [0, g], \mathcal{W} = \mathcal{B}_{[0,g]} = \text{ the collection of Borel subsets of } W,$$

$$X = \{2, 3, \dots\} \times \{-1, 1\}, \mathcal{X} = \text{ the collection of all subsets of } X,$$

$$u(w, (i, -1)) = \begin{cases} \frac{1}{i-w}, & \text{if } w \in [0, g^2) = W_1 \\ \frac{1}{i-g^2}, & \text{if } w \in [g^2, g] = W_2 \end{cases}$$

$$u(w, (i, 1)) = \frac{1}{i + w}, \, w \in W,$$

$$P(w, (i, \epsilon)) = \begin{cases} \frac{(4-w^2)(1-\delta(i,2)/2)}{4(\epsilon w + i - (1-\delta(i,2))/2)(\epsilon w + i + 1/2)}, & \text{if } w \in W_1 \\ \frac{(w+2)(1-\delta(i,2)/2)\delta(\epsilon,1)}{(w+i-(1-\delta(i,2))/2)(w+i+1/2)}, & \text{if } w \in W_2 \end{cases}$$

for all $i \geq 2$ and $\epsilon = \pm 1$. It is very important in what follows that

$$u(W, (2, \epsilon)) \subset W_2, \quad u(W, (i, \epsilon)) \subset \overline{W_1} \tag{5}$$

for all $\epsilon = \pm 1$ and $i \geq 3$.

According to the general theory (see Iosifescu and Grigorescu, 1990, Section 1.1), for any $w \epsilon W$ there exist a probability $\mathbf{P}_w$ on $\mathcal{X}^{\mathbf{N}^*}$, X-valued random variables ξ_n, $n \in \mathbf{N}^*$, and W-valued random variables $\zeta_n, n \in \mathbf{N}$, on $(X^{\mathbf{N}^*}, \mathcal{X}^{\mathbf{N}^*}, \mathbf{P}_w)$ such that $\zeta_0 = w$, $\zeta_n = u(\xi_n, \zeta_{n-1})$, $n \in \mathbf{N}^*$, and

$$\mathbf{P}_w(\xi_1 = (i, \epsilon)) = P(w, (i, \epsilon)),$$

$$\mathbf{P}_w(\xi_{n+1} = (i, \epsilon) | \xi_\ell, \ 1 \le \ell \le n) = P(\zeta_n, (i, \epsilon)), \ n \in \mathbf{N}^*,$$

for all $(i, \epsilon) \in X$. Similarly to the case of the continued fraction expansion, we can assert that

(i) for $w = 0$, the sequences $(\xi_n)_{n \in \mathbf{N}^*}$ and $(\zeta_n)_{n \in \mathbf{N}}$ associated with RSCC(4) under $\mathbf{P}_0$ are equivalent to the sequences $(b_n)_{n \in \mathbf{N}^*}$ and $(s_n)_{n \in \mathbf{N}}$, respectively, under λ;

(ii) for a rational $0 \ne w = \begin{bmatrix} 1, & j_r, \dots, j_2 \\ i_r, & i_{r-1}, \dots, i_1 \end{bmatrix} \in W$, $r \in \mathbf{N}^*$, the sequences $(\xi_n)_{n \in \mathbf{N}^*}$ and $(\zeta_n)_{n \in \mathbf{N}}$ associated with RSCC(4) under $\mathbf{P}_w$ are equivalent to the sequences $(b_{r+n})_{n \in \mathbf{N}^*}$ and $(s_{r+n})_{n \in \mathbf{N}}$, respectively, under $\lambda(\cdot | E_{i_1 \cdots i_r}^{j_1 \cdots j_r})$, with either $j_1 = 1$ or $j_1 = -1$.

Now, the sequence $(\zeta_n)_{n \in \mathbf{N}}$ is a W-valued Markov chain with transition probability function Q defined as

$$Q(w, B) = \sum_{\{(i, \epsilon) : u(w, (i, \epsilon)) \in B\}} P(w, (i, \epsilon)), \quad w \in W, \ B \in \mathcal{W}.$$

We shall prove that a stationary probability Q^∞ for this Markov chain is given by

$$Q^\infty(B) = \int_B dF(w), \quad B \in \mathcal{W},$$

where

$$F(w) = \frac{1}{\log G} \times \begin{cases} \log \frac{2+w}{2-w}, & \text{if } w \in W_1, \\ \log \frac{2+w}{2-g^2}, & \text{if } w \in W_2. \end{cases}$$

(cf. Jagers, 1985, Theorem 1), that is the following result holds.

Proposition 1 *For any $B \in \mathcal{B}_{[0,g]}$ we have*

$$\int_0^g Q^\infty(dw) Q(w, B) = Q^\infty(B).$$

Proof. Since the intervals $[0, x) \subset [0, g]$ generate $\mathcal{B}_{[0,g]}$ it is sufficient to check the above equation just for $B = [0, x)$, $0 < x \le g$. For any given

$w \in W_1$ we have $u(w, (i, -1)) \in [0, x)$ iff $i \geq [x^{-1} + w] + 1$, while for any given $w \in W$ we have $u(w, (i, 1)) \in [0, x)$ iff $i \geq [x^{-1} - w] + 1$. Thus

$$A = (\log G) \int_0^g Q^\infty(dw) Q(w, [0, x))$$

$$= \int_0^{g^2} \frac{dw}{-w + [x^{-1} + w] + 1 - (1 - \delta([x^{-1} + w] + 1, 2))/2}$$

$$+ \int_0^g \frac{dw}{w + [x^{-1} - w] + 1 - (1 - \delta([x^{-1} - w] + 1, 2))/2}.$$

Now, we distinguish three cases. **Case 1:** $x < g^2$. Letting $\{\cdot\}$ denote fractionary part, we have

$$A = \int_0^{g^2} \frac{dw}{-w + [x^{-1} + w] + 1/2} + \int_0^g \frac{dw}{w + [x^{-1} - w] + 1/2}$$

$$= \int_0^{\min(g^2, 1-\{x^{-1}\})} \frac{dw}{-w + [x^{-1}] + 1/2} + \int_{\min(g^2, 1-\{x^{-1}\})}^{g^2} \frac{dw}{-w + [x^{-1}] + 3/2}$$

$$+ \int_0^{\min(g, \{x^{-1}\})} \frac{dw}{w + [x^{-1}] + 1/2} + \int_{\min(g, \{x^{-1}\})}^{g} \frac{dw}{w + [x^{-1}] - 1/2}$$

$$= \log \frac{2+x}{2-x} = (\log G) F(x) = (\log G) Q^\infty([0, x)).$$

[Let us note that for computing the last four integrals above we should distinguish two subcases, namely

1a: $1 - \{x^{-1}\} < g^2$ (equivalent to $\{x^{-1}\} > 1 - g^2 = g$), and

1b: $\{x^{-1}\} < g$ (equivalent to $1 - \{x^{-1}\} > 1 - g = g^2$). Also, we make use of the identity
$x^{-1} = [x^{-1}] + \{x^{-1}\}.$]

Case 2: $g^2 \leq x \leq 1/2$. In this case we have $2 \leq x^{-1} \leq G^2 = g + 2$, whence $[x^{-1}] = 2$, $\{x^{-1}\} \leq g$, and $1 - \{x^{-1}\} \geq g^2$. Then noting that $2 + \{x^{-1}\} = x^{-1}$, we have

$$A = \int_0^{g^2} \frac{dw}{-w + 2 + 1/2} + \int_0^{\{x^{-1}\}} \frac{dw}{w + 2 + 1/2} + \int_{\{x^{-1}\}}^{g} \frac{dw}{w + 2}$$

$$= -\log(-g^2 + \frac{5}{2}) + \log \frac{5}{2} + \log(x^{-1} + \frac{1}{2}) - \log \frac{5}{2} + \log(g + 2) - \log x^{-1}$$

$$= \log \frac{2+x}{2(5/2 - g^2)/(g+2)} = \log \frac{2+x}{2 - g^2} = (\log G) F(x) = (\log G) Q^\infty([0, x)).$$

Case 3: $1/2 < x \leq g$. In this case we have $g + 1 = G \leq x^{-1} < 2$, whence $[x^{-1}] = 1, \{x^{-1}\} \geq g$, and $1 - \{x^{-1}\} \leq g^2$. Then, noting that $1 + \{x^{-1}\} = x^{-1}$, we have

$$A = \int_0^{1-\{x^{-1}\}} \frac{\mathrm{d}w}{-w+2} + \int_{1-\{x^{-1}\}}^{g^2} \frac{\mathrm{d}w}{-w+3-1/2} + \int_0^g \frac{\mathrm{d}w}{w+2}$$

$$= -\log x^{-1} + \log 2 - \log(-g^2 + \frac{5}{2}) + \log(x^{-1} + \frac{1}{2}) + \log(g+2) - \log 2$$

$$= \log \frac{2+x}{2(5/2 - g^2)/(g+2)} = (\log G)Q^\infty([0,x)).$$

3 THE MAIN RESULT

Let $L(W_1, W_2)$ denote the collection of all bounded complex-valued functions f on W such that

$$s(f_k) = \sup_{w_1 \neq w_2 \in W_k} \frac{|f_k(w_1) - f_k(w_2)|}{|w_1 - w_2|} < \infty,$$

where f_k is the restriction of f to W_k, $k = 1, 2$. Let also $C(W_1, W_2)$ denote the collection of all bounded complex-valued functions f on W such that f_k is continuous on W_k, $k = 1, 2$, and f_1 extends to a continuous function on the closure $\overline{W_1} = [0, g^2]$. It is easy to see that $C(W_1, W_2)$ is a Banach space under the supremum norm $|f| = \sup_{w \in W} |f(w)|$, while $L(W_1, W_2)$ is a Banach space under the norm $\|f\| = |f| + s(f)$, where $s(f) = \max_{1 \leq k \leq 2} s(f_k)$. Moreover, this pair of Banach spaces satisfies Condition (ITM$_1$) in the Ionescu Tulcea-Marinescu ergodic theorem (see Iosifescu, 1992). Cf. also Problems 28 and 29 at the end of Chapter 3 in Iosifescu and Grigorescu (1990).

Now consider the transition operator U associated with RSCC(4), which is defined as

$$Uf(w) = \sum_{(i,\epsilon) \in X} P(w, (i, \epsilon)) f(u(w, (i, \epsilon))), \quad w \in W,$$

for any $f \in B(W) =$ the collection of all bounded measurable complex-valued functions on W. Note that U is also the transition operator of the Markov chain $(\zeta_n)_{n \in \mathbb{N}}$ and we have

$$Uf(w) = \int_W Q(w, \mathrm{d}w') f(w'),$$

which implies that

$$U^n f(w) = \int_W Q^n(w, dw') f(w'), \quad w \in W, \ n \in \mathbb{N}^*, \tag{6}$$

where Q^n is the n-step transition probability function constructed from $Q = Q^1$. Clearly, we have

$$|Uf| \leq |f| \tag{7}$$

for any $f \in L(W_1, W_2)$—actually for any $f \in B(W) \supset L(W_1, W_2)$. Next, it is easy to see that for some constants $C > 0$ and $0 < q < 1$ we have

$$\sup_{w \in W_k} \left| \frac{d}{dw} \sum_{(i,\epsilon) \in A} P(w, (i, \epsilon)) \right| \leq C, \, k = 1, 2,$$

for all $A \subset X$, and

$$\frac{|u(w_1, (i, \epsilon)) - u(w_2, (i, \epsilon))|}{|w_1 - w_2|} \leq q$$

for all $w_1 \neq w_2 \in W$ and $(i, \epsilon) \in X$. Then, from the equation

$$Uf(w_1) - Uf(w_2) = \sum_{(i,\epsilon) \in X} P(w_1, (i, \epsilon))(f(u(w_1, (i, \epsilon))) - f(u(w_2, (i, \epsilon))))$$

$$+ \sum_{(i,\epsilon) \in X} (P(w_1, (i, \epsilon)) - P(w_2, (i, \epsilon))) f(u(w_2, (i, \epsilon))),$$

by (5) we obtain

$$s(Uf) \leq qs(f) + C|f|, \quad f \in L(W_1, W_2). \tag{8}$$

Inequalities (7) and (8) show that U takes boundedly $L(W_1, W_2)$ into itself, and Conditions (ITM$_2$) and (ITM$_3$) in the Ionescu Tulcea-Marinescu ergodic theorem hold.

Now, we are prepared to prove our main result.

Theorem 2 *There exist positive constants $\theta < 1$ and c such that for all $f \in L(W_1, W_2)$ and $n \in \mathbb{N}^*$ we have*

$$\left\| U^n f - \int_0^g f(w) Q^\infty(dw) \right\| \leq c\theta^n \|f\|.$$

Proof. We have already seen that Condition (ITM$_1$) holds for the pair of Banach spaces $(C(W_1, W_2), L(W_1, W_2))$, and that Conditions (ITM$_2$) and (ITM$_3$) hold for the transition operator U as an operator on $L(W_1, W_2)$.

Next, checking Condition(ITM$_4$) is a very simple exercise. We can there-fore apply the Ionescu Tulcea-Marinescu ergodic theorem. The only thing which remains to be proved is that 1 is the unique eigenvalue of U of modulus 1, and this eigenvalue is simple. This follows from the uniform convergence of $U^n f$ as $n \to \infty$ to a constant function $U^\infty f$ depending on f for any $f \in L(W_1, W_2)$, which is a consequence of (5) and of The-orem 2.4.2 and Proposition 2.2.8 in Iosifescu and Grigorescu (1990). In conjunction with (6) and our Proposition 1, that convergence also implies that

$$U^\infty f = \int_0^g f(w) Q^\infty(\mathrm{d}w)(= U_1 f).$$

The proof is complete.

We note that the optimal value of θ in Theorem 2 is not known.

4 A Few Consequences

Theorem 2 allows us to solve a Gauss-type problem (cf. Iosifescu and Grigorescu, 1990, Subsection 5.5.2), namely to obtain the asymptotic be-haviour as $n \to \infty$ of the conditional probability $\lambda(r_{n+m} > t, \epsilon_{n+m} = \epsilon | E_{i_1 \cdots i_m}^{j_1 \cdots j_m}), t \geq 2, \epsilon = \pm 1, n, m \in \mathbb{N}^*$.

To proceed, for any real number $t \geq 2$ and any $\epsilon = \pm 1$, consider the function

$$f_{t,\epsilon}(w) = \begin{cases} \frac{4-w^2}{4(\epsilon w + t)}, & \text{if } w \in W_1 \\ \frac{(w+2)\delta(\epsilon, 1)}{w+t}, & \text{if } w \in W_2 \end{cases}$$

which belongs to $L(W_1, W_2)$. It is easy to prove that

$$\lambda(r_n > t, \epsilon_n = \epsilon) = U^{n-1} f_{t,\epsilon}(0),$$

$$\lambda(r_{n+m} > t, \epsilon_{n+m} = \epsilon | E_{i_1 \cdots i_m}^{j_1 \cdots j_m}) = U^{n-1} f_{t,\epsilon}(s_m),$$

where $s_m = \begin{bmatrix} 1, & j_m, \ldots, j_2 \\ i_m, & i_{m-1}, \ldots, i_1 \end{bmatrix}$, for any $t \geq 2, \epsilon = \pm 1$ and $n, m \in \mathbb{N}^*$. Indeed, for $n = 1$ the above equations reduce to formulas established in Section 2. The general case is obtained by induction with respect to n and using

(2). We have

$$U^\infty f_{t,\epsilon} = \int_0^g f_{t,\epsilon}(w) Q^\infty(\mathrm{d}w)$$

$$= \frac{1}{\log G}\left(\int_0^{g^2} \frac{\mathrm{d}w}{\epsilon w + t} + \int_{g^2}^g \frac{\delta(\epsilon,1)\mathrm{d}w}{w+t}\right) = \frac{1}{\log G} \times \begin{cases} \log(1 + \frac{g}{t}), & \text{if } \epsilon = 1 \\[2mm] \log \frac{t}{t-g^2}, & \text{if } \epsilon = -1 \end{cases}$$

for all $t \geq 2$. By a reasoning similar to that in Iosifescu and Grigorescu (1990, Subsection 5.5.2), we can deduce from Theorem 2 that there exists some positive constant c' such that

$$|U^n(f_{s,\epsilon} - f_{t,\epsilon}) - U^\infty(f_{s,\epsilon} - f_{t,\epsilon})| \leq c'\theta^n U^\infty(f_{s,\epsilon} - f_{t,\epsilon}),$$

$$2 \leq s < t, \ \epsilon = \pm 1, \ n \in \mathbb{N}^*.$$

Thus we can state the following corollary to Theorem 2.

Corollary 3 *(Solution of Gauss' problem). There exist positive constants $\theta < 1$ and c such that*

$$\lambda(s < r_n \leq t, \epsilon_n = \epsilon) =$$

$$\frac{1}{\log G}(1 + \alpha_0\theta^n)(\delta(\epsilon,1)\log\frac{1+g/s}{1+g/t} + \delta(\epsilon,-1)\log\frac{1-g^2/t}{1-g^2/s}),$$

$$\lambda(s < r_{n+m} \leq t, \epsilon_{n+m} = \epsilon | E^{j_1 \cdots j_m}_{i_1 \cdots i_m}) =$$

$$\frac{1}{\log G}(1 + \alpha_m\theta^n)(\delta(\epsilon,1)\log\frac{1+g/s}{1+g/t} + \delta(\epsilon,-1)\frac{1-g^2/t}{1-g^2/s})$$

for all $2 \leq s < t$, $\epsilon = \pm 1$, and $m, n \in \mathbb{N}^*$, where $\alpha_0 = \alpha_0(n,s,t,\epsilon)$, $\alpha_m = \alpha_m(n,s,t,\epsilon, \begin{bmatrix} 1, & j_m,\ldots,j_2 \\ i_m, & i_{m-1},\ldots,i_1 \end{bmatrix})$, with $|\alpha_0|, |\alpha_m| \leq c$, $m \in \mathbb{N}^*$.

This can be immediately rephrased in terms of $\tau^n(y) = \epsilon_{n+1}(y)/r_{n+1}(y)$, $n \in \mathbb{N}$.

Corollary 4 *There exist positive constants $\theta < 1$ and c such that*

$$\lambda(y : 0 < \tau^n(y) < x) = \frac{1}{\log G}(1 + \alpha_0^{(1)}\theta^n)\log\frac{x+G}{G}, \ \textit{if } 0 \leq x \leq \tfrac{1}{2},$$

$$\lambda(y : -\frac{1}{2} < \tau^n(y) < x) = \frac{1}{\log G}(1+\alpha_0^{(-1)}\theta^n)\log\frac{x+G+1}{G+1/2}, \ \textit{if } -\tfrac{1}{2} \leq x \leq 0,$$

$$\lambda(y : 0 < \tau^{n+m}(y) < x | E^{j_1 \cdots j_m}_{i_1 \cdots i_m})$$

$$= \frac{1}{\log G}(1+\alpha_m^{(1)}\theta^n)\log\frac{x+G}{G}, \quad \textit{if } 0 \leq x \leq \tfrac{1}{2},$$

$$\lambda(y: -\frac{1}{2} < \tau^{n+m}(y) < x | E^{j_1 \ldots j_m}_{i_1 \ldots i_m})$$

$$= \frac{1}{\log G}(1+\alpha_m^{(-1)}\theta^n)\log \frac{x+G+1}{G+1/2}, \ if -\frac{1}{2} \le x \le 0,$$

for all $n \in \mathbb{N}$ and $m \in \mathbb{N}^*$, where

$$\alpha_0^{(\epsilon)} = \alpha_0^{(\epsilon)}(n,x), \alpha_m^{(\epsilon)} = \alpha_m^{(\epsilon)}\left(n,x,\left[\begin{smallmatrix} 1, & j_m,\ldots,j_2 \\ i_m, & i_{m-1},\ldots,i_1 \end{smallmatrix}\right]\right),$$

with $|\alpha_0^{(\epsilon)}|, |\alpha_m^{(\epsilon)}| \le c, \epsilon = \pm 1, m \in \mathbb{N}^*$.

It should be noted that Corollary 4 emphasizes Rieger's measure ρ on $\mathcal{B}_I$ defined as

$$\rho(B) = \frac{1}{\log G}\int_B r(x)\mathrm{d}x, \ B \in \mathcal{B}_I,$$

where

$$r(x) = \begin{cases} \frac{1}{x+G+1}, & if -\frac{1}{2} \le x < 0 \\ \frac{1}{x+G}, & if\ 0 \le x \le \frac{1}{2}. \end{cases}$$

As is well known (see, e.g., Rieger, 1979), τ is ρ-preserving, and this amounts to the strict stationarity of the sequence $(b_n, \epsilon_n)_{n \in \mathbb{N}^*}$ under ρ.

Corollary 3 implies the ψ-mixing of $(b_n, \epsilon_n)_{n \in \mathbb{N}^*}$ under both λ and ρ. The proof is similar to that of the corresponding result for the continued fraction expansion. See Iosifescu (1989) and Iosifescu and Grigorescu (1990, pp. 182-183). In turn, ψ-mixing implies lots of limit theorems in both classical and functional versions. To form an idea of the results to be expected it is sufficient to look again at the corresponding results for the continued fraction expansion.

To conclude with, we wish to note that our approach is suitable for studying the metric properties of many other algorithms for representing real numbers. See, e.g., Kalpazidou (1986b,c,1987), where a different approach is used.

REFERENCES

DOEBLIN, W. (1940) Remarques sur la théorie métrique des fractions continues. *Compositio Math.* 7, 353-371.

IOSIFESCU, M. (1974) On the applications of random systems with complete connections to the theory of f-expansions. In: J. Gani et al, Eds. *Progress in Statistics* (European Meeting of Statisticians, Budapest 1972), Colloq. Math. Soc. János Bolyai Vol. 9, 335-365. North-Holland, Amsterdam.

IOSIFESCU, M. (1989) On mixing coefficients for the continued fraction expansion. *Stud. Cerc. Mat.* 41, 491-499.

IOSIFESCU, M. (1990) A survey of the metric theory of continued fractions, fifty years after Doeblin's 1940 paper. In: B. Grigelionis et al, Eds. *Probability Theory and Mathematical Statistics* (Proc. V Internat. Conf. Vilnius, Lithuania, 1989), Vol. 1, 550-572. VSP, Zeist, The Netherlands.

IOSIFESCU, M. (1992) A basic tool in mathematical chaos theory: Doeblin and Fortet's ergodic theorem and Ionescu Tulcea and Marinescu's generalization. In: these Proceedings.

IOSIFESCU, M. & GRIGORESCU, S. (1990) *Dependence with Complete Connections and Its Applications.* Cambridge Univ. Press, Cambridge.

JAGER, H. (1985) Metrical results for the nearest integer continued fraction. *Indag. Math.* 47, 417-427.

KALPAZIDOU, S. (1983) *Contributions to probabilistic number theory.* Ph.D. Thesis, Dept. of Mathematics, Univ. of Bucharest. (Romanian)

KALPAZIDOU, S. (1985) On a random system with complete connections associated with the continued fraction to the nearer integer expansion. *Rev. Roumaine Math. Pures Appl.* 30,527-537.

KALPAZIDOU, S. (1986a) A class of Markov chains arising in the metrical theory of the continued fraction to the nearer integer expansion. *Rev. Roumaine Math. Pures Appl.* 31, 877-890.

KALPAZIDOU, S. (1986b) A Gaussian measure for certain continued fractions. *Proc.Amer. Math. Soc.* 96, 629-635.

KALPAZIDOU, S. (1986c) On a problem of Gauss-Kuzmin type for continued fractions with odd partial quotients. *Pacific J. Math.* 123, 103-114.

KALPAZIDOU, S. (1987) On the application of dependence with complete connections to the metrical theory of *G*-continued fractions. *Lithuanian Math. J.* 27(1), 32-40.

PERRON, O. (1977) *Die Lehre von den Kettenbrüchen.* Band I, 3. Aufl. Teubner, Stuttgart.

RIEGER, G.J. (1979) Mischung und Ergodizität bei Kettenbrüchen nach nächsten Ganzen. *J. Reine Angew. Math.* 310, 171-181.

ROCKETT, A.M. (1980) The metrical theory of continued fractions to the nearer integer. *Acta Arith.* 38, 97-103.

ROMANIAN ACADEMY, CENTRE FOR MATHEMATICAL STATISTICS, BD. MAGHERU 22, RO-70158, BUCHAREST, ROMANIA.

ARISTOTLE UNIVERSITY OF THESSALONIKI, DEPARTMENT OF MATHEMATICS, 54006, THESSALONIKI, GREECE.

Contemporary Mathematics
Volume **149**, 1993

ON THE WEIGHTED ASYMPTOTICS OF PARTIAL SUMS AND EMPIRICAL PROCESSES OF INDEPENDENT RANDOM VARIABLES

M. Csörgő, L. Horváth, Q-M. Shao and B. Szyszkowicz

1. Introduction

This exposition summarizes some of the results which were presented by the first named author at the Doeblin Memorial Conference in Blaubeuren, Germany. These results are concerned with the weighted asymptotic behaviour of partial sums, empirical and quantile processes. In the Doeblin-spirit of the occasion, our emphasis will be on partial sums, as summarized in Section 3. In Section 2 we present the results on empirical and quantile processes which preceded and inspired the ones described in Section 3. Some dichotomy theorems for random integrals will play a crucial role in both sections.

Doeblin's main fundamental contributions to the asymptotic theory of sums of independent random variables, to central limit theorems, infinitely divisible distributions and domains of attraction, are contained in his papers [13] and [14].

Doeblin (1939) studies arrays of independent random variables satisfying a condition of uniform asymptotic negligibility and establishes the first necessary and sufficient conditions for the existence of normalizing constants to make the sequence of row sums in such an array convergent in distribution. Writing about the general theory of stable and infinitely divisible distributions, concerning this Doeblin work [13], Feller (1966, p.169, **Note on history**) writes: "The interest in the theory was stimulated by Doeblin's masterful analysis of the domains of attraction (1939). His criteria was the first to involve regularly varying functions. The modern theory still carries the imprint of this pioneer work...".

Doeblin (1940) is also concerned with the problem of domains of attraction and, among others, it contains his ever fascinating universal laws that belong to the domain of partial attraction of every infinitely divisible distribution. That is to

1991 *Mathematics Subject classification.* Primary 60F17, 60G50,60F25.

Research supported by NSERC Canada grants at Carleton University, Ottawa.

This paper is in final form and no version of it will be submitted for publication elsewhere

say, these universal laws are distributions $\mathcal{D}(X)$ such that for sums $S_n = X_1 + \cdots + X_n$ of independent random variables each having the distribution $\mathcal{D}(X)$, there are normalizing constants A_n, B_n such that for Y having any infinitely divisible distribution, there is a subsequence $\{n_k\}$ with $S_{n_k}/A_{n_k} - B_{n_k} \xrightarrow{\mathcal{D}} Y$.

Lindvall (1991) emphasizes quite rightly that the paper [12] should also be remembered for a central limit theorem for sums $\sum_1^n f(X_i)$, where Doeblin for the first time makes use of the fact that the segments of a recurrent Markov chain between successive visits to a reference state constitute i.i.d. random elements. For example, in the case of a recurrent random walk $\{S_i\}_{i=1}^\infty$ of an integer valued i.i.d. sequence $\{X_i\}_{i=1}^\infty$, define $\rho_0 = 0$ and the n-th return to zero, ρ_n, by $\rho_n = \min\{i : \rho_{n-1} < i, S_i = 0\}$, $n = 1, 2, \ldots$. Let $\xi(x, n)$ be the local time of this random walk, defined by $\xi(x, n) = \#\{k : 0 < k \le n, S_k = x\}$, $x \in \mathbf{Z}$, $n = 1, 2, \ldots$. Then, studying general additive functionals of the form $D_N = \sum_{k=1}^N f(S_k)$, where f is assumed to be summable over $\mathbf{Z}$ with $\bar{f} = \sum_{x=-\infty}^\infty f(x)$, D_N can be expressed by local times as

$$D_N = \sum_{x=-\infty}^\infty f(x)\xi(x, N).$$

The properties of R_N can be conveniently studied by putting $N = \rho_n$, for in this case the "Doeblin trick" yields the sum of i.i.d. random variables

$$D_{\rho_n} = \sum_{i=1}^n \sum_{x=-\infty}^\infty f(x)\left(\xi(x, \rho_i) - \xi(x, \rho_{i-1})\right)$$

with $E \sum_{x=-\infty}^\infty f(x)\xi(x, \rho_1) = \bar{f}$, and hence, by the law of large numbers,

$$\lim_{n\to\infty} D_{\rho_n}/n = \bar{f} \quad \text{a.s.}$$

Consequently, by putting $n = \xi(0, N)$, we obtain

$$(1.1) \qquad \lim_{N\to\infty} D_N/\xi(0, N) = \bar{f} \quad \text{a.s.}$$

This, in turn, implies for example that, on assuming $\bar{f} \ne 0$, $D_N/\bar{f}$ has the same limit distribution and the same law of the iterated logarithm as $\xi(0, N)$ does. In the literature of additive functionals (1.1) and its consequences are called *first order limit theorems* for D_N.

In the case of $\bar{f} = 0$, or by considering $D_N - \bar{f}\xi(0, N)$, the conclusions change drastically, and one obtains the so–called *second order limit theorems* for D_N.

For an insightful review of, and references on, these manifold consequences of the "Doeblin trick" we refer to Csáki (1991) and, for related results on strong approximations of additive functionals, to Csáki, Csörgő, Földes and Révész (1991).

Back to our task at hand, however, we now turn to reviewing the results on empirical and quantile processes, which preceded and inspired our results on partial sums presented in Section 3.

2. Weighted Convergence of Uniform Empirical and Quantile Processes

Let $U_1, U_2, \ldots$ be independent r.v.'s uniformly distributed on $[0, 1]$. The uniform empirical process $e_n(t)$ of $U_1, U_2, \ldots, U_n$ is defined by

$$e_n(t) = n^{1/2}(E_n(t) - t), \quad 0 \le t \le 1,$$

where

$$E_n(t) = \frac{1}{n} \sum_{1 \le i \le n} I\{U_i \le t\}$$

denotes the uniform empirical distribution function. If $U_{1,n} \le U_{2,n} \le \ldots \le U_{n,n}$ denote the order statistics of $U_1, U_2, \ldots, U_n$, then we define the uniform empirical quantile function $U_n(t)$ by

$$U_n(t) = U_{k,n}, \quad k/(n+2) < t \le (k+1)/(n+2), \quad k = 0, \ldots, n+1,$$

where $U_{0,n} = 0$ and $U_{n+1,n} = 1$ and the uniform quantile process $\tilde{u}_n(t)$ by

$$\tilde{u}_n(t) = n^{1/2}(t - U_n(t)), \quad 0 \le t \le 1.$$

Starting off with Chibisov (1964) and O'Reilly (1974), there has been considerable interest in the asymptotic behaviour of weighted uniform empirical and quantile processes. For an insightful treatise of this subject we refer to Shorack and Wellner (1986), and to M. Csörgő, S. Csörgő, Horváth and Mason (1986a), as well as to the references in these works.

Throughout this section we assume that q, our weight function, is a positive function on $(0, 1)$, i.e. $\inf_{\delta \le t \le 1-\delta} q(t) > 0$ for all $0 < \delta < 1$. There are now complete characterizations available for describing the asymptotic behaviour of the weighted uniform empirical and quantile processes in supremum norm. The proofs of the following two theorems can be found in M. Csörgő, S. Csörgő, Horváth and Mason (1986a).

Theorem A. *We assume that q is positive on $(0, 1)$, and is nondecreasing in a neighbourhood of 0 and nonincreasing in a neighbourhood of 1.*

(i)   *We can define a sequence of Brownian bridges $\{B_n(t), \ 0 \le t \le 1\}$ such that*

(2.1) $$\sup_{0 < t < 1} |e_n(t) - B_n(t)|/q(t) = o_P(1)$$

if and only if

$$I(q, \lambda) = \int_0^1 \frac{1}{t(1-t)} \exp(-\lambda q^2(t)/(t(1-t))) dt < \infty$$

for all $\lambda > 0$.

(ii)   *We have*

(2.2) $$\sup_{0 < t < 1} |e_n(t)|/q(t) \xrightarrow{\mathcal{D}} \sup_{0 < t < 1} |B(t)|/q(t),$$

where $\{B(t), \ 0 \le t \le 1\}$ is a Brownian bridge, if and only if $I(q, \lambda) < \infty$ for some $\lambda > 0$.

Similar results hold true for the uniform quantile process $\tilde{u}_n$.

Theorem B. *We assume that q is positive on $(0,1)$ and is nondecreasing in a neighbourhood of 0 and nonincreasing in a neighbourhood of 1.*

(i)   *We can define a sequence of Brownian bridges $\{B_n(t),\ 0 \le t \le 1\}$ such that*

$$(2.3) \qquad \sup_{0<t<1} |\tilde{u}_n(t) - B_n(t)|/q(t) = o_P(1)$$

if and only if $I(q, \lambda) < \infty$ for all $\lambda > 0$.

(ii)   *We have*

$$(2.4) \qquad \sup_{0<t<1} |\tilde{u}_n(t)|/q(t) \xrightarrow{\mathcal{D}} \sup_{0<t<1} |B(t)|/q(t),$$

where $\{B(t), 0 \le t \le 1\}$ is a Brownian bridge, if and only if $I(q, \lambda) < \infty$ for some $\lambda > 0$.

It is interesting to note that (2.1) does not imply (2.2) for all possible weight functions q of interest. For example, choosing $q(t) = (t(1-t) \log \log(1/(t(1-t))))^{1/2}$, we have (2.2) and (2.4), but (2.1) and (2.3) do not hold true.

For a discussion of the relationship of the integral $I(q, \lambda)$ to that used in the classical Kolmogorov test for upper and lower class functions of a Wiener process (cf., e.g., Itô and McKean (1965)), we refer to Csörgő, Shao and Szyszkowicz (1991).

Concerning now L_p–functionals of e_n/q and u_n/q, we quote the following two theorems from Csörgő, Horváth and Shao (1991a).

Theorem C. *We assume that $0 < p < \infty$ and q is positive on $(0,1)$. Then the following statements are equivalent:*

(i)   *We have*

$$(2.5) \qquad \int_0^1 (t(1-t))^{p/2}/q(t)dt < \infty,$$

(ii)   *There is a sequence of Brownian bridges $\{B_n(t),\ 0 \le t \le 1\}$ such that*

$$(2.6) \qquad \int_0^1 |e_n(t) - B_n(t)|^p/q(t)dt = o_P(1),$$

(iii)   *We have*

$$(2.7) \qquad \int_0^1 |e_n(t)|^p/q(t)dt \xrightarrow{\mathcal{D}} \int_0^1 |B(t)|^p/q(t)dt,$$

where $\{B(t),\ 0 \le t \le 1\}$ is a Brownian bridge.

Theorem D. *We assume that $0 < p < \infty$ and q is positive on $(0,1)$. Then the following statements are equivalent:*

(i)   *We have*

$$\text{(2.8)} \qquad \int_0^1 (t(1-t))^{p/2}/q(t)dt < \infty,$$

(ii)   *There is a sequence of Brownian bridges $\{B_n(t),\ 0 \le t \le 1\}$ such that*

$$\text{(2.9)} \qquad \int_0^1 |\tilde{u}_n(t) - B_n(t)|^p/q(t)dt = o_P(1),$$

(iii)   *We have*

$$\text{(2.10)} \qquad \int_0^1 |\tilde{u}_n(t)|^p/q(t) \overset{\mathcal{D}}{\to} \int_0^1 |B(t)|^p/q(t)dt,$$

where $\{B(t),\ 0 \le t \le 1\}$ is a Brownian bridge.

Csörgő and Horváth (1988a) showed that (2.5) implies (2.7) and, similarly, that (2.8) yields (2.10) if $1 \le p < \infty$. Shorack and Wellner (1986, p.470) proved that (2.5) implies (2.6) and, similarly, (2.8) gives (2.9) if $0 < p \le 2$. Shorack and Wellner (1986, p.471) and Csörgő and Horváth (1988a) also showed that (2.5) is necessary and sufficient for (2.7), if $p = 2$. The necessary part for $p = 2$ followed from a dichotomy theorem for a Brownian bridge $\{B(t),\ 0 \le t \le 1\}$ in Shepp (1966). Namely, Shepp (1966) proved that

$$\text{(2.11)} \qquad \int_0^1 B^2(t)/q(t)dt < \infty \quad \text{a.s.}$$

if and only if

$$\text{(2.12)} \qquad \int_0^1 t(1-t)/q(t)dt < \infty.$$

Shepp's proof of the equivalence of (2.11) and (2.12) is based on computation of Radon–Nikodym derivatives of Gaussian processes and may not be carried over to cover the general case of $0 < p < \infty$.

Using a different, simpler method, Csörgő, Horváth and Shao (1991a) show that Shepp's result is true for all $0 < p < \infty$. Namely, we have

Theorem E. *Let $\{B(t),\ 0 \le t \le 1\}$ be a Brownian bridge, $0 < p < \infty$ and $-\infty < \beta < \infty$. Then, with a positive function q on $(0,1)$, we have*

$$\text{(2.13)} \qquad \int_0^1 |B(t) + \beta t(1-t)|^p/q(t)dt < \infty \quad \text{a.s.}$$

if and only if

$$\text{(2.14)} \qquad \int_0^1 (t(1-t))^{p/2}/q(t)dt < \infty.$$

3. Weighted Convergence and Almost Sure Summability of Partial Sums

The just quoted Theorem E is a consequence of a more general result, Corollary 2.1, of Csörgő, Horváth and Shao (1991a). The same corollary also yields the following dichotomies.

Theorem F.   *Let $\{W(t),\ 0 \le t < \infty\}$ be a standard Wiener process, $0 < p < \infty$. Then,*
(i)   *With a positive function q on $(0, 1]$ we have*

$$(3.1) \qquad \int_0^1 |W(t)|^p/q(t)dt < \infty \quad a.s.$$

if and only if

$$(3.2) \qquad \int_0^1 t^{p/2}/q(t)dt < \infty,$$

(ii)   *With a positive function on $[1, \infty)$ we have*

$$(3.3) \qquad \int_1^\infty |W(t)|^p/q(t)dt < \infty \quad a.s.$$

if and only if

$$(3.4) \qquad \int_1^\infty t^{p/2}/q(t)dt < \infty.$$

The equivalence of (3.1) and (3.2) poses the question of possible analogs of Theorems C and D for partial sums of independent r.v.'s, while that of (3.3) and (3.4) makes one think about necessary and sufficient conditions for (3.3) when $W(t)$ in it is replaced by partial sums of independent r.v.'s. In this section we present results along these lines, as well as analogs of Theorems A and B for partial sums.

Let $X_1, X_2, \ldots$ be independent, identically distributed random variables with $EX_1 = 0$ and $EX_1^2 = 1$. The weak convergence of the weighted partial sum process $n^{-1/2}S_{[nt]}/q(t),\ 0 < t \le 1$, where $S_{[nt]} = X_1 + \ldots + X_{[nt]}$, in supremum norm was proved by O'Reilly (1974) for continuous weight functions under the assumption of $E|X_1|^3 < \infty$. He asserted also, without proof, that the third moment condition could be dropped to two, and his theorem would remain true.

For an extension of the Komlós, Major and Tusnády (1975, 1976) approximation of partial sums to weighted supremum norm approximations which improve also the just mentioned result of O'Reilly (1974) in terms of the optimal class of weight functions as in M. Csörgő, S. Csörgő, Horváth and Mason (1986a), we refer to Csörgő and Horváth (1988b) and the references given there.

Assuming the existence of two moments only, Szyszkowicz (1991, 1992a) obtained the following weighted supremum norm approximation (cf. Theorem 2.1 and Remark 2.2 there).

Let Q be the class of positive functions on $(0, 1]$, i.e., $\inf_{\delta \le t \le 1} q(t) > 0$ for all $0 < \delta < 1$, which are nondecreasing in a neighbourhood of zero. Let also

$$I^*(q, c) = \int_0^1 t^{-1} \exp(-ct^{-1}q^2(t))dt, \quad c > 0.$$

Theorem G. *Let* $X_1, X_2, \ldots$ *be i.i.d.r.v.'s such that* $EX_1 = 0$ *and* $EX_1^2 = 1$. *Then a standard Wiener process* $\{W(t),\ 0 \le t < \infty\}$ *can be constructed in such a way that the following hold true.*
(a) *Let* $q \in Q$. *Then*

$$\sup_{0 < t \le 1} |n^{-1/2}\left(S_{[nt]} - W(nt)\right)|/q(t) = o_P(1)$$

if and only if $I^*(q, c) < \infty$ *for all* $c > 0$.
(b) *Let* $q \in Q$. *Then*

$$\sup_{0 < t \le 1} |n^{-1/2}(S_{[nt]} - W(nt))|/q(t) = \mathcal{O}_P(1)$$

if and only if $I^*(q, c) < \infty$ *for some* $c > 0$.

Concerning the asymptotic behaviour of L_p–approximations and functionals and weighted partial sum processes, Szyszkowicz (1992b,c) obtained the following result.

Theorem H. *Let* $X_1, X_2, \ldots$ *be i.i.d.r.v.'s such that* $EX_1 = 0$ *and* $EX_1^2 = 1$. *We assume that* $0 < p < \infty$ *and* $q \in Q$.
(a) A standard Wiener process $\{W(t),\ 0 \le t < \infty\}$ *can be constructed in such a way that*

$$\int_0^1 |n^{-1/2}(S_{[nt]} - W(nt))|^p/q(t)dt = o_P(1)$$

if and only if (3.2) *holds.*
(b) Let $\{W(t),\ 0 \le t < \infty\}$ *be a standard Wiener process. Then*

$$\int_0^1 |n^{-1/2}S_{[nt]}|^p/q(t)dt \xrightarrow{\mathcal{D}} \int_0^1 |W(t)|^p/q(t)dt$$

if and only if (3.2) *holds.*

Throughout the rest of this paper we assume that

(3.5) $\{X_n,\ n \ge 1\}$ is a sequence of independent r.v.'s

and impose some regularity conditions on the positive functions q of (ii) of Theorem F as follows:

(3.6) $q(n) > 0$ for all $n \ge 1$,

(3.7)   there is a constant C_1 such that $\max_{1 \le i \le n} q(i) \le C_1 q(n)$ for all $n \ge 1$,

(3.8)   there is a constant C_2 such that $q(2n) \le C_2 q(n)$ for all $n \ge 1$.

We note that if $q(n)$ is a positive regularly varying sequence with positive exponent, then (3.6)–(3.8) hold true.

Concerning almost sure weighted summability of partial sums, we quote a few results from Csörgő, Horváth and Shao (1991b). The first of these concludes that (ii) of Theorem F remains true if the Wiener process in it is replaced by partial sums of i.i.d.r.v.'s having a finite variance.

Theorem I. *Let $0 < p < \infty$. We assume that $(3.6) - (3.8)$ hold true and that $\{X_n,\ n \geq 1\}$ are i.i.d.r.v.'s with $EX_1 = 0$ and $0 < EX_1^2 < \infty$. Then the statements*

$$(3.9) \qquad \sum_{1 \leq n < \infty} |S_n|^p / q(n) < \infty \quad a.s.$$

$$(3.10) \qquad \sum_{1 \leq n < \infty} \max_{1 \leq i \leq n} |S_i|^p / q(n) < \infty \quad a.s.$$

and

$$(3.11) \qquad \sum_{1 \leq n < \infty} n^{p/2} / q(n) < \infty$$

are equivalent.

A version of Theorem I for not necessarily identically distributed r.v.'s reads as follows.

Theorem J. *Let $0 < p < \infty$. We assume that $(3.5) - (3.8)$ holds true and that $EX_n = 0$, $EX_n^2 < \infty$. Set $\sigma_n^2 = \sum_{1 \leq i \leq n} EX_i^2$. Then*

$$\sum_{1 \leq n < \infty} \sigma_n^p / q(n) < \infty$$

and

$$\sum_{1 \leq n < \infty} \frac{1}{n} \sum_{1 \leq i \leq n} P\{|X_i| \geq \sigma_n\} < \infty$$

imply (3.10).

If $EX_1^2 = \infty$, then, in general, condition (3.11) is not enough for having also (3.9).

Proposition A. *Let $g(x)$ be a non-decreasing continuous function satisfying $\lim_{x \to \infty} g(x) = \infty$. For any $0 < p < \infty$ we can find a sequence $\{X_n,\ n \geq 1\}$ of i.i.d.r.v.'s with $EX_1 = 0$, $EX_1^2 / g(|X_1|) < \infty$ and a sequence $\{q(n), n \geq 1\}$ satisfying $(3.5) - (3.8)$ such that (3.11) holds true but (3.9) does not.*

On the other hand, if $EX_1^2 = \infty$, then the condition

$$(3.12) \qquad \sum_{1 \leq n < \infty} E \frac{\max_{1 \leq i \leq n} |X_i|^p}{q(n) + n \max_{1 \leq i \leq n} |X_i|^p} < \infty$$

is necessary and sufficient for (3.9) and (3.10), as in the following, Doeblin–type, theorem.

Theorem K. *Let $0 < p < \infty$. We assume that $(3.6) - (3.8)$ hold true and that $\{X_n,\ n \geq 1\}$ are i.i.d.r.v.'s. We assume also that there is a sequence $\{a(n),\ n \geq 1\}$ such that*

$$S_n / a(n) \xrightarrow{\mathcal{D}} Y,$$

where Y is a non–degenerate stable r.v. with exponent $0 < \nu < 2$. Then (3.9), (3.10) and (3.12) are equivalent statements.

It is of interest to note here that, if $\{X_n, \ n \geq 1\}$ are i.i.d.r.v.'s in the domain of attraction of a stable law, then the necessary and sufficient condition (3.12) for having (3.9) and (3.10) is given in terms of $\max_{1 \leq i \leq n} |X_i|$. This observation rhymes well with the main results in M. Csörgő, S. Csörgő, Horváth and Mason (1986b), and S. Csörgő, Horváth and Mason (1986). Namely, with G^{-1} standing for the inverse of $G(t) = P\{|X_1| \geq t\}$, they show that $S_n/G^{-1}(1/n)$ is asymptotically stable if $\max_{1 \leq i \leq n} |X_i|$ is so large that its asymptotic order is the same as that of S_n.

REFERENCES

1. Chibisov, D., *Some theorems on the limiting behaviour of empirical distribution functions.*, Selected Transl. Math. Statist. Probab. **6** (1964), 147–156.
2. Csáki, E., *On the local time of Wiener process and random walk*, Lecture Notes, 7th Int. Summer School, Varna (1991) (to appear).
3. Csáki, E., Csörgő, M., Földes, A. and Révész, P., *Strong approximations of additive functionals*, J. Theor. Prob. (1991) (to appear).
4. Csörgő, M., Csörgő, S., Horváth, L. and Mason, D.M., *Weighted empirical and quantile processes*, Ann. Prob. **14** (1986a), 31–85.
5. Csörgő, M., Csörgő, S., Horváth, L. and Mason, D.M., *Normal and stable convergence of integral functionals of the empirical distribution function*, Ann. Prob. **14** (1986b), 86–118.
6. Csörgő, M. and Horváth, L., *On the distributions of L_p–norms of weighted uniform empirical and quantile processes*, Ann. Prob. **16** (1988a), 142–161.
7. Csörgő, M. and Horváth, L., *Nonparametric methods for changepoint problems*, Handbook of Statistics **7** (1988b), Elsevier Science Publishers B.V., North Holland, 403-425.
8. Csörgő, M., Horváth, L. and Shao, Q.M., *Convergence of integrals of uniform empirical and quantile processes*, Tech. Rep. Ser. Lab. Res. Stat. Prob., Carleton U. – U. of Ottawa **168** (1991); To appear in *Stochastic Process. Appl.*.
9. Csörgő, M., Horváth, L. and Shao, Q.M., *Almost sure summability of partial sums*, Tech. Rep. Ser. Lab. Res. Stat. Prob., Carleton U – U. of Ottawa **168** (1991).
10. Csörgő, M., Shao, Q.-M. and Szyszkowicz, B., *A note on local and global functions of a Wiener process and some Rényi–type statistics*, Studia Sci. Math. Hung (1991) (to appear).
11. Csörgő, S., Horváth, L. and Mason, D.M., *What portion of the sample makes a partial sum asymptotically stable or normal?*, Probab. Theory Rel. Fields **82** (1986), 1–16.
12. Doeblin, W., *Sur deux problèmes de M. Kolmogoroff concernant les chaînes dénombrables*, Bull. Soc. Math. France **66** (1938), 210–220.
13. Doeblin, W., *Sur les sommes d'un grand nombre de variables indépendantes*, Bull. Sci. Math. **63** (1939), 23–32, 35–64.
14. Doeblin, W., *Sur l'ensemble des puissances d'une loi de probabilité*, Studia Math. **9** (1940), 71–96; Reprinted with a complement in Ann. École Norm. Sup. **63** (1947), 317–350.
15. Feller, W., *An Introduction to Probability Theory and Its Applications II*, Wiley, New York (1966).
16. Itô, K. and McKean, H.P. Jr., *Diffusion Processes and their Sample Paths*, Springer–Verlag, Berlin (1965).
17. Komlós, J., Major, P. and Tusnády, G., *An approximation of partial sums of independent R.V.'s and the sample DF. I.*, Z. Wahrsch. verw. Gebiete **32** (1975), 111–131.
18. Komlós, J., Major, P. and Tusnády, G., *An approximation of partial sums of independent R.V.'s and the sample DF. II.*, Z. Wahrsch. verw. Gebiete **34** (1976), 33–58.
19. Lindwall, T., *W. Doeblin 1915–1940*, Ann. Prob. **19** (1991), 929–934.
20. O'Reilly, N., *On the weak convergence of empirical processes in sup–norm metrics*, Ann. Prob. **2** (1974), 642–651.

21. Shepp, L.A., *Radon–Nikodym derivatives of Gaussian measures*, Ann. Math. Statist. **37** (1966), 321–354.

22. Shorack, G.R. and Wellner, J.A., *Empirical Processes with Applications to Statistics*, Wiley, New York. (1986).

23. Szyszkowicz, B., *Weighted stochastic processes under contiguous alternatives*, C.R. Math. Rep. Acad. Sci. Canada XIII **5** (1991), 161–166.

24. Szyszkowicz, B., *On $\| \cdot /q \|$–metric convergence and contiguous alternatives.*, Tech. Rep. Ser. Lab. Res. Stat. Prob. Carleton U. – U. Ottawa. **191** (1992a).

25. Szyszkowicz, B., *L_p–approximations of weighted partial sum processes*, Tech. Rep. Ser. Lab. Res. Stat. Prob., Carleton U. – U. Ottawa. **191** To appear in *Stochastic Process. Appl.*.

26. Szyszkowicz, B., *L_p–functionals of weighted partial sum processes*, C.R.Math. Rep. Acad. Sci. Canada **XIV** (1992c), 31–36.

DEPARTMENT OF MATHEMATICS AND STATISTICS, CARLETON UNIVERSITY, OTTAWA, ONTARIO, CANADA, K1S 5B6

DEPARTMENT OF MATHEMATICS, UNIVERSITY OF UTAH, SALT LAKE CITY, UT 84112, U.S.A.

DEPARTMENT OF MATHEMATICS, HANGZHOU UNIVERSITY, HANGZHOU, PEOPLE'S REPUBLIC OF CHINA

DEPARTMENT OF MATHEMATICS AND STATISTICS, CARLETON UNIVERSITY, OTTAWA, ONTARIO, CANADA, K1S 5B6

Contemporary Mathematics
Volume **149**, 1993

Homoclinic approach to the central limit theorem for dynamical systems

MIKHAIL GORDIN

ABSTRACT. A concept of weak dependence for stationary sequences which is alternative to mixing conditions is introduced and discussed. A general central limit theorem (CLT) result is stated in terms of so called homoclinic Laplace operator. The proof uses Ch. Stein's approach. As an application a new proof of the CLT for hyperbolic toral automorphisms is given.

1. Introduction

Let $\{X_n, n \in \mathbb{Z}\}$ be a sequence of random variables defined on a probability space $(\mathcal{X}, \mathcal{F}, \mathcal{P})$. One of possible ways for the sequence $\{X_n, n \in \mathbb{Z}\}$ to demonstrate the asymptotic independence property is to admit some perturbations which asymptotically vanish whenever $|n| \to \infty$. More exactly, one can expect that for such an asymptotically independent sequence there exists another sequence $\{X'_n, n \in \mathbb{Z}\}$ such that $\{X_n\}$ and $\{X'_n\}$ are defined on a common probability space and for some metrics d $d(X_n, X'_n) \to 0$ whenever $|n| \to \infty$. Certainly, some nondegeneracy conditions are necessary to ensure that the perturbated variable X'_n differs essentially from X_n if n is small.

Probably W. Doeblin was the first mathematician who used similar ideas to deduce some consequences which are usually associated to certain forms of weak dependence. This approach has given rise to the coupling theory which was developed very far (first of all for Markov processes) and has now many important applications.

Actually we use in the present paper some form of (deterministic) coupling to prove a version of the Central Limit Theorem (CLT) for a certain class of dynamical systems. As an application we get a new proof of the CLT for hyperbolic toral automorphisms. It would be reasonable to mention that the known proofs

1991 *Mathematics Subject Classification.* Primary 60F05, 60G10.

The final version of this paper will be submitted for publication elsewhere.

of this result hardly depend either on Markov partitions and mixing conditions technique [1] or on harmonic analysis [2].

The present paper follows the line of the author's paper [7] but gives more tractable conditions for the CLT. We use in our proof one of the basic ideas from Ch. Stein's paper [3]. It was the Stein's approach what inspired the author's attempts to find out a way to prove the CLT directly, without any use of mixing conditions. Instead of these conditions we apply in our proof a kind of transformations which can be defined in terms of the dynamical system under consideration. By the reason of the direct connection with the H. Poincare's concept of homoclinic point they were called homoclinic transformations (further discussion can be found in the author's paper [7]). After the paper [7] has been published the author has discovered that essentially the same notion was introduced and studied (outside the CLT context) by Capocaccia [4] and Ruelle [5] (see also Krieger's paper [6]) under the name of conjugating transformations. Nevertheless, the author continues to apply his "homoclinic" terminology to emphasize the connection with Poincare's ideas and to avoid some terminological confusions in continuous time case (for example, for geodesic flows).

To make the conditions for the CLT to hold more tractable we introduce and use in the present paper an invariant with respect to dynamical system unbounded difference operator defined in terms of homoclinic transformations. It has many features reminding the Laplace operator on a compact Riemannian manifold but some other properties are essentially different. This operator gives a kind of lacunary decomposition for functions on the space of dynamical system and can be used to express some invariant with respect to dynamics regularity conditions imposed on such functions.

2. Statement of result

Let T be an automorphism of a probability space $(\mathcal{X}, \mathcal{F}, \mathcal{P})$. By definition, it means that T is an invertible bimeasurable probability preserving transformation.

For every function f with domain $\mathcal{X}$ we write Uf instead of $f \circ T$. If B is a Banach space and $p \in [1, \infty]$ then U acts in invertible and isometric way on the space $L_p(B)$ of measurable functions taking their values in B and having finite L_p-norm $|\cdot|_{p,E}$. The character B will be omitted when $B = \mathbb{R}^1$. In what follows brackets $(\cdot, \cdot)$ denote both the inner product on L_2 and the coupling form on $L_p \times L_q$ $(p + q = 1)$, constant functions have the same denotations as their values and, by definition, $L_p^0 = \{ f \mid f \in L_p, (f, 1) = 0 \}$. Suppose H is an unitary operator generated by an automorphism R of $(\mathcal{X}, \mathcal{F}, \mathcal{P})$ according to relation $Hf = f \circ R$, $f \in L_2$.

DEFINITION. R is said to be a (measure preserving) homoclinic transformation for T and H is said to be a homoclinic operator for U (or for T) iff

$$U^n H U^{-n} \to I$$

as $|n| \to \infty$. All operators here are considered as acting on L_2, I is the identity operator and convergence is meant in the sense of the strong operator topology.

Very simple proposition which follows can illustrate some connections between the properties of T and the existence of an ergodic set of homoclinic transformations.

PROPOSITION. *Let $\{ R_\gamma, \gamma \in \Gamma \}$ be a set of homoclinic transformations (and $\{ H_\gamma, \gamma \in \Gamma \}$ be corresponding homoclinic operators) for an automorphism T of a probability space $(\mathcal{X}, \mathcal{F}, \mathcal{P})$. If the set $\{ R_\gamma, \gamma \in \Gamma \}$ acts ergodically on $(\mathcal{X}, \mathcal{F}, \mathcal{P})$ then T has the mixing property.*

PROOF. It is sufficient to prove that $(U^n f, h) \to 0$ as $n \to \infty$ for every pair f, g, where $g \in L_2^0$ and f belongs to a dense subset of L_2^0.

Let us remark that the closed linear span of the set

$$C = \{ f \mid f \in L_2, f = g - H_\gamma^* g \text{ for some } g \in L_2, \gamma \in \Gamma \}$$

coincides with the orthogonal complement to the subspace

$$\{ f \mid f \in L_2, H_\gamma f = f \text{ for every } \gamma \in \Gamma \}$$

which is, by the reason of ergodicity, the subspace of constants. So the closed linear span of C is L_2^0, the orthogonal complement in L_2 to the subspace of constants.

Then we can see that $(U^n f, h) \to 0$ as $|n| \to \infty$ for every $f \in C$, $h \in L_2$. This statement is a consequence of representation $f = g - H_\gamma^* g$, $g \in L_2$, and homoclinicity of H_γ, $\gamma \in \Gamma$, because

$$\begin{aligned}
\left| (U^n f, h) \right| &= \left| (U^n g, h) - (U^n H_\gamma^* g, h) \right| \\
&= \left| (U^n g, h) - (g, H_\gamma U^{-n} h) \right| \\
&= \left| (U^n g, h - U^n H_\gamma U^{-n} h) \right| \\
&\leqslant |g|_2 |h - U^n H_\gamma U^{-n} h|_2 \to 0
\end{aligned}$$

as $|n| \to \infty$. $\square$

Now we need to fix some objects and notations to state our main result. Let $H_1, \ldots, H_r$ be a set of homoclinic operators. Let us define

$$S_n = \sum_{k=0}^{n-1} U^k, \qquad n \in \mathbb{Z},$$

$$D_{p,l} = U^p(H_l - I)U^{-p}, \qquad p \in \mathbb{Z}, \, l \in \{1, \ldots, r\},$$

$$M_p = \sum_{l=1}^{r} D_{p,l}^* D_{p,l}, \qquad p \in \mathbb{Z},$$

$$\Delta = \sum_{p \in \mathbb{Z}} M_p.$$

All $D_{p,l}$ and M_p are bounded operators but it is not the case for Δ. We will use the notation Δf only for situations where either the series $\sum_{p\in\mathbb{Z}}|M_pf|_2$ converges or some closure of Δ is fixed. It should be mentioned that at any rate on the formal level Δ and U commute.

THEOREM. *Suppose the automorphism T is ergodic and $f \in L_2$ is such that:*

1) $\sum_{p\in\mathbb{Z}}\sum_{l=1}^{r}|D_{p,l}f|_2 < \infty$,
2) *for some $F \in L_2^0$ satisfying the condition*

$$\sum_{p\in\mathbb{Z}}\sum_{l=1}^{r}|D_{p,l}F|_2 < \infty$$

f can be represented in the form

$$f = \Delta F.$$

Then the sequence $\{n^{-\frac{1}{2}}\sum_{k=0}^{n-1}U^kf\}$ converges in distribution to the normal law $N(0,\sigma^2)$, where

$$\sigma^2 = \sum_{l=1}^{r}\left(\sum_{p\in\mathbb{Z}}D_{0,l}U^pF, \sum_{p\in\mathbb{Z}}D_{0,l}U^pf\right).$$

COROLLARY. *Let us denote $\mathcal{D}$ the space of all $f \in L_2^0$ satisfying the condition*

$$\sum_{p\in\mathbb{Z}}\sum_{l=1}^{r}|D_{p,l}f|_2 < \infty.$$

Suppose that there exists such $c > 0$ that

$$(\Delta f, f) \geqslant c(f,f),$$

$\mathcal{D}$ is a dense subspace of L_2^0, and the operators in the collection $\{D_{p,l}, D_{p,l}^*, p \in \mathbb{Z}, l = 1,\ldots,r\}$ pairwise commute. Then for every $f \in \mathcal{D}$ the conclusion of the theorem is valid.

3. Proofs

First we are going to derive the corollary from the main theorem. Let us mention that Δ sends $\mathcal{D} \subset L_2^0$ into L_2^0 because it is the case for all $D_{p,l}$, $D_{p,l}^*$. Let Δ_0 be Δ considered as an unbounded operator from $\mathcal{D} \subset L_2^0$ into L_2^0. Under our hypothesis Δ_0 has Friedrichs' extension (we will also denote it Δ_0) having the range L_2^0 and the right inverse Δ_0^{-1} with the norm bounded by c^{-1}.

Moreover, Δ_0^{-1} commutes with all $D_{p,l}$, $D_{p,l}^*$. By this reason we can set $F = \Delta_0^{-1}f$ and, taking into consideration the relation

$$|D_{p,l}F|_2 = |D_{p,l}\Delta_0^{-1}f|_2 \leqslant c^{-1}|D_{p,l}f|_2,$$

apply the theorem to these f and F. $\square$

We need some notations and lemmas before to prove the theorem stated in the preceding section.

For $\Lambda = \mathbb{Z} \times \{1, \ldots, r\}$ let us introduce the Hilbert space (over $\mathbb{R}^1$) $E = \ell_2(\Lambda)$ of functions (sequences) on Λ taking values in $\mathbb{R}^1$. E is equipped with the inner product $[\cdot, \cdot]$ and the norm $\|\cdot\|$. We need also the Hilbert space $L_2(E)$ of functions on $\mathcal{X}$ taking their values in E. $L_2(E)$ has the inner product $(\cdot, \cdot)_{2,E}$ and the norm $|\cdot|_{2,E}$. We can describe a function $h \in L_2(E)$ as a family $\{\, h_{(p,l)}, \, (p,l) \in \Lambda \,\}$ where every $h_{(p,l)}$ is a function belonging to L_2. We have for such a function h

$$|h|_{2,E}^2 = \sum_{(p,l) \in \Lambda} |h_{(p,l)}|_2^2.$$

Let grad and div be partial mappings from L_2 to $L_2(E)$ and from $L_2(E)$ to L_2, correspondingly, defined by the formulas:

$$(\operatorname{grad} f)(x) = \{\, (D_{p,l} f)(x), \, (p,l) \in \Lambda \,\}, \qquad x \in \mathcal{X},$$

$$(\operatorname{div} h)(x) = \sum_{p \in z} \sum_{l=1}^{r} (D_{p,l}^* h_{(p,l)})(x), \, x \in \mathcal{X},$$

for $f \in L_2^0$ and $h = \{\, h_{(p,l)}, \, (p,l) \in \Lambda \,\} \in L_2(E)$ such that

(1)
$$\sum_{p \in \mathbb{Z}} \sum_{l=1}^{r} |D_{p,l} f|_2 < \infty,$$

(2)
$$\sum_{p \in \mathbb{Z}} \sum_{l=1}^{r} |h_{(p,l)}|_2 < \infty.$$

The following formulas should be mentioned:

$$\Delta f = \operatorname{div} \operatorname{grad} f,$$
$$(\Delta f, g) = (\operatorname{grad} f, \operatorname{grad} g)_{2,E} = ([\operatorname{grad} f, \operatorname{grad} g], 1),$$
$$\operatorname{grad} U f = V \operatorname{grad} f = U(s \operatorname{grad} f),$$
$$U(\operatorname{div} h) = \operatorname{div}(V h) = \operatorname{div}(s U h),$$

where $h = \{\, h_{(p,l)}, \, (p,l) \in \Lambda \,\}$ and the operators U, s (the shift operator) and V are defined by the following relations:

$$U h = \{\, U h_{(p,l)}, \, (p,l) \in \Lambda \,\},$$
$$s h = \{\, h_{(p-1,l)}, \, (p,l) \in \Lambda \,\},$$
$$V h = s U h = \{\, U h_{(p-1,l)}, \, (p,l) \in \Lambda \,\}.$$

Let us denote A the subspace of $L_2(E)$ consisting of all such $h \in L_2(E)$, which satisfy the condition

$$\sum_{(p,l) \in \Lambda} |h_{(p,l)}|_2 < \infty.$$

LEMMA 1. *If $h \in A$ then the limit*

$$\lim n^{-1} \left| \sum_{k=0}^{n-1} V^k h \right|_{2,E}^2 = v(h)$$

exists and

$$v(h) = \sum_{l=1}^{r} \left| \sum_{p \in \mathbb{Z}} U^{-n} h_{(p,l)} \right|_2^2 ,$$

where the series converges absolutely.

PROOF. One has

$$\sum_{k \in \mathbb{Z}} \left| (V^k h, h)_{2,E} \right|$$

$$= \sum_{k \in \mathbb{Z}} \left| \sum_{(p,l) \in \Lambda} \left(U^k h_{(p-k,l)}, h_{(p,l)} \right) \right|$$

$$\leqslant \sum_{k \in \mathbb{Z}} \sum_{(p,l) \in \Lambda} |h_{(p-k,l)}|_2 |h_{(p,l)}|_2$$

$$\leqslant \left(\sum_{(p,l) \in \Lambda} |h_{(p,l)}|_2 \right)^2 < \infty .$$

Hence the series

$$\sum_{k \in \mathbb{Z}} (V^k h, h)_{2,E} = \sum_{l=1}^{r} \sum_{p \in \mathbb{Z}} \sum_{q \in \mathbb{Z}} (U^{-p} h_{(p,l)}, U^{-q} h_{(q,l)})$$

converges absolutely and by this reason

$$\lim n^{-1} \left| \sum_{k=0}^{n-1} V^k h \right|_{2,E}^2 = \lim \sum_{k=-(n-1)}^{(n-1)} (1 - |k|/n)(V^k h, h)_{2,E}$$

exists as $n \to \infty$ and equals to $v(h)$. $\square$

LEMMA 2. *Let $F \in L_2^0$ and satisfy the condition* (1). *Let us define a function $B(F) \in L_2(E)$ by the expression*

$$B_{(0,l)}(F) = \sum_{p \in \mathbb{Z}} U^{-p} D_{p,l} F,$$

$$B_{(s,l)}(F) = 0, \qquad s \neq 0, \quad l = 1, \dots, r.$$

Then

$$n^{-1} \left| \operatorname{grad} S_n F - \sum_{k=0}^{n-1} V^k B(F) \right|_{2,E} \to 0 \quad as \quad n \to \infty .$$

PROOF. If h is defined by the expression $h = \operatorname{grad} F - B(F) \in A$ then for $l \in \{1, \ldots, r\}$

$$h_{(0,l)} = D_{0,l}F - \sum_{q \in \mathbb{Z}} U^{-q} D_{q,l}F,$$

$$h_{(p,l)} = D_{p,l}F, \qquad p \in \mathbb{Z} \setminus \{0\},$$

and

$$\sum_{p \in \mathbb{Z}} U^{-p} h_{(p,l)} = 0.$$

Taking into consideration the identity

$$\operatorname{grad} S_n F - \sum_{k=0}^{n-1} V^k B(F) = \sum_{k=0}^{n-1} V^k \big(\operatorname{grad} F - B(F) \big)$$

one has, according to Lemma 1, that

$$\lim_{n \to \infty} n^{-\frac{1}{2}} \left| \operatorname{grad} S_n F - \sum V^k B(F) \right|_{2,E} = \sum_{l=1}^{r} \left| \sum_{p \in \mathbb{Z}} U^{-p} h_{(p,l)} \right|_2^2 = 0. \quad \square$$

LEMMA 3. *If $f, g \in L_2^0$ are such that*

$$\sum_{l=1}^{r} \sum_{p \in \mathbb{Z}} |D_{p,l}f|_2 < \infty, \qquad \sum_{l=1}^{r} \sum_{p \in \mathbb{Z}} |D_{p,l}g|_2 < \infty.$$

and, by definition,

$$\langle f, g \rangle = \sum_{l=1}^{r} \left(\sum_{p \in \mathbb{Z}} U^{-p} D_{p,l}f \right) \left(\sum_{q \in \mathbb{Z}} U^{-q} D_{q,l}g \right)$$

$$= \sum_{l=1}^{r} \left(\sum_{p \in \mathbb{Z}} D_{0,l} U^{-p} f \right) \left(\sum_{q \in \mathbb{Z}} D_{0,l} U^{-q} g \right),$$

then the series representing $\langle f, g \rangle$ converges in the L_1-norm and

$$n^{-1} \big| [\operatorname{grad} S_n f, \operatorname{grad} S_n g] - S_n \langle f, g \rangle \big|_1 \to 0$$

as $n \to \infty$.

PROOF. The L_1-convergence of the series is a consequence of the estimate

$$|U^{-p} D_{p,l}f, U^{-q} D_{q,l}g|_1 < |D_{p,l}f|_2 |D_{q,l}g|_2.$$

By means of the standard polarization arguments it is possible to reduce the statement of Lemma to the case $f = g$. Then, one can see that $S \langle f, f \rangle = [\sum_{k=0}^{n-1} V^k B(f), \sum_{k=0}^{n-1} V^k B(f)]$ where $B(f)$ is a function defined in the statement of Lemma 2.

It is not difficult to derive from Lemmas 1 and 2 that

$$\lim_{n \to \infty} n^{-\frac{1}{2}} \left| \operatorname{grad} S_n f - \sum_{k=0}^{n-1} V^k B(f) \right|_{2,E} = 0$$

and

$$\lim n^{-\frac{1}{2}} |\operatorname{grad} S_n f|_{2,E} = \lim n^{-\frac{1}{2}} \left| \sum_{k=0}^{n-1} V^k B(f) \right|_{2,E} < \infty.$$

Then

$$n^{-1} \left| [\operatorname{grad} S_n f, \operatorname{grad} S_n f] - S_n \langle f, f \rangle \right|_1$$

$$= n^{-1} \left| \| \operatorname{grad} S_n f \|^2 - \left\| \sum_{k=0}^{n-1} V^k B(f) \right\|^2 \right|_1$$

$$\leqslant n^{-1} \left| \left\| \operatorname{grad} S_n f - \sum_{k=0}^{n-1} V^k B(f) \right\| \cdot \left(\| \operatorname{grad} S_n f \| + \left\| \sum_{k=0}^{n-1} V^k B(f) \right\| \right) \right|_1$$

$$\leqslant n^{-1} \left| \operatorname{grad} S_n f - \sum_{k=0}^{n-1} V^k B(f) \right|_{2,E} \cdot \left(2 \| \operatorname{grad} S_n f \|^2 + 2 \left\| \sum_{k=0}^{n-1} V^k B(f) \right\|^2 \right)^{\frac{1}{2}}_1$$

$$\leqslant 2^{\frac{1}{2}} n^{-\frac{1}{2}} \left| \operatorname{grad} S_n f - \sum_{k=0}^{n-1} V^k B(f) \right|_{2,E}$$

$$\times n^{-\frac{1}{2}} \left(|\operatorname{grad} S_n f|^2_{2,E} + \left| \sum_{k=0}^{n-1} V^k B(f) \right|^2_{2,E} \right)^{\frac{1}{2}} \to 0$$

as $n \to \infty$. $\square$

LEMMA 4. *Let $\varphi \colon \mathbb{R}^1 \to \mathbb{R}^1$ be a bounded continuously differentiable function with uniformly continuous derivative. If $f \in L_2^0$ satisfies the condition (1) then we have the relation*

$$\left| \operatorname{grad} \varphi(n^{-\frac{1}{2}} S_n f) - \varphi'(n^{-\frac{1}{2}} S_n f) \operatorname{grad}(n^{-\frac{1}{2}} S_n f) \right|_{2,E} \to 0$$

as $n \to \infty$.

PROOF. To reduce subscripts we will assume that $r = 1$ and thus the character l can be omitted in indices. The proof for the general case is essentially the same.

We have

$$\left| \operatorname{grad} \varphi(n^{-\frac{1}{2}} S_n f) - \varphi'(n^{-\frac{1}{2}} S_n f) \operatorname{grad}(n^{-\frac{1}{2}} S_n f) \right|^2_{2,E}$$

$$= \sum_{p \in \mathbb{Z}} \left| D_p \varphi(n^{-\frac{1}{2}} S_n f) - \varphi'(n^{-\frac{1}{2}} S_n f) D_p n^{-\frac{1}{2}} S_n f \right|^2_2$$

$$= \sum_{p \in \mathbb{Z}} \left| D_0 \varphi(n^{-\frac{1}{2}} U^{-p} S_n f) - \varphi'(n^{-\frac{1}{2}} U^{-p} S_n f) D_0 n^{-\frac{1}{2}} U^{-p} S_n f \right|^2_2.$$

Since

$$|D_0 U^{-p} S_n f| \leqslant \sum_{q \in \mathbb{Z}} |D_0(U^q f)| \in L_2$$

and φ' is a uniformly continuous function then

$$\delta_n = \sup_{p\in\mathbb{Z}} \left(\left| D_0\varphi(n^{-\frac{1}{2}}U^{-p}S_nf) - \varphi'(n^{-\frac{1}{2}}U^{-p}S_nf)D_0 n^{-\frac{1}{2}}U^{-p}S_nf \right| \right)$$

$$= \sup_{p\in\mathbb{Z}} \left| \varphi'(n^{-\frac{1}{2}}U^{-p}S_nf + \theta_{p,n}D_0 n^{-\frac{1}{2}}U^{-p}S_nf) - \varphi'(n^{-\frac{1}{2}}U^{-p}S_nf) \right| \xrightarrow[n\to\infty]{} 0$$

where $\theta_{p,n}$, $|\theta_{p,n}| \leqslant 1$, are multipliers from the Lagrange formula and $0/0 = 0$ by definition. As a consequence of this we have that $\delta_n \to 0$ in probability as $n \to \infty$. Besides we have $|\delta_n|_\infty \leqslant 2\sup|\varphi'|$.

Furthermore,

$$\sum_{p\in\mathbb{Z}}(D_0 n^{-\frac{1}{2}}U^{-p}S_nf)^2 = \frac{1}{n}\sum_{m=0}^{n-1}\sum_{k=0}^{n-1}\sum_{p\in\mathbb{Z}}(D_0 U^{p+m}f)(D_0 U^{p+k}f)$$

$$= \frac{1}{n}\sum_{m=0}^{n-1}\sum_{k=-m}^{m}\sum_{p\in\mathbb{Z}}(D_0 U^{p+k}f)(D_0 U^p f)$$

$$\leqslant \left(\sum_{p\in\mathbb{Z}}|D_0 U^p f|\right)^2 \in L_1$$

because

$$\sum_{p\in\mathbb{Z}}|D_0 U^p f|_2 = \sum_{p\in\mathbb{Z}}|D_p f|_2 < \infty.$$

Together with convergence and boundedness properties of $\{\delta_n\}$ it gives us that

$$\left| \operatorname{grad}\varphi(n^{-\frac{1}{2}}S_nf) - \varphi'(n^{-\frac{1}{2}}S_nf)\operatorname{grad}(n^{-\frac{1}{2}}S_nf) \right|_{2,E}^2$$

$$\leqslant \left| \delta_n \sum_{p\in\mathbb{Z}}|D_o U^p f| \right|_2^2 \to 0$$

as $n \to \infty$, and Lemma 4 is proved. $\square$

PROOF OF THEOREM. We will use Ch. Stein's approach (see [3] or [7] for details). According to this approach, to prove the convergence in distribution of a sequence $\{W_n\}$ of random variables to the normal law $N(0,\sigma^2)$ it is sufficient to establish that $(W_n, \varphi(W_n)) - \sigma^2(\varphi'(W_n), 1) \to 0$ as $n \to \infty$ for every continuously differentiable function $\varphi\colon \mathbb{R}^1 \to \mathbb{R}^1$ such that $\varphi(x)$ and $\varphi'(x)$ converges as $|x| \to \infty$.

We have

$$
\begin{aligned}
\left(W_n, \varphi(W_n)\right) &= \left(n^{-\frac{1}{2}} S_n f, \varphi(W_n)\right) \\
&= \left(n^{-\frac{1}{2}} S_n \Delta F, \varphi(W_n)\right) \\
&= \left(n^{-\frac{1}{2}} \Delta S_n F, \varphi(W_n)\right) \\
&= \left(n^{-\frac{1}{2}} \operatorname{div} \operatorname{grad}(S_n F), \varphi(W_n)\right) \\
&= \left(n^{-\frac{1}{2}} \operatorname{grad} S_n F, \operatorname{grad} \varphi(W_n)\right)_{2,E} \\
&= \left(n^{-\frac{1}{2}} \sum_{k=0}^{n-1} V^k \operatorname{grad} F, \operatorname{grad} \varphi(W_n)\right)_{2,E} \\
&= \left(n^{-\frac{1}{2}} \operatorname{grad} S_n F, \varphi'(W_n) \operatorname{grad} W_n\right)_{2,E} + C_{1,n} \\
&= n^{-1}\left([\operatorname{grad} S_n F, \operatorname{grad} S_n f], \varphi'(W_n)\right)_2 + C_{1,n} \\
&= n^{-1}\left(S_n \langle F, f \rangle, \varphi'(W_n)\right)_2 + C_{1,n} + C_{2,n} \\
&= \left(\langle F, f \rangle, 1\right)\left(\varphi'(W_n), 1\right) + C_{1,n} + C_{2,n} + C_{3,n},
\end{aligned}
$$

where

$$
\begin{aligned}
C_{1,n} &= \left(n^{-\frac{1}{2}} \operatorname{grad} S_n F, \operatorname{grad} \varphi(W_n) - \varphi'(W_n) \operatorname{grad} W_n\right)_{2,E}, \\
C_{2,n} &= \left(n^{-1}[\operatorname{grad} S_n F, \operatorname{grad} S_n f] - n^{-1} S_n \langle F, f \rangle, \varphi'(W_n)\right), \\
C_{3,n} &= \left(n^{-1} S_n \langle F, f \rangle, \varphi'(W_n)\right) - \left(\langle F, f \rangle, 1\right)\left(\varphi'(W_n), 1\right)
\end{aligned}
$$

and to complete the proof it is sufficient to check that $C_{i,n} \to 0$ as $n \to \infty$ for $i = 1, 2, 3$.

Indeed, it is the case for $C_{1,n}$, because the left multiplier of the inner product is bounded (that is a consequence of Lemma 1 and the identity $\operatorname{grad} S_n F = \sum_{k=0}^{n-1} V^k \operatorname{grad} F$) and the right one tends to 0 according to Lemma 4. Further, it follows, correspondingly, from Lemma 3 and from the Ergodic Theorem that $C_{2,n}$ and $C_{3,n}$ go to 0. The relation

$$
\left(\langle F, f \rangle, 1\right) = \sum_{l=1}^{r}\left(\sum_{p \in \mathbb{Z}} D_{0,l} U^p F, \sum_q D_{0,l} U^q f\right)
$$

takes place according to Lemma 3. $\square$

4. The CLT for hyperbolic toral automorphisms

Let A be an algebraic automorphism of the d-dimensional torus $\mathbb{T}^d = \mathbb{R}^d / \mathbb{Z}^d$. A is defined by the $(d \times d)$ matrix M, $|\det M| = 1$, with integer entries. A is covered by the linear transformation $\tilde{A}$ of $\mathbb{R}^d$ described by the matrix M in the canonical basis of $\mathbb{R}^d$. A preserves the usual Lebesgue (= Haar) measure on $\mathbb{T}^d$. We will consider the case when A is a hyperbolic automorphism. By definition, it means that the spectrum of M has the empty intersection with the subset $\{z \mid |z| = 1\}$ of $\mathbb{C}^1$.

We are going to prove that the corollary of the main theorem of the present paper is applicable to the case of hyperbolic toral automorphism if we take $r = d$.

To establish it, first we should construct a finite set of homoclinic transformations. There is a (uniquely determined) splitting of $\mathbb{R}^d$ into the direct sum of the stable and the unstable A-invariant subspaces, say $\widetilde{\Gamma}_s$ and $\widetilde{\Gamma}_u$ (they correspond to the splitting of the spectrum of M by the unit circle). The canonical map $\pi \colon \mathbb{R}^d \to \mathbb{T}^d$ injectively sends each of them to the corresponding subgroup (Γ_s or Γ_u) of $\mathbb{T}^d$ (the injectivity for the case of $\widetilde{\Gamma}_s$ is a consequence of the contractivity of A restricted to the intersection $\widetilde{\Gamma}_s \cap \mathbb{Z}^d$: this implies that the intersection just mentioned is trivial; the case of $\widetilde{\Gamma}_u$ can be considered in the similar way by substituting A^{-1} and $\widetilde{\Gamma}_u$ instead of A and $\widetilde{\Gamma}_s$).

The splitting $\mathbb{R}^d = \widetilde{\Gamma}_s + \widetilde{\Gamma}_u$ gives rise to two projections p_s and p_u commuting with A and having ranges $\widetilde{\Gamma}_s$ and $\widetilde{\Gamma}_u$, correspondingly. Since $\operatorname{Ker} p_s = \widetilde{\Gamma}_u$, $\operatorname{Ker} p_u = \widetilde{\Gamma}_s$ and $\widetilde{\Gamma}_u \cap \mathbb{Z}^d = \widetilde{\Gamma}_s \cap \mathbb{Z}^d = 0$, p_s and p_u map injectively $\mathbb{Z}^d$ to Γ_s and Γ_u. As a result, we have that $\pi \circ p_s$ and $\pi \circ p_u$ send injectively $\mathbb{Z}^d$ into $\mathbb{T}^d$.

Let us define the homoclinic subgroup Γ of $\mathbb{T}^d$ by the formula $\Gamma = \Gamma_s \cap \Gamma_u$. As it will be explained below, Γ is a countable A-invariant dense subgroup of $\mathbb{T}^d$. It can be characterized by the homoclinic property of its elements: $\Gamma = \{\gamma \mid \gamma \in \mathbb{T}^d, A^n \gamma \to 0 \text{ as } |n| \to \infty\}$. Let us mention that the convergence in this characterization is exponentially fast with respect to the standard distance on $\mathbb{T}^d$. As it will be proven Γ is a free abelian group on d generators.

The dual automorphism to A has the dual splitting with the same properties. It implies that both subgroups $\widetilde{\Gamma}_s + \mathbb{Z}^d$ and $\widetilde{\Gamma}_u + \mathbb{Z}^d$ are dense in $\mathbb{R}^d$. Hence $p_s(\mathbb{Z}^d) = p_s(\widetilde{\Gamma}_u + \mathbb{Z}^d)$ is dense in $\widetilde{\Gamma}_s$ and $p_u(\mathbb{Z}^d) = p_u(\widetilde{\Gamma}_s + \mathbb{Z}^d)$ is dense in $\widetilde{\Gamma}_u$. Furthermore, $\Gamma_s = \pi(\widetilde{\Gamma}_s) = \pi(\widetilde{\Gamma}_s + \mathbb{Z}^d)$ and $\Gamma_u = \pi(\widetilde{\Gamma}_u) = \pi(\widetilde{\Gamma}_u + \mathbb{Z}^d)$ are dense in $\mathbb{T}^d$ and one can conclude that $\pi \circ p_s$ and $\pi \circ p_u$ map $\mathbb{Z}^d$ isomorphically (in the algebraic sense) on some countable dense subgroups of $\mathbb{T}^d$.

We are going to prove now that these subgroups coincide. More exactly, it will be proven that
$$\Gamma = \pi(p_s(\mathbb{Z}^d)) = \pi(p_u(\mathbb{Z}^d)).$$

Indeed, since
$$p_s(\mathbb{Z}^d) = p_s(\mathbb{Z}^d) \cap (I - p_u)\mathbb{Z}^d \subset \widetilde{\Gamma}_s \cap (\widetilde{\Gamma}_u + \mathbb{Z}^d)$$

and
$$\begin{aligned} \widetilde{\Gamma}_s \cap (\widetilde{\Gamma}_u + \mathbb{Z}^d) &= p_s(\widetilde{\Gamma}_s \cap (\widetilde{\Gamma}_u + \mathbb{Z}^d)) \\ &\subset p_s(\widetilde{\Gamma}_u) \cap p_s(\widetilde{\Gamma}_s + \mathbb{Z}^d) \\ &= \Gamma_s \cap p_s(\mathbb{Z}^d) = p_s(\mathbb{Z}^d) \end{aligned}$$

we have
$$p_s(\mathbb{Z}^d) = \widetilde{\Gamma}_s \cap (\widetilde{\Gamma}_u + \mathbb{Z}^d)$$

and in the same way
$$p_u(\mathbb{Z}^d) = (\widetilde{\Gamma}_s + \mathbb{Z}^d) \cap \widetilde{\Gamma}_u.$$

Applying π to these two identities one can see that

$$\pi p_s(\mathbb{Z}^d) = \Gamma_s \cap \Gamma_u = \Gamma$$

and

$$\pi p_u(\mathbb{Z}^d) = \Gamma_s \cap \Gamma_u = \Gamma.$$

Let us fix the set $\{\gamma, \ldots, \gamma_d\}$ of generators for Γ coming from the canonical basis of $\mathbb{Z}^n$ by means of the map $\pi \circ p_s$. The automorphism A restricted to Γ has the same matrix M with respect to the basis $\{\gamma, \ldots, \gamma_d\}$ as A has with respect to the canonical basis of $\mathbb{Z}^d$ (or $\mathbb{R}^d$).

To every γ_l corresponds a (measure preserving) translation R_l of the torus $\mathbb{T}^d$. Such transformations and corresponding unitary operators H_l on L_2 have the homoclinic property in the sense of the definition from Section 2 and the inequality

$$(1') \qquad \sum_{p\in\mathbb{Z}}\sum_{l=1}^{d} |U^p H_l U^{-p} f - f|_2 = \sum_{p\in\mathbb{Z}}\sum_{l=1}^{d} |D_{p,l} f|_2 < \infty$$

holds for $f \in L_2$ under weak regularity conditions imposed on f (for example, the Hoelder or some versions of the Dini property in L_2 are sufficient).

So, if $H_1, \ldots, H_l$ denote the homoclinic operators corresponding to the generators fixed above the operator Δ is defined on a dense subset of L_2. To complete the proof that the CLT holds for every function with the property $(1')$ it is sufficient to check for Δ the semiboundedness inequality from the statement of the corollary from Section 2.

To prove this inequality we have to check, since all characters of $\mathbb{T}^d$ are eigenfunctions of Δ, that there exists such a positive constant c that for every character χ of $\mathbb{T}^d$ distinct from χ_0 (where χ_0 is the unity character) we have

$$(\Delta\chi, \chi) = \sum_{p\in\mathbb{Z}}\sum_{l=1}^{d} |D_{p,l}\chi|_2^2 = \sum_{p\in\mathbb{Z}}\sum_{l=1}^{d} |\chi(A^p\gamma_l) - 1|^2 \geqslant c.$$

An equivalent form of this inequality is

$$\sum_{p\in\mathbb{Z}}\sum_{l=1}^{d} |\langle \gamma_l, ((A^\perp)^*)^p\chi - \chi_0\rangle|^2 \geqslant c$$

where $\langle \cdot, \cdot \rangle$ is the duality form between the countable group Γ and its (compact) character group $\widehat{\Gamma}$ ($\widehat{\Gamma}$ is isomorphic to $\mathbb{T}^d$) and $A^\perp$ is an automorphism A restricted to Γ. Since $(A^\perp)^*$ is defined by the matrix conjugate to M it is also hyperbolic and the last inequality can be easily deduced from the property

$$\inf_{\substack{\chi\in\widehat{\Gamma} \\ \chi\neq\chi_0 \\ p\in\mathbb{Z}}} \mathrm{dist}(((A^\perp)^*)^p\chi, \chi_0) > 0$$

(where dist is a translation invariant distance on $\widehat{\Gamma}$), which is a consequence of the well known Bowen expansiveness property [1] of the (hyperbolic) automorphism $(A^{\perp})^*$ of $\widehat{\Gamma}$.

5. Concluding remarks

There exist many various modifications and generalizations of the definition given in Section 2. For example, nonsingular homoclinic transformations, homoclinic operators acting on L_2 and homoclinic homeomorphisms were defined in [7].

Random homoclinic transformations generating homoclinic transition operators could give a common framework for the study of some hyperbolic dynamical systems and more traditional objects of application of coupling such as Markov processes. We restrict ourselves within the present paper to the case of deterministic invertible P-preserving homoclinic transformations to avoid some technical problems and to minimize the volume of the paper.

The author would like to remark that there exists an alternative proof of the semiboundedness property of the operator Δ similar to the proof of the classical Poincare inequality.

A natural sphere of application of the homoclinic approach to the CLT could be the CLT for random fields. A version of the CLT for random fields based on Stein's technique can be found in [8]. The statement of this result is given in terms of mixing condition. It would be interesting to compare this result with some form of extension of homoclinic approach to the case of fields.

Another important problem is to extend this approach to the case of general ergodic toral automorphisms [2].

Acknowledgements. A part of this paper was done in October 1991 during the author's stay at Goettingen as a guest of SFB–170. The author is very grateful to M. Denker for his hospitality and a number of interesting discussions.

A special thanks to the Organizing Committee of Doeblin Conference at Blaubeuren, especially to J. Gani and H. Hering for their kind attention to the participants during the conference and to H. Cohn whose support and patience helped the author to prepare this paper.

The author is very indebted to E. Antonovskaya for typing the final version of this paper in TEX.

References

1. Bowen R., *Equilibrum states and the ergodic theory of Anosov diffeomorphisms*, Lect. Notes in Math., Springer-Verlag **470** (1975).
2. Leonov V.P., *Some applications of the high order cumulants to the theory of stationary random processes*, Moscow, 1964. (Russian)
3. Stein Ch., *A bound for the error in the normal approximation to the distribution a sum of dependent random variables*, Proc. Sixth Berkeley Symp. Math. Statist. and Prob., vol. 2, 1977, pp. 583–602.

4. Capocaccia D., *A definition of Gibbs state for a compact set with $\mathbb{Z}^d$ action*, Commun. Math. Phys. **48** (1976), 85–88.

5. Ruelle D., *Thermodynamic formalism*, Addison-Wesley Publ. Comp., 1978.

6. Krieger W., *On dimension function and topological Markov chains*, Invent. Math. **56** (1980), 239–250.

7. Gordin M. I., *Homoclinic version of the central limit theorem*, Notes of Scientific Seminars of the Leningrad branch of Steklov Math. Institute **184** (1990), 80–91 (Russian); English translation in Journal of Soviet Mathematics (under preparation).

8. Bolthausen E., *On the central limit theorem for stationary mixing random fields*, Ann. of Prob. **10** (1982), 1047–1050.

KUIBYSHEV ST. 1/5, APT. 112, 197046 ST. PETERSBURG, RUSSIA

Contemporary Mathematics
Volume **149**, 1993

Asymptotic results for φ-mixing sequences

Magda Peligrad[1]

0   Introduction

In order to study the asymptotical properties of Markov chains, Doeblin (1937) introduced a condition known as condition (D) or Doeblin's hypothesis. We give it here as it appears in Doob (1953), page 192.

Let $(X, \mathcal{P}_X)$ be a measurable space and let $P(x; A)$, $x \in X$, $A \in \mathcal{P}_X$, be a transition probability function.

Doeblin's Hypothesis: There is a (finite-valued) measure μ on $\mathcal{P}_X$ with $\mu(X) > 0$, an integer $n \geq 1$ and a positive ε, such that

$$P^n(x, A) \leq 1 - \varepsilon \ \text{ if } \ \mu(A) \leq \varepsilon.$$

This means, roughly speaking, that whenever A is small, $P^n(x, A)$ is not big and this happens uniformly in $x \in X$.

For stationary Markov chains Doeblin's condition is equivalent to the following condition known as φ-mixing (for a proof see Rosenblatt (1971), page 212, eqn. (18)).

Definition. [Ibragimov (1962)]. Let $\{X_n\}_{n \in Z}$ be a strictly stationary sequence of random variables on $(\Omega, \mathcal{K}, P)$. Denoting $\mathcal{F}_n^m = \sigma(X_k : n \leq k \leq m)$, we say that $\{X_n\}_{n \in Z}$ is φ-mixing if $\varphi_n \to 0$, where

$$\varphi_n = \sup\{|P(B|A) - P(B)|; \ A \in \mathcal{F}_{-\infty}^0, \ B \in \mathcal{F}_n^\infty, \ P(A) \neq 0\}.$$

Put $S_n = \sum_{i=1}^n X_i$, $\sigma_n^2 = \mathrm{var}(S_n)$, $W_n = \{S_{[nt]}/\sigma_n : 0 \leq t \leq 1\}$, $M_n = \max_{1 \leq i \leq n} |S_i|$, and $N_n = \max_{1 \leq i \leq n} |X_i|$; here $[\cdot]$ is the greatest integer function.

In 1962 Ibragimov proved that a centered strictly stationary ϕ-mixing sequence of random variables satisfying $\sigma_n^2 \to \infty$ and $E|X_1|^{2+\delta} < \infty$ obeys the

[1]1991 Mathematics Subject Classification. Primary 60F05
Key Words: Doeblin's condition, mixing sequences, central limit theorem.
Partially supported by an N.S.F. Grant and a Taft Grant.
This paper is in final form and no version of it will be submitted for publication elsewhere.

CLT. In 1975 he proved that the result remains valid if one replaces $E|X_1|^{2+\delta} < \infty$ by the conditions $EX_1^2 < \infty$ and $\sum \phi^{1/2}(2^i) < \infty$. He also showed that both these results can be extended to weak invariance principles. These results are considered only steps in establishing the truth of the following conjectures.

CONJECTURE 1 *(Ibragimov (1962)). Let $\{X_n\}_{n\in Z}$ be a strictly stationary, centered ϕ-mixing sequence of random variables such that $EX_1^2 < \infty$ and $\sigma_n^2 \to \infty$. Then S_n/σ_n converges weakly to the normal distribution.*

CONJECTURE 2 *(Iosifescu (1977)). Let $\{X_n\}_{n\in Z}$ be as in Conjecture 1. Then W_n converges weakly to the standard Brownian motion on $[0,1]$.*

Herrndorf (1983) showed that if there is a strictly stationary ϕ-mixing sequence such that $\sigma_n^2 \to \infty$ and $\liminf \sigma_n^2/n = 0$ then Conjecture 2 is not true. Peligrad (1985) proved that both conjectures are true under the assumption $\liminf \sigma_n^2/n \neq 0$, that reduces the study of the above conjectures to a study of the variance of the partial sum. In Peligrad (1990) the CLT and its weak invariance principle are established for partial sums of strictly stationary sequences having the marginal distribution of $|X_1|$ regularly varying with exponent -2, and $\phi_1 < 1$.

An important condition for φ-mixing sequences appears to be:

(C) $\lim_{n\to\infty} nP(|X_1| > \varepsilon\sigma_n) = 0$ for every $\varepsilon > 0$,

(C) was used as a condition in many papers in connection with CLT and its weak invariance principle for stationary ϕ-mixing sequences.

Herrndorf (1983) proved that (C) is the condition that added to the CLT implies the weak invariance principle, while Samur (1984, Theorem 4.1) assumed (C) together with the convergence in distribution of sums of ϕ-mixing triangular arrays in order to characterize the limiting distribution that appeared to be Gaussian. Assuming (C) Jakubowski (1988, Theorem 2) proved tightness of W_n and Hahn, Kuelbs and Samur (1987, Corollary 3) established that under this condition S_n/σ_n is tight with only centered Gaussian limits.

In this note we prove

THEOREM 1 *Let $\{X_n\}_{n\in Z}$ be a strictly stationary φ-mixing sequence of random variables such that $EX_1 = 0$, $EX_1^2 < \infty$, $\sigma_n^2 \to \infty$ and $\varphi_1 < 1$. Then there is a subsequence Q, $Q \subset N$ such that $\lim_{\substack{m\to\infty \\ m\in Q}} mP(|X_1| > \varepsilon\sigma_m) = 0$ for every $\varepsilon > 0$.*

COROLLARY 1 *Under the conditions of Theorem 1, $\{S_n/\sigma_n\}$ contains a subsequence attracted to a normal distribution, (possibly degenerate).*

1 Preliminaries

We group here different results that will be used later on.

Let $\{X_n^*\}$ be the i.i.d. associated sequence of $\{X_n\}$. That is, X_1^* has the same distribution function as X_1.

It is easy to see that for every positive x we have

(1.1) $$P(\max_{1\leq i\leq n}|X_i^*| > x) \geq nP(|X_1| > x)P(\max_{1\leq i\leq n}|X_i^*| \leq x)$$

(for a similar relation see Lai, (1977), (3.28)).

(Peligrad (1990), Proposition 3.1). For every x and every $n \geq 1$ we have

(1.2) $$\begin{aligned}(1 - \phi_1)P(\max_{1\leq i\leq n} X_i^* > x) &\leq P(\max_{1\leq i\leq n} X_i > x) \\ &\leq (1 + \phi_1)P(\max_{1\leq i\leq n} X_i^* > x).\end{aligned}$$

It is easy to see that for every positive x we have

(1.3) $$P(\max_{1\leq i\leq n}|X_i| \geq x) \leq P(\max_{1\leq i\leq n}|S_i| \geq x/2).$$

(Iosifescu-Theodorescu (1969), Lemma 1.1.6). If $\{X_n\}$ is strictly stationary and $\varphi_1 < 1$, then $P(|S_k| < a) > \varphi_1$ for $1 \leq k < n$ implies

(1.4) $$P(\max_{1\leq i\leq n}|S_i| > a + x) \leq \frac{P(|S_n| > x)}{\min_{1\leq i<n} P(|S_i| < a) - \varphi_1}.$$

For similar results see also Cohn (1965), (1966).

(Ibragimov-Linnik, (1971)). Under the conditions of Theorem 1, σ_n^2 has the representation

(1.5) $$\sigma_n^2 = nh(n)$$

with $h(n)$ a slowly varying function at infinity.

(Peligrad (1990), Lemma 3.3.)

(1.6) $$\max_{1\leq i\leq n} \sigma_i^2/\sigma_n^2 \text{ is a bounded sequence}$$

LEMMA 1 *Let $\{X_n\}_{n\in Z}$ be as in Theorem 1 and let $m \in Q \subset N$ and $p = p(m) = \sigma(m)$ as $m \to \infty$, $k = \left[\frac{m}{p}\right] + 1$. Then there is a constant C, such that*

$$mP(|X| > \varepsilon\sigma_m) \leq CkP(|S_p| > \varepsilon\sigma_m/4)$$

for every $m \in Q$.

Proof: By (1.3) and (1.4) with $a = x = \varepsilon\sigma_m/2$, and Tchebyshev's inequality

$$
\begin{aligned}
P(N_p > \varepsilon\sigma_m) &\leq\ P(M_p > \varepsilon\sigma_m/2) \\
&\leq\ \frac{P(|S_p| > \varepsilon\sigma_m/4)}{1 - 4\min_{1\leq i\leq p}\sigma_i^2/\varepsilon^2\sigma_m^2 - \varphi_1}
\end{aligned}
$$

Because of (1.5) and (1.6) and because $p = o(m)$ as $m \to \infty$, $m \in Q$, we have $\min_{1\leq i\leq p}\sigma_i^2 = o(\sigma_m^2)$ and we can find a constant C_1 such that

$$(1.7) \qquad\qquad P(N_p > \varepsilon\sigma_m) < C_1 P(|S_p| > \varepsilon\sigma_m/4).$$

Now by (1.1) and (1.2) after an easy computation we have

$$pP(|X_1| > \varepsilon\sigma_m) \leq P(N_p > \varepsilon\sigma_m)\frac{1}{1 - \varphi_1 - P(N_p > \varepsilon\sigma_m)}$$

Hence,

$$pP(|X_1| > \varepsilon\sigma_m) \leq C_1 P(|S_p| > \varepsilon\sigma_m/4)\frac{1}{1 - \varphi_1 - C_1 P(|S_p| > \varepsilon\sigma_m/4)}$$

Now, by applying Tchebyshev's inequality and taking into account that $\sigma_p = o(\sigma_m)$ as $m \to \infty$, $m \in Q$ we can find a positive constant C_2 such that for every $m \in Q$ and $p = o(m)$ we have

$$pP(|X_1| > \varepsilon\sigma_m) \leq C_2 P(|S_p| > \varepsilon\sigma_m/4)$$

The result follows by multiplying with k.

2   Proof of Theorem 1

The proof is divided into two parts, according to the behavior of $E|S_n|/\sigma_n$. First we assume that

$$(2.1) \qquad\qquad \liminf \frac{E|S_n|}{\sigma_n} = a > 0$$

By Theorem 1 in Dehling, Denker and Philipp (1986), there is a subsequence $Q \subset N$ such that $S_m/\sqrt{\frac{\pi}{2}}E|S_m| \xrightarrow{\mathcal{D}} N(0,1)$ along Q. Put $b_n = \sqrt{\frac{\pi}{2}}E|S_n|$. The proof uses now the small-big block argument. For $p = p(m)$, $q = q(m)$, $i = 0, ..., k - 1$ where $k := k(m)$ satisfies $(k-1)(p+2) \leq m \leq k(p+q)$, define

$$Y_i = Y_i(m) = X_{i(p+q)+1} + ... + X_{i(p+q)+p}$$

$$Z_i = Z_i(m) = X_{i(p+q)+p+1} + ... + X_{(i+1)(p+q)}$$

$$Z_k = X_{k(p+q)+1} + ... + X_n.$$

Because $\varphi_n \to 0$ and σ_n^2 satisfies (1.5) we can select p and q with the following properties:

$$(2.2) \qquad k\varphi(q) \to 0 \;\; \text{as} \;\; m \to \infty, \; m \in Q$$

$$(2.3) \qquad \frac{\text{var}(\sum_{i=0}^k Z_i)}{\sigma_m^2} \to 0 \;\; \text{as} \;\; m \to \infty, \; m \in Q$$

$$(2.4) \qquad \frac{k\sigma_p^2}{\sigma_m^2} \to 1 \;\; \text{as} \;\; m \to \infty, \; m \in Q.$$

By (2.1) and (2.3) we see that the behaviour in distribution of $\{S_m/b_m\}$ is the same as the behavior of $\{\sum_{i=0}^{k-1} Y_i/b_m\}$. But by Ibragimov-Linnik (1971; page 338),

$$E(\exp\left(\frac{it\sum_{j=0}^{k-1} Y_j}{b_m}\right) - \Pi_{j=0}^{k-1} E \exp\left(\frac{itY_j}{b_m}\right)) \leq k\varphi(q)$$

which by (2.2) is going to 0 as $m \to \infty$, $m \in Q$. Therefore if $\{Y_i^*\}_{0 \leq i \leq k-1}$ is an independent sequence associated with $\{Y_i\}_{0 \leq i \leq k-1}$, the asymptotic distribution of $\sum_{i=0}^{k-1} Y_i/b_m$ will be the same as that of $\sum_{i=0}^{k-1} Y_i^*/b_m$. But, a necessary and sufficient condition for the CLT for $\sum_{i=0}^{k-1} Y_i^*/b_m$ is the Lindeberg condition which in this case is

$$(2.5) \qquad \frac{kEY_0^2 I(|Y_0| \geq \varepsilon b_m)}{b_m^2} \to 0 \;\; \text{as} \;\; m \to \infty, \;\; m \in Q$$

for every $\varepsilon > 0$, which by (2.1) is equivalent to

$$\frac{kEY_0^2 I(|Y_0| > \varepsilon \sigma_m)}{\sigma_m^2} \to 0 \;\; \text{as} \;\; m \to \infty, \; m \in Q,$$

for every $\varepsilon > 0$.

This relation implies

$$kP(|Y_p| > \varepsilon \sigma_m) \to 0 \;\; \text{as} \;\; m \to \infty, \; m \in Q$$

for every $\varepsilon > 0$, which by Lemma 1 implies

$$mP(|X_1| > \varepsilon \sigma_m) \to 0 \;\; m \to \infty, \;\; \text{as} \;\; m \in Q$$

for every $\varepsilon > 0$. This gives the desired result under (2.1).

Assume now that (2.1) does not hold, i.e.

$$(2.6) \qquad \frac{E|S_p|}{\sigma_p} \to 0$$

along a subsequence $p \in Q'$. Take $k = k_p = \left[\frac{\sigma_p}{E|S_p|}\right] + 1$ and notice that $k \to \infty$. Put $m = k_p$. According to Lemma 1, Tchebyshev's inequality and (1.5) we have

$$
\begin{aligned}
mP(|X_1| > \varepsilon\sigma_m) &\leq CkP(|S_p| > \varepsilon\sigma_m/4) \\
&\leq Ck\frac{4E|S_p|}{\varepsilon\sigma_m}O\left(\frac{\sigma_p}{\sigma_m}\right) \to 0 \text{ as } m \to \infty, \ p \in Q'.
\end{aligned}
$$

This concludes the proof of Theorem 1.

I wish to thank the referee for the careful reading of the manuscript which improved on the presentationn of this paper.

References

[1] Billingsley, P., *Convergence of Probability Measures*, Wiley, New York, (1968).

[2] Cohn, H., On a class of dependent random variables, Rev. Roum. Math. Pures et Appl., **10**,*10* (1965) 1593-1606.

[3] Cohn, H., Teoreme limită pentru procese stochastice, Studii şi Cercetări Matematice, **18**,*7*(1966) 993-1027.

[4] Dehling, H., Denker, M., and Philipp, W., Central limit theorems for mixing sequences of random variables under minimal conditions, Ann. Probability, **14**(1986) 1359-1370.

[5] Doeblin, W., Sur les propriétés asymptotiques de mouvements régis par certains types de chanes simples, Bull. Math. Soc. Roumaine Sci. **39**, **1**(1937) 57-115; **39**, **2**(1937) 3-61.

[6] Doob, J.L., *Stochastic Processes*, Wiley, New York, (1953).

[7] Hahn, M.G., Kuelbs, J., and Samur, J.D., Asymptotic normality of trimmed sums of ϕ-mixing random variables, Ann. Probab. 5(1987) 1395-1418.

[8] Herrndorf, N., The invariance principle for ϕ-mixing sequences, Z. Wahrsch. Verw. Gebiete **63**, **1**(1983) 97-109.

[9] Ibragimov, I.A., Some limit theorems for stationary processes, Theory Prob. Appl. 7(1962) 349-382.

[10] Ibragimov, I.A., Linnik, Yu, V., *Independent and Stationary Sequences of Random Variables*, Wolters-Noordhoff, Groningen, (1971).

[11] Iosifescu, M., Limit theorems for ϕ-mixing sequences: a survey, *Proc. 5th Conf. Probab. Theory (Brasov, 1974)*, pp. 51-57. Ed. Academiei R.S., Romania, Bucuresti, (1977).

[12] Iosifescu, M., Theodorescu, R., *Random Processes and Learning*, Springer-Verlag, New York, (1969).

[13] Jakubowski, A., A note on the invariance principle for stationary ϕ-mixing sequences: tightness via stopping times, Revue Roumaine Math. Pures Appl. **33**(1988) 407-412.

[14] Lai, T.L., Convergence rates and r-quick version of the strong law for stationary mixing sequences, Ann. Probab. **15**(1987) 693-706.

[15] Peligrad, M., An invariance principle for ϕ-mixing sequences, Ann. Probab. **13**(1985) 1304-1313.

[16] Peligrad, M., On Ibragimov-Iosifescu conjecture for ϕ-mixing sequences, Stochastic Process Appl. **35**(1990) 293-308.

[17] Rosenblatt, M., *Markov Processes, Structure and Asymptotic Behavior*, Springer-Verlag, Berlin, (1971).

[18] Samur, J., Convergence of sums of mixing triangular arrays of random vectors with stationary sums, Ann. Probab. **12**(1984) 390-426.

Department of Mathematical Sciences, University of Cincinnati, Cincinnati, Ohio 45221-0025
Current address: Department of Mathematical Sciences, University of Cincinnati, Cincinnati, Ohio 45221-0025
E-mail address: peligrad@ucbeh.san.uc.edu

Contemporary Mathematics
Volume **149**, 1993

The Central Limit Theorem and Markov Sequences

MURRAY ROSENBLATT

ABSTRACT. Remarks are made about the Doeblin condition, its relation to certain mixing conditions and the central limit theorem. Related conditions for random fields are discussed. A representation problem proposed by Wiener is also considered.

Mixing Conditions. Let Ω be a state space of points with $\mathcal{A}$ a Borel field of subsets of Ω. $P(\cdot, \cdot)$ is a transition probability function if (i) for each $x \in \Omega$, $P(x, \cdot)$ is a probability measure on $\mathcal{A}$ and (ii) for each $A \in \mathcal{A}$, $P(\cdot, A)$ is an $\mathcal{A}$ measurable function of x on Ω. Consider a Markov sequence with one step transition function $P(\cdot, \cdot)$ that has a stationary transition function. The higher step transition probabilities $P_n(\cdot, \cdot)$, $n > 1$, are given recursively by

$$(1) \qquad P_1(x, A) = P(x, A) \,,$$

$$P_{n+1}(x, A) = \int P_n(x, dy) P(y, A) \,, \quad n = 1, 2, \dots \,.$$

A probability measure μ on $\mathcal{A}$ is invariant with respect to $P(\cdot, \cdot)$ if

$$(2) \qquad \int \mu(dx) P(x, A) = \mu(A) \,.$$

One can then introduce a Markov sequence that is stationary with instantaneous distribution μ and 1 step transition function $P(\cdot, \cdot)$. Generally an invariant probablity measure μ with respect to $P(\cdot, \cdot)$ needn't exist. An early sufficient condition for the existence of such an invariant probability measure is the condition D introduced by Doeblin [5] (also see Doob [7]). This is the condition that there is a finite measure ϕ on $\mathcal{A}(0 < \phi(\Omega) < \infty)$, an integer $n \geq 1$, and an $\epsilon > 0$ such that

$$(3) \qquad P_n(x, A) \leq 1 - \epsilon \quad \text{if} \quad \phi(A) \leq \epsilon, \quad A \in \mathcal{A}$$

This research was partially supported by the Office of Naval Research through grant N00014-90-J1371.

This paper is in final form and no version of it will be submitted for publication elsewhere.

AMS 1991 subject classifications: 60J05, 60F05, 60G60.

for all $x \in \Omega$.

The Doeblin condition D_0 is the Doeblin condition D with the added requirement that the stationary Markov process generated by $P(\cdot, \cdot)$ and a unique invariant probability measure μ be purely nondeterministic. Consider the condition

$$(4) \qquad \sup_{B \in \mathcal{A}} |P_n(x, B) - \mu(B)| \le \alpha(n) \to 0$$

for almost all $x(d\mu)$ as $n \to \infty$. One can show that this condition (4) is equivalent to the Doeblin condition D_0 with $\phi = \mu$. Let $\{X_n, n = \ldots, -1, 0, 1, \ldots\}$ be the stationary Markov sequence generated by μ and $P(\cdot, \cdot)$. Let $\mathcal{B}_n$ and $\mathcal{F}_n$ be the Borel fields generated by the random variables in the brackets

$$\mathcal{B}_n = \mathcal{B}(X_k, k \le n), \quad \mathcal{F}_n = \mathcal{B}(X_k, k \ge n).$$

$P(\cdot)$ will denote the probability measure of the sequence $\{X_n, n = \ldots, -1, 0, 1, \ldots\}$. Then one can show (see [10]) that (4) is equivalent to

$$(5) \qquad \sup_{B \in \mathcal{F}_n} |P(AB) - P(A)P(B)| \le \phi(n)P(A)$$

for all $A \in \mathcal{B}_0$.

The transition probability function $P(\cdot, \cdot)$ determines a bounded operator T acting on the bounded $\mathcal{A}$ measurable functions f

$$(Tf)(x) = \int P(x, dy)f(y)$$

with

$$\sup_x |(Tf)(x)| \le \sup_x |f(x)|.$$

The operator T can be extended naturally so that it acts as a bounded operator on

$$L^p(d\mu) = \left\{ f \colon \|f\|_p = \left(\int |f(x)|^p \mu(dx) \right)^{1/p} < \infty \right\}, \quad 1 \le p \le \infty,$$

where μ is the invariant probability measure relative to $P(\cdot, \cdot)$. Call the condition

$$(6) \qquad \lim_{n \to \infty} \langle T^n \rangle_p = \lim_{n \to \infty} \sup_{f \perp 1} \frac{\|T^n f\|_p}{\|f\|_p} \to 0,$$

$1 \le p \le \infty$, the L^p norm condition. The mixing condition (4) can be shown to be equivalent to the L^∞ norm condition. The Riesz convexity theorem shows that the L^p norm conditions, $1 < p < \infty$, are all equivalent to each other and that the L^1 and L^∞ norm conditions are stronger than the L^p norm conditions, $1 < p < \infty$. The condition (6) implies that $\langle T^n \rangle_p$ decreases to zero exponentially fast.

Let $\{X_n\}$ be a stationary Markov process satisfying an L^p norm condition (6), $1 \leq p \leq \infty$. Let f be a Borel function on Ω. If $Ef(X)^2 < \infty$ $Ef(X) = 0$,

$$\sigma_n^2 = E \left| \sum_{j=1}^{n} f(X_j) \right|^2 \to \infty$$

as $n \to \infty$ one can show that

$$\sigma_n^{-1} \sum_{j=1}^{n} f(X_j)$$

is asymptotically $N(0, 1)$ as $n \to \infty$.

Some results of Doeblin were extended by T. Harris [9] who introduced the following condition: the Borel σ-field $\mathcal{A}$ is separable and there is a σ-finite measure ϕ on $(\Omega, \mathcal{A})$ with the property that $P[\text{entering} \quad A \quad \text{at some time} \mid X_0 = x] = 1$ for all $x \in \Omega$ and all $A \in \mathcal{A}$ with $\phi(A) > 0$. Harris showed that his condition implies the existence of a unique (up to a constant factor) σ-finite measure μ that is stationary in that $\mu(A) = \int P(x, A)\mu(dx)$. If the stationary measure μ is a probability measure and the Markov process $\{X_n\}$ is purely nondeterministic, the Harris condition implies that the total variation

$$\|P_n(x, \cdot) - \mu(\cdot)\| \to 0$$

as $n \to \infty$ (see Orey [11]). One can construct Markov sequences $\{X_n\}$ that satisfy the L^2 norm condition but not the Harris condition. One can achieve this by letting the process have a transition function $P(\cdot, \cdot)$ such that $P_n(x, \cdot)$ is singular with respect to the invariant probability measure μ for each integer $n \geq 1$ and each $x \in \Omega$.

It is natural to try to extend a condition like the L^2 norm condition in a form that would be appropriate for a stationary possibly non-Markovian sequence. Let us consider a stationary process $\{X_n, \ n = \ldots, -1, 0, 1, \ldots\}$. Consider $L^2(\mathcal{B}_n)$, $L^2(\mathcal{F}_n)$ the collection of random variables with finite second moments measurable with respect to $\mathcal{B}_n$ and $\mathcal{F}_n$ respectively. Set

(7)
$$\alpha(k) = \sup_{\substack{U \in L^2(\mathcal{B}_0) \\ V \in L^2(\mathcal{F}_k)}} |\text{corr}\,(U, V)| \, .$$

We shall say that the process $\{X_n\}$ has asymptotically correlation zero if $\alpha(k) \to 0$ as $k \to \infty$. This is a natural extension of the L^2 norm condition for Markov processes adapted so that it can be applied to a stationary process.

The condition that

(8)
$$\beta(k) = \sup_{\substack{A \in \mathcal{B}_0 \\ B \in \mathcal{F}_k}} |P(A\,B) - P(A)P(B)|$$

approach zero as $k \to \infty$ is often referred to as strong mixing. In the case of stationary Gaussian processes, asymptotic correlation zero and strong mixing are equivalent to each other. However, generally asymptotic correlation zero is a stronger condition than strong mixing. This can be seen since a stationary Markov process satisfies the strong mixing condition if and only if

$$(9) \qquad \sup_{f \perp 1} \frac{\|T^n f\|_1}{\|f\|_\infty} \to 0$$

as $n \to \infty$.

Cogburn [3] introduced the condition of uniform ergodicity

$$(10) \qquad \sup_{A \in \mathcal{B}_0, B \in \mathcal{F}_0} \left| \frac{1}{n} \sum_{k=1}^{n} P(\tau^k B \cap A) - P(A)P(B) \right| \to 0$$

as $n \to \infty$ where τ is the shift transformation. It has been shown that if a stationary process $\{X_n\}$ is uniformly ergodic and mixing, then it must be strongly mixing (see [12]). Cogburn has shown that the only limit laws of stationary asymptotically negligible instantaneous real valued functions of a stationary Markov sequence are infinitely divisible if and only if the sequence is uniformly mixing. Denker [4] has shown that if one has a strongly mixing stationary sequence $\{X_n\}$ with $E X_k \equiv 0$,

$$S_n = \sum_{k=1}^{n} X_k \,,$$

$$\sigma_n^2 = E|S_n|^2 \to \infty \quad \text{as} \quad n \to \infty \,,$$

the sequence

$$\sigma_n^{-1} S_n$$

is asymptotically normal if and only if the $\sigma_n^{-2} S_n^2$ are uniformly integrable.

Consider now the case of stationary random fields $\{X_t, \ t \in Z^d\}$ with $Z = \{\ldots, -1, 0, 1, \ldots\}$ and $d \geq 2$. The concepts of asymptotic correlation zero and strong mixing can be extended in a number of different ways to the case $d \geq 2$. We shall choose one of these. Let S and T be the disjoint sets of Z^d. Set

$$(11) \qquad \alpha(S,T) = \sup \ |\text{corr}\,(U,V)| , \ U \in L^2(\mathcal{B}(X_t, \ t \in S)) ,$$
$$V \in L^2(\mathcal{B}(X_t, \ t \in T))$$

and

$$(12) \qquad \beta(S,T) = \sup \ |P(A\,B) - P(A)P(B)| , \ A \in \mathcal{B}(X_t, t \in S) ,$$
$$B \in \mathcal{B}(X_t, t \in T) \,.$$

Further let

$$(13) \qquad \alpha^*(r) = \sup \ \alpha(S,T)$$
$$\beta^*(r) = \sup \ \beta(S,T)$$

where the sup is taken over all pairs of nonempty sets $S, T \subset Z^d$ with distance $\text{dist}(S, T) \geq r$. We shall say the random field $\{X_t, \ t \in Z^d\}$ has asymptotic correlation zero if $\alpha^*(r) \to 0$ as $r \to \infty$ and is strongly mixing if $\beta^*(r) \to 0$ as $r \to \infty$. Bradley ([1] and [2]) has shown that these versions of asymptotic correlation zero and strong mixing are equivalent for $d \geq 2$. It should be noted that if we use these versions of asymptotic correlation zero and strong mixing in the one-dimensional case $d = 1$, they are stronger than the versions we originally introduced there.

A Representation Problem. N. Wiener considered a structural problem in his book on nonlinear problems in random theory [14]. Given a stationary process $\{X_n\}$ the object was to determine conditions under which one could generate a stationary sequence of independent random variables ξ_n (each uniformly distributed on $[0, 1]$) such that ξ_n is measurable with respect to $\mathcal{B}_n = \mathcal{B}(X_s, \ s \leq n)$ and independent of $\mathcal{B}_{n-1}$

$$(14) \qquad\qquad \xi_n = g(\tau^n X) \,.$$

Here g is a fixed function, X the process $\{X_n\}$ and τ the shift operator. Also this representation is to be invertible in the sense that X_n is measurable with respect to the Borel field $\mathcal{B}(\xi_k, \ k \leq n)$. Wiener suggested conditions that might insure the existence of such a mapping and its construction. Let the conditional distribution of X_n given the past be

$$(15) \qquad F(x_n | x_k, \ k \leq n - 1) = P\{X_n \leq x_n | X_k = x_k, \ k \leq n - 1\}$$
$$= P\{X_n \leq x_n | \mathcal{B}_{n-1}\} \,.$$

The conditions suggested were (i) $\{X_n\}$ is purely nondeterministic in the sense that $\mathcal{B}_{-\infty} = \bigcap_n \mathcal{B}_n$ is the trivial σ-field.

(ii) $F(x_n | x_k, \ k \leq n - 1)$ is a strictly increasing continuous function of x_n for almost every past.

Assume that $F(x_n | x_k, \ k \leq n - 1)$ is a Borel function of its arguments x_k, $k \leq n$. One can then construct the independent uniformly distributed random variables by setting

$$(16) \qquad\qquad \xi_n = F(X_n | X_k, \ k \leq n - 1) \,.$$

It is clear that X_n is determined by $\xi_n, \ldots, \xi_{n-k+1}, X_{n-k-1}, \ldots$ if $k > 0$. It seemed plausible to Wiener that one should be able to let $k \to \infty$ and obtain a representation of X_n in terms of $\xi_k, \ k \leq n$. However, one can construct examples showing that it is not possible generally with this construction suggested by Wiener.

A related problem can be phrased as follows. Consider a stationary process $\{X_n\}$. Let $\{\eta_n\}$ be a sequence of independent uniformly distributed random variables. Under what conditions can one find a one-sided function f of the η sequence

$$f(\eta_0, \eta_{-1}, \eta_{-2}, \ldots) = f(\eta)$$

such that the process $\{Y_n\}$

$$Y_n = f(\tau^n \eta)$$

has the same probability structure as the process $\{X_n\}$? If $\{X_n\}$ is a countable state Markov chain, one can show that this is possible if and only if the chain is mixing [13]. D. Hanson [8] obtained the following interesting extension of this result. Let $\{Y_n\}$ be a real-valued strictly stationary Markov sequence with

(i) trivial tail field $\mathcal{B}_{-\infty}$
(ii) Borel sets $A, B \in \mathcal{A}$ on the real line and a nonnegative measure φ such that $P(B)$, $\varphi(A) > 0$ and for all $x \in B$ and $A' \subset A$ one has $P(x, A') \geq \varphi(A')$.

If then $\{\xi_n\}$ is a sequence of independent uniformly distributed random variables on $[0, 1]$, there is a one-sided function $f(\xi_0, \xi_{-1}, \ldots)$ such that $\{Y_n\}$ and $\{f(\tau^n \xi)\}$ have the same probability structure. Hanson's result is obtained by a coupling argument, a device that has become popular in recent years and that was used by Doeblin. One should add that if the conditional distribution function $F(x_n | x_{n-1})$ is continuous in x_n for almost every past x_{n-1} and Hanson's condition is satisfied, an invertible one-sided representation of the type that Wiener desired in terms of independent uniformly distributed random variables can be constructed for the Markov sequence.

REFERENCES

[1] R. C. Bradley, "On the spectral density and asymptotic normality of weakly dependent random fields." #332, March 1991, Center for Stochastic Processes, University of North Carolina, Chapel Hill.

[2] R. C. Bradley, "Equivalent mixing conditions for random fields." Technical Report #336, March 1991, Center for Stochastic Processes, University of North Carolina, Chapel Hill.

[3] R. Cogburn, "Conditional probability operators," Ann. Math. Statist., 33 (1962).

[4] M. Denker, "Uniform integrability and the central limit theorem for strongly mixing processes," in *Dependence in Probability and Statistics* (editors, E. Eberlein and M. Taqqu), 269–274, Birkhäuser 1986.

[5] W. Doeblin, "Sur les propriétés asymptotiques de mouvement régis par certains types de chaines simples," Bull. Math. Soc. Roum. Sci., 39, no. 1, 57–115, no. 2, 3–61 (1937).

[6] W. Doeblin, "Eléments d'une théorie générale des chaines simple constantes de Markoff," Ann. Sci., École Norm. Sup. (3), 57 (1940), 61–111.

[7] J. L. Doob, *Stochastic Processes*. John Wiley, 1953.

[8] D. L. Hanson, "On the representation theorem for stationary stochastic processes with trivial tail field," J. Math. Mech., 12 (1963), 293–301.

[9] T. E. Harris, "The existence of stationary measures for certain Markov processes," Third Berkeley Symposium on Mathematical Statistics and Probability, vol. II, 1956. Berkeley, 113–124.

[10] I. Ibragimov and Yu. Linnik, *Independent and Stationary Sequences of Random Variables*, Wolters-Noordhoff, 1971.

[11] S. Orey, "Recurrent Markov chains," Pacific J. Math., **9** (1959), 805–827.

[12] M. Rosenblatt, "Uniform ergodicity and strong mixing," Zeit. Warschein., **24** (1972), 79–84.

[13] M. Rosenblatt, "Stationary Markov chains and independent random variables," J. Math. Mech., **9** (1960), 945–950.

[14] N. Wiener, *Nonlinear Problems in Random Theory*, John Wiley, 1958.

Department of Mathematics, University of California, San Diego, La Jolla, California 92093-0012

Current address: Department of Mathematics, University of California, San Diego, La Jolla, California 92093-0012

E-mail address: mrosenblatt@ucsd.edu

Contemporary Mathematics
Volume **149**, 1993

Behaviour of Infinite Products
with Applications to
Non-Homogeneous Markov Chains

I. FLEISCHER AND A. JOFFE

ABSTRACT. We want to review, in the huge literature devoted to the study of non-homogeneous Markov Chains, the results relying on the general theory of perturbation of infinite products of Lipschitz operators acting on metric spaces.This approach unifies many known results, some of which have been and are being rediscovered many times. In the case of finite Markov Chains important results have been obtained which up to now have not been derived from the general theory: we will discuss several generalisations of these results to more general state spaces.

0. Introduction

The only limits of powers of real numbers are 1 or 0: a^n converges if and only if $|a| < 1$ (and the limit is 0) or $a = 1$ (and the limit is 1). The behaviour of these products under perturbation is quite different: The product $\prod_1^\infty (1 + e_n)$ is convergent if $\sum_1^\infty |e_n|$ is convergent ("small perturbation") while for $|a| < 1$ the product $\prod_1^\infty (a + e_n)$ is convergent if e_n converges to 0 ("large perturbation"). The reason for this different behaviour is that under multiplication 0 is absorbent $(x \cdot 0 = 0)$.

The proof of the first statement follows from a simple algebraic identity which does not use exponentials or logarithms; it extends to Banach algebras (not necessarily commutative) and even to more general situations such as Lipschitz operators acting on metric spaces; this was the point of view of Fleischer and

1991 *Mathematics Subject Classification.* Primary 40A10, 47N30, 60J10, 60J35; Secondary 47B38, 47H10, 54E40.

Key words and phrases. Metric spaces, infinite products, perturbations, contractions, non-homogeneous Markov Chains, weak and strong ergodicity, Birkhoff metric.

Support by NSERC Rosenberg is gratefully acknowledged by the first author, research of the second author was supported by an NSERC grant.

This paper is in final form and no version of it will be submitted for publication elsewhere.

Joffe [7]. Unfortunately, the treatment in that paper is highly abstract and it has been generally overlooked that the specialization e.g. to Banach algebras leads to new results; therefore we present a short exposition, restricted to Banach algebras just to avoid too many definitions, in Section 2.

While our proof of the first statement presents an interest even in the case of real numbers, that of the second is so trivial that it does not enlighten the general situation. For finite stochastic matrices it was proved by J. L. Mott [15]. We generalized this result in Fleischer and Joffe [8] and we will present the part pertinent to probabilists in Section 3. Section 1 contains well known definitions and a development of the theory in the setting of metric spaces; as we hope will be apparent to the reader, our point of view, based on contraction maps on metric spaces, unifies many results, simplifies proofs and is very natural. (We do not claim originality here: this point of view belongs to the folklore of the subject and was already in essence adopted in the well known work of R. L. Dobrushin [5]. For a thorough survey in the case of stochastic matrices the interested reader should consult [16].) Finally in Section 4 we review other results in the literature for which our methods seem well adapted.

As in many other branches of probability, Dœblin was a pioneer in the study of non-homogeneous Markov chains; in particular our Proposition 5 is an abstract version of his treatment in [6].

1. Metric Spaces

Lipschitz constant, for an operator T between (possibly different) metric spaces is

$$|T| := \sup \frac{(fT, gT)}{(f, g)} \qquad \text{for } f \neq g,$$

where $(\cdot, \cdot)$ denotes the distance.

If $T_n \to T$ pointwise, $\liminf |T_n| \geq |T|$ (since $(fT, gT) > (|T| - \epsilon)(f, g)$ yields $|T_n|(f, g) \geq (fT_n, gT_n) > (|T| - \epsilon)(f, g)$ for sufficiently large n), we thus obtain:

PROPOSITION 1. *The Lipschitz constant is lower semicontinuous under pointwise convergence.*

T_n is *asymptotically close* if

$$\lim_{n \to \infty} (fT_n, gT_n) = 0 \qquad \text{for all } f, g.$$

This follows, uniformly in (f, g) bounded, from $|T_n| \to 0$.

For T_n selfmaps of a complete metric space, $|T_n| < 1$ entails that T_n has a unique fixpoint to which its powers converge at every element (Banach)—call it f_n. The following proposition (continuity of the fixpoint) is already in [3]; for the sake of completeness we present it with a slightly different proof.

PROPOSITION 2. *Let T_n, with fixed point f_n, be a sequence of selfmaps converging pointwise to T_∞; then if $\sup |T_n| < 1$, the sequence f_n converges to the fixpoint of T_∞.*

PROOF. First observe that if $|T_n| \le K < 1$ then T_∞ has a Lipschitz constant bounded by K: let f_∞ be its fixed point; then (f_m, f_∞) is the limit of

$$(f_\infty T_m^n, f_\infty) \le \sum_{i=1}^n (f_\infty T_m^i, f_\infty T_m^{i-1}) \le \frac{1}{1 - |T_m|}(f_\infty T_m, f_\infty)$$

$$\le \frac{1}{1 - K}(f_\infty T_m, f_\infty T_\infty) \to 0.$$

REMARK 1. Uniform convergence of selfmaps on a set B can be metrized by

$$(T_1, T_2)_B = \sup_{f \in B}(fT_1, fT_2);$$

when T is a linear operator on a Banach space, its operator norm is its Lipschitz constant $|T|$ which induces this metric for B the unit ball.

Asymptotic closeness also follows from $fT_n \to f_\infty$ for all f. If the convergence is uniform on the f, the latter implies $f_n = f_n T_n \to f_\infty$ (continuity of the fixed point); more generally we have:

PROPOSITION 3. *Let T_n, with fixed point f_n, be a sequence of selfmaps asymptotically close uniformly on the f_n, with $fT_n \to f_\infty$ for some f, then $f_n = f_n T_n \to f_\infty$.*

The proof follows immediately fom the triangle inequality:

$$(f_n T_n, f_\infty) \le (f_n T_n, fT_n) + (fT_n, f_\infty).$$

Conversely:

PROPOSITION 4. *Let T_n, with fixed point f_n, be an asymptotically close sequence of selfmaps; then either of the following conditions*

(i) *$fT_n \to f_\infty$ for some f; or*
(ii) *$f_n \to f_\infty$ and the T_n are uniformly asymptotically close*
    implies $fT_n \to f_\infty$ for all f.

The proof follows again from an obvious use of the triangle inequality.

Consider forward products (goy products in the terminology of [7]) $T_k \cdots T_{k+n}$: the T_n are *weakly ergodic* if $\{T_k \cdot T_{k+1} \cdots T_{k+n}\}$ is asymptotically close for each k; *properly convergent* if the product converges pointwise as $n \to \infty$ for each k; and *strongly ergodic* if $fT_k \cdots T_{k+n} \to f_k$, independent of f (Cf. [16], p.136); this is equivalent to weak ergodicity plus convergence at some f. For forward products it would suffice to have convergence $fT_1 \cdots T_n$ for some f since the limit f_k is independent of k: $(fT_k)T_{k+1} \cdots \to f_{k+1}$; therefore in the case of forward products this is the same as: $T_k \cdots T_{k+n}$ convergent to a constant map*; but this need no longer hold for backward products. The following is obvious, but is so fundamental that we state it as our next proposition:

---

*In [7] we defined strong ergodicity as the latter stronger condition, which caused some formal incorrectness in the immediately following statements.

PROPOSITION 5. *A sufficient condition that weak ergodicity hold uniformly on bounded sets is that* $|T_n \cdot T_{n+1} \cdots T_{n+p}| \to 0$ *for every n; which, if* $|T_n| \leq 1$, *is the same as: there exists a subsequence* n_k *such that* $\sum(1 - |T_{n_k} \cdots T_{n_{k+1}-1}|) = \infty$; *this is implied by proper convergence to zero of the product of the* $|T_n|$. *This condition is necessary in normed vector spaces.*

REMARK 2. The above goes over mutatis mutandis for backward products (hebrew products in the terminology of [7]) or for any order (anarchic products) as in [13]; the latter will be denoted by $\prod^A$.

Finally we have

THEOREM. *For backward products in a complete metric space, uniform weak ergodicity implies uniform strong ergodicity.*

PROOF. Since the hypothesis implies that $(xT_nT_{n-1} \cdots T_1, yT_nT_{n-1} \cdots T_1) \to 0$ uniformly in x and y, it suffices to replace x by $yT_{n+p}T_{n+p-1} \cdots T_{n+1}$ to have by the uniformity that $yT_nT_{n-1} \cdots T_1$ is a Cauchy sequence.

In the special case of finite stochastic matrices this theorem is due to Chatterjee and Seneta (see theorem 4.17, p. 154 in [16]) and for infinite stochastic matrices Seneta [18]. Motivated by the work of [14] we give in the next proposition some sufficient conditions for strong ergodicity; this is the abstract version of theorems 2.1 and 2.2 of [14]. First we need the following lemma:

LEMMA. *Let* T_n *be a weakly ergodic sequence such that for some* f_∞

$$\lim_{n \to \infty} (f_\infty T_n \cdots T_{n+p}, f_\infty) = 0, \quad uniformly\ in\ p;$$

then strong ergodicity obtains with limit f_∞.

PROOF. Again by the triangle inequality we have:

$$(fT_1 \cdots T_{n+p}, f_\infty) \leq (fT_1 \cdots T_{n+p}, f_\infty T_n \cdots T_{n+p}) + (f_\infty T_n \cdots T_{n+p}, f_\infty);$$

the last term can be made arbitrarily small for all p by choosing n large enough and then the preceding term is small for large p by weak ergodicity.

PROPOSITION 6. *Let* T_n *be a weakly ergodic sequence of selfmaps with fixpoint* f_n *converging to* f_∞; *then either of the following conditions implies strong ergodicity with limit* f_∞.

  (i) $\limsup_n |T_n| = C < 1$;

  (ii) *For large enough n,* $|T_n \cdot T_{n+1} \cdots T_{n+p}| \leq 1$, *and*

$$\sum_{n=1}^{\infty} (f_n, f_\infty) < \infty.$$

PROOF. By the triangle inequality we have:

$$(f_nT_n \cdot T_{n+1} \cdots T_{n+p}, f_\infty)$$
$$\leq (f_nT_{n+1} \cdots T_{n+p}, f_{n+1}T_{n+1} \cdots T_{n+p}) + \cdots + (f_{n+p}, f_\infty).$$

This expression is majorized in the first case by $\epsilon/(1-C)$, where n is chosen in such a way that $\sup_p(f_{n+p}, f_n) < \epsilon$; while in the second case, it is majorized by $\sum_{j=0}^{p-1}(f_{n+j}, f_{n+j+1}) + (f_{n+p}, f_\infty)$; the lemma then yields the result.

Applications to non-homogeneous Markov chains. The above presentation is well adapted to the study of non-homogeneous Markov chains in general measurable state spaces $(F, \mathfrak{F})$, the object of interest being the set $\mathcal{P}$ of transition operators: $P(x, A)$, acting from the left on the space $\mathcal{L}^\infty$ of bounded functions and from the right on the space $\mathcal{M}$ of probability measures, (the "probas" in [8]). There are many ways to endow these spaces with different metrics so that one can apply our set up. Two choices are very important and a vast literature has been devoted to their study. With the $\mathcal{L}^\infty$ norm, the space of transition operators $(\mathcal{P}, var)$ is endowed with the metric of total variation, the Lipschitz constant τ_D is one minus Dobrushin's coefficient of ergodicity, and the corresponding notions of weak and strong ergodicity correspond to "subtractive ergodicity"—since in the case of finite matrices it is equivalent to the difference of rows of the products converging to zero. (There are many good expositions of this theory: let us mention [5], [11], [12], and [16]); while if we use Birkhoff's metric on the cone of non-negative measure we are led to the notion of "ratio ergodicity"—since in the case of matrices it corresponds to the ratio of elements in the same column of the products converging to one—here the Lipschitz constant is Birkhoff's coefficient of ergodicity τ_B. In the case of finite and even infinite matrices the inequality: $\tau_D \leq \tau_B$ yields that convergence of Birkhoff's coefficient entails both ratio ergodicity and subtractive ergodicity (cf. [18]); such a result holds in the general case $\mathcal{P}$, and will be presented in a forthcoming paper. The general Birkhoff theory [2] is not easily accessible, but in the special case of a finite state space we can refer to Seneta [16] and [17].

2. Small perturbations

Let $(\mathcal{A}, \| \ \|)$ be a Banach algebra, $a_n, b_n \in \mathcal{A}$.

LEMMA. *If* $\Sigma_1^\infty \| a_n - b_n \| < \infty$ *and* $\sup_j \sup_{j<p} \| \prod_{k=j}^p a_{n+k} \| < \infty$ *then* $\sup_p \| \prod_{j=0}^p b_{n+j} \| < \infty$.

PROOF. The following elementary identity:

$$\prod_{j=0}^p a_{n+j} - \prod_{j=0}^p b_{n+j} = \sum_{j=0}^p \prod_{k=0}^{j-1} b_{n+k} \, (a_{n+j} - b_{n+j}) \prod_{k=j+1}^p a_{n+k}$$

yields:

$$\| \prod_{j=0}^p b_{n+j} \| \leq C + C_p \cdot \epsilon \cdot C,$$

where C_p is $\sup_{j\leq p} \| \prod_{k=0}^j b_{n+k} \|$, C is a bound for the norms of the products of the a's and n is chosen such that $\sum_{j=o}^{\infty} \|a_{n+j} - b_{n+j}\| < \epsilon$, $\epsilon < \frac{1}{C}$; thus:

$$C_p \leq \frac{C}{1 - \epsilon\, C}.$$

From the lemma follows:

THEOREM*. *Under the assumptions of the lemma we have*

$$\lim_{n \to \infty} \sup_p \| a_n \cdot a_{n+1} \cdots a_{n+p} - b_n \cdot b_{n+1} \cdots b_{n+p} \| = 0.$$

In particular, convergence, weak and strong ergodicity are preserved under small perturbations.

An important case of this set up is the Banach algebra of linear operators on a Banach space; as mentioned in the introduction we refer to [7] for a much more general approach, but it is important to notice that the above theorem has nothing to do with linear operators or finite dimensionality of the underlying space.

This theorem can be used in the context of non-homogeneous Markov chains in general measurable state spaces $(F, \mathfrak{F})$ with the Banach algebra of transition operators.

3. Large perturbations: Variations on Mott's Theorem

The original Mott theorem, [15], for finite matrices has been extended to infinite matrices in [11] and to an abstact setting in [8]; here we state a special case of the latter which suffices for the probabilistic applications. The setup is a metric space $E, (\ , \ )$ with a familly Θ of selfmaps whose Lipschitz constants are bounded by 1. We assume moreover that for any $T_1, T_2 \in \Theta$, $\sup_x (xT_1, xT_2) < \infty$ and we still denote by $(\ , \ )$ this sup; the space Θ becomes a metric space (with the topology of uniform convergence). The following obvious fact is useful: $(xT_1, yT_2) \leq (x, y) + (T_1, T_2)$.

LEMMA.

$$(T_1 \cdot T_2 \cdots T_n,\ T_1' \cdot T_2' \cdots T_n') \leq \sum_{k=1}^{n} (T_k\ ,\ T_k').$$

PROOF. From the triangle inequality we have:

$$(T_1 \cdot T_2 \cdots T_n\ ,\ T_1' \cdot T_2' \cdots T_n') \leq (T_1 \cdot T_2 \cdots T_n\ ,\ T_1' \cdot T_2' \cdots T_{n-1}' \cdot T_n)$$
$$+ (T_1' \cdot T_2' \cdots T_{n-1}' T_n\ ,\ T_1' \cdot T_2' \cdots T_{n-1}' T_n').$$

The first term is dominated by $(T_1 \cdot T_2 \cdots T_{n-1}, T_1' \cdot T_2' \cdots T_{n-1}')$ since $|T_n| \leq 1$; the last term is dominated by (T_n, T_n'), so the lemma follows by iteration.

---

*We adapt the proof of the lemma of [7] to the particular case of Banach algebras. It is surprising that special cases of these results still appear in the literature; Theorem 2.2 of [19] is the special case in which all the a's are identical to e, the unit of the Banach algebra (a result which is already in [4]) and Theorem 2.1 of [1] is the special case for the algebra of finite matrices in which all the a's have norm 1!

THEOREM. *Let $T_1, T_2, \ldots, T_n, \ldots$ be a sequence of selfmaps, strongly ergodic uniformly in n and on bounded sets, with limit f, and let $T_1', T_2', \ldots, T_n', \ldots$ be a perturbed sequence, $\lim(T_n, T_n') = 0$; then the perturbed sequence is also uniformly strongly ergodic, with the same limit f.*

PROOF. From the lemma we have that:

$$\lim_n \left(y T_n \cdot T_{n+1} \cdots T_{n+p}, y T_n' \cdot T_{n+1}' \cdots T_{n+p}'\right) = 0$$

for every p, uniformly for y bounded. Since:

$$(f, y T_n' \cdots T_{n+p}') \leq (f, y T_n \cdots T_{n+p}) + (y T_n \cdots T_{n+p}, y T_n' \cdots T_{n+p}')$$

it follows that for any ϵ and sufficiently large j there is a k_0 such that

$$(y T_{k+1}' \cdots T_{k+j}', f) < \epsilon,$$

uniformly on bounded sets, for all $k > k_0$; putting $y = x T_n' \cdot T_{n+1}' \cdots T_k'$ yields the result.

In the special case of Markov transition operators acting on probability measures on $(F, \mathfrak{F})$ the above theorem is already new even in the case of finite state space; Mott's theorem, [15], deals with the case where all the T_n are identical to T, a regular matrix (Theorem 4.14, p. 150 in [16], which has been extended to infinite stochastic matrices, Theorem V.4.5. p 170 in [11], whose exercise 20 p. 182 has inspired our proof of the above theorem); observe that for a regular matrix, the Lipschitz constant τ_D is less than one.

Of course when one perturbs a strongly ergodic sequence, this theorem subsumes the theorem of Section 2; however that theorem is still useful for probabilists since it allows one to preserve, under perturbation, all the asymptotic properties of the tails of the original sequence; in particular convergence to a not necessarily constant transition operator will be preserved.

If one uses the metric induced by the Birkhoff metric on the cone of probability measures [2] we have a new formulation of Mott's theorem and its generalisation which might be useful. The use of this metric has been exploited in [10], [9], and [18].

4. Weak and strong ergodicity

Recently Leizarowitz [13] considered the matrix case of the following situation: Let $\mathcal{S}$ be a semigroup of bounded linear operators on a closed subset of a Banach space, whose Lipschitz constants are bounded, $C := \sup\{|T| : T \in S\}$, and assume that $\mathcal{S}$ contains the identity, so that $C \geq 1$. For a given sequence $T_n \in \mathcal{S}$ we denote by $\mathcal{L}$ the set of limit points (i.e. limits of subsequences) of the sequence under norm convergence and we assume in all this section that under this topology $\mathcal{S}$ is compact.

THEOREM 1. (i) *Suppose that there exists at least one $T \in \mathcal{L}$ which satisfies*

$$|T| < \frac{1}{C}; \tag{1}$$

then weak ergodicity obtains for infinite products of the $\{T_n\}_{n=0}^{\infty}$ taken in arbitrary order of multiplication, and strong ergodicity obtains for the backward products.

(ii) *If in addition we require*

$$|T| < 1 \qquad \text{for every } T \in \mathcal{L}, \tag{2}$$

then the convergence has geometric rate.

PROOF.

(i) Let T_{n_k} be a subsequence which converges to T; two non-adjacent consecutive elements of that subsequence are separated by a block whose product has Lipschitz constant at most C. For a small enough ϵ and large n_k: $|T_{n_k}| < \frac{1}{C} - \epsilon$; it follows that the infinite product can be decomposed into products of blocks whose Lipschitz constant is bounded by $C(\frac{1}{C} - \epsilon) = 1 - C\epsilon$, thus the Lipschitz constants of the partial products go to zero; this implies weak ergodicity by Section 1; strong ergodicity for the backward product follows from the theorem of Section 1.

(ii) It suffices to notice by the continuity of the Lipschitz constant that $\limsup\{ |T| : T \in \mathcal{L}\} < 1$; therefore for all large n, $|T_n| < 1 - \epsilon$.

In the case of finite matrices this is theorem A, p. 190 of [13] whose proof is on pp. 192-94.

THEOREM 2. *Assume that*

$$|T| < 1 \qquad \text{for every } T \in \mathcal{L},$$

and suppose that all members of $\mathcal{L}$ have the same fixed point f_{∞}; then provided there are an infinite number of terms on the right, the product taken in arbitrary order of multiplication is strongly ergodic with

$$\lim_{n \to \infty} \prod^A T_n = R, \qquad \text{independent of the order,}$$

and

$$fR = f_{\infty}.$$

PROOF. As in the proof of Theorem 1 we have that $\limsup_n |T_n| < C$, for some $C < 1$. For the "goy" products the theorem is now a direct consequence of part (i) of Proposition 6 of Section 1, since by the continuity of the fixpoint we have that $\lim f_n = f_{\infty}$. Finally it suffices to observe that the proof of Proposition 6 still stands regardless of the order of multiplication, provided that there are an infinite number of terms on the right.

For the case of finite matrices this is essentially theorem 3.2 of [13] which is stated without any restriction on the order of multiplication: that statement is not true in its full generality as is seen by taking the "hebrew" product $\cdots R \cdot R \cdots RP$, where R is a constant stochastic matrix and P a stochastic matrix with a different fixpoint. Theorem 2, as noted in [13] yields another proof of Mott's theorem; as shown by the above example, this restriction on the order of multiplication has to be assumed also in Theorem 3.4 of [13].

Acknowledgment

We thank Richard Duncan for some helpful comments and Ghislain Léveillé for having discovered a few errors in the previous version of this article.

REFERENCES

1. A. Artzrouni, *On the convergence of infinite products of matrices.*, Linear Algebra Appl. **74** (1986), 11–21.
2. G. Birkhoff, *Lattice theory*, Amer. Math. Soc. Colloq. Publ., XXV, 1967.
3. F. F. Bonsall, *Lectures on some fixed point theorems of functional analysis*, Tata Institute, 1962.
4. N. Bourbaki, *Topologie générale, Ch IX. Utilisation des nombres réels en topologie générale*, Hermann, Paris, 1948.
5. R. L. Dobrushin, *Central limit theorem for nonstationnary Markov Chains 1, 2*, Theory Probab. Appl. **1** (1956), 65–80, 329–384.
6. W. Dœblin, *Le cas discontinu des probabilités en chaîne.*, Publ. Fac. Sci. Univ. Masaryk (Brno) **236** (1937), 2–18.
7. I. Fleischer and A. Joffe, *Perturbation des produits infinis et applications*, Journal d' Analyse Math. **31** (1977), 69–75.
8. I. Fleischer and A. Joffe, *Preservation of ergodicity under perturbation*, Submitted for publication.
9. S. Gibert and A. Mukherjea, *Products of infinite-dimensional non-negative matrices; extension of Hajnal's results*, Math Z. **196** (1987), 485–490.
10. J. Hajnal, *On products of non-negative matrices*, Math. Proc. Camb. Phil. Soc. **79** (1976), 521–530.
11. D. L. Isaacson and R. W. Madsen, *Markov chains theory and applications*, Wiley, 1976.
12. M. Iosifescu and R. Theodorescu, *Random processes and learning*, Springer-Verlag, 1969.
13. A. Leizarowitz, *On infinite products of stochastic matrices*, Linear Algebra Appl. **168** (1992), 189–219.
14. R. W. Madsen and D. L. Isaacson, *Strongly ergodic behavior for non-stationary Markov processes*, The Annals of Probability **1** (1973), 329–335.
15. J. L. Mott, *Conditions for the ergodicity of non-homogeneous finite Markov chains*, Proc. Roy. Soc. Edinburgh, Section A **64** (1957), 369–380.
16. E. Seneta, *Non-negative matrices and Markov chains*, Springer Verlag, Second ed., 1981.
17. E. Seneta, *Coefficients of ergodicity: structure and applications*, Adv. Appl. Prob. **11** (1979), 576–590.
18. E Seneta, *Ergodicity for products of infinite stochastic matrices*, This volume.
19. S.T. Welstead, *Infinite products in a Banach algebra*, J. Math. Anal. Appl. **105** (1985), 523– 532.

DEPARTMENT OF MATHEMATICS AND STATISTICS, UNIVERSITY OF WINDSOR, WINDSOR (ONTARIO) N9B 3P4, CANADA

DÉPARTEMENT DE MATHÉMATIQUES ET STATISTIQUE, UNIVERSITÉ DE MONTRÉAL.C.P. 6128, SUCC. A. MONTRÉAL (QUÉBEC) H3C 3J7, CANADA

Contemporary Mathematics
Volume **149**, 1993

APPLICATIONS OF ERGODICITY COEFFICIENTS TO HOMOGENEOUS MARKOV CHAINS

E. SENETA

ABSTRACT. Coefficients of ergodicity have been used primarily as tools for investigating weak convergence in finite inhomogeneous Markov chains. The first application of this kind is probably due to Doeblin (1937), with significant contributions in the 1950's by Dobrushin, Hajnal and Sarymsakov. For a finite square matrix with constant row sums, a (e.g. a stochastic matrix), the Markov–Dobrushin coefficient of ergodocity bounds eigenvalues other than a . On account of this key property, this coefficient now finds application also to homogeneous Markov chains; the purpose of this paper is to illustrate how a number of results may be synthesized using such a coefficient.

1. Basic Concepts

For a real matrix $A = \{a_{ij}\}$, $i,j = 1,...,n$, satisfying

$A\underset{\sim}{1} = a\underset{\sim}{1}$ for some real a , we shall be concerned with the quantity

$$(1.1) \qquad \tau_1(A) = \tfrac{1}{2} \max_{i,j} \sum_{s=1}^{n} |a_{is} - a_{js}| \equiv a - \min_{i,j} \sum_{s=1}^{n} \min(a_{is}, a_{js}) .$$

Historically, this has been applied to the case where A is a stochastic matrix P (that is $P \geq 0$, $P\underset{\sim}{1} = \underset{\sim}{1}$) in probabilistic contexts.

In such contexts, $\tau_1(P)$ is a "coefficient of ergodicity".

1991 AMS Classifications. Primary 60J05, 15A51.
This paper is in final form and no version of it will be submitted for publication elsewhere.

An exposition on such coefficients in the probabilistic setting may be found now in several books, amongst them Iosifescu (1980, Chapter 7), and Seneta (1981, Chapter 4). This section is a sketch of properties and history relevant to this paper and to the occasion which this conference commemorates.

The ergodicity coefficient τ_1 was used, probably for the first time in a probabilistic context by A.A. Markov (1906) to express the contractive property of a stochastic matrix P : If $\underset{\sim}{z} = P \underset{\sim}{w}$ then

$$(1.2) \qquad \max_{h,h'} |z_h - z_{h'}| \leq \tau_1(P) \{\max_{j,j'} |w_j - w_{j'}|\}$$

where $\underset{\sim}{w} = \{w_i\}$ is an arbitrary real vector, and $\underset{\sim}{z} = \{z_i\}$. This result is clearly of most interest when $\tau_1(P) < 1$, and this holds if and only if any two rows of P "intersect" (both have strictly positive elements in at least one common position). In particular this is true if $P > 0$; and Markov's intention was to use (1.1) to prove, for a finite homogeneous Markov chain with such a strictly positive transition matrix the ergodic theorem: as $k \to \infty$,

$$P^k \to \underset{\sim}{1} \underset{\sim}{\pi}' \text{ for some } \underset{\sim}{\pi}' \text{ satisfying } \underset{\sim}{\pi} > \underset{\sim}{0} , \underset{\sim}{\pi}'\underset{\sim}{1} = 1 .$$

The main usage of coefficients of ergodicity such as τ_1 has been in the study of weak ergodicity of inhomogeneous products:

$$(1.3) \qquad P_{m,r} = P_{m+1} P_{m+2} \cdots P_{m+r} , \quad m \geq 1$$

formed from a sequence $\{P_i\}$ of $(n \times n)$ stochastic matrices, as $r \to \infty$. In this setting, writing $P_{m,r} = \{p_{is}^{(m,r)}\}$, the definition of weak ergodicity of Kolmogorov (1931) reduces to

$$p_{is}^{(m,r)} - p_{js}^{(m,r)} \overset{r \to \infty}{\to} 0 , \text{ for all } i , j , s , m .$$

This work flourished in the 1950's in the writings of Dobrushin, Hajnal and Sarymsakov. It was discovered later (Seneta, 1973) that Doeblin (1937), in his very first paper, had already established a necessary and sufficient condition for the sequence $\{P_i\}$, $i \geq 1$, to be weakly ergodic using the ergodicity coefficient

$$\delta(A) = a - \sum_{s=1}^{n} (\min_i a_{is})$$

(for any A with $A \underset{\sim}{1} = a \underset{\sim}{1}$) .

The 1950's work focussed on the case when, for stochastic P, $\tau_1(P) < 1$ (i.e. when P is "scrambling"), and on certain inequalities satisfied by τ_1. It turns out that for any $(n \times n)$ real A with equal row sums a,

(1.4)
$$\tau_1(A) = \sup_{\substack{\|\underset{\sim}{\delta}'\|_1 = 1 \\ \underset{\sim}{\delta}'\underset{\sim}{1} = 0}} \|\underset{\sim}{\delta}'A\|_1 .$$

Hence if A_1 and A_2 have row sums respectively a_1 and a_2 then the submultiplicative property $\tau_1(A_1 A_2) \leq \tau_1(A_1)\tau_1(A_2)$ holds. If P, P_1, P_2 are stochastic, then, in totality, using also (1.1).

(i) $0 \leq \tau_1(P) \leq 1$; (ii) $\tau_1(P_1 P_2) \leq \tau_1(P_1)\tau_1(P_2)$ [Dobrushin, 1956];

(iii) $\tau_1(P) = 0$ iff $P = \underset{\sim}{1}\,\underset{\sim}{v}'$, where $\underset{\sim}{v}' \geq \underset{\sim}{0}'$, $\underset{\sim}{v}'\underset{\sim}{1} = 1$.

It is reasonably obvious now how the scalar functional τ_1 satisfying these properties can be used as a scalar measure of the tendency to coincidence of rows of $P_{m,r}$, defined by (1.3), as $r \to \infty$; and this is essentially what weak ergodicity amounts to.

Of more interest in the setting of homogeneous Markov chains (i.e. powers of a <u>single</u> stochastic matrix P), is the property that for any real matrix A with equal row sums a , if λ is any eigenvalue of A other than a , (i.e. $\lambda \neq a$) then

(1.5)
$$|\lambda| \leq \tau_1(A) .$$

This result is due to E . Deutsch in 1969 (see Seneta, 1981, p.64. It is curious that the weaker result $|\lambda| \leq \delta(A)$ was still being rediscovered in 1973). Notice in particular that if $A = P$ is stochastic and scrambling, $\tau_1(P)$ provides a non–trivial bound on the largest non–unit eigenvalue, and we may thus expect it to be a useful measure of quantities such as convergence rate and numerical stability, in the setting of homogeneous finite Markov chains.

2. Finite Regular Markov Chains

Finite regular Markov chains consist of a single closed aperiodic set of states, as well as, possibly some inessential states. Regularity is necessary and sufficient for the ergodic property:

(2.1) $P^k \to \underset{\sim}{1}\ \underset{\sim}{\pi}'\ ,\quad \underset{\sim}{\pi}' \geq \underset{\sim}{0}'\ ,\quad \underset{\sim}{\pi}'\underset{\sim}{1} = 1$

as $k \to \infty$. $\underset{\sim}{\pi}'$ is the unique solution of $\pi'(I - P) = \underset{\sim}{0}'$, $\underset{\sim}{\pi}'\underset{\sim}{1} = 1$. It is

clear from the limit property (2.1) that for such a chain P^k has a strictly

positive column, so for $k \geq \gamma$, P^k is scrambling for some γ . In fact

(Paz, 1971) we may always take $\gamma = n(n - 1)/2$, which depends only on the

size n of the regular P . For $k \geq \gamma$ for such a matrix, the contractive

property comes into force for P^k . In particular $\tau_1(P^\gamma) < 1$. If P is

itself scrambling, we may take $\gamma = 1$. The following results illustrate the

applications of the ergodicity coefficient foreshadowed at the end of Section 1.

THEOREM 2.1 . For regular P , and $k \geq \gamma$,

$$\left\| P^k - \underset{\sim}{1}\ \underset{\sim}{\pi}' \right\|_1 \leq C(\tau_1^{1/\gamma}(P^\gamma))^k$$

where C is a constant independent of k .

PROOF: $\left\| P^k - \underset{\sim}{1}\ \underset{\sim}{\pi}' \right\|_1 = \left\| (I - \underset{\sim}{1}\ \underset{\sim}{\pi}')P^k \right\|_1 = \underset{\|\underset{\sim}{x}'\|=1}{\sup}\ \left\| \underset{\sim}{x}'(I - \underset{\sim}{1}\ \underset{\sim}{\pi}')P^k \right\|_1$

$$\leq \left\| \underset{\sim}{x}'(I - \underset{\sim}{1}\ \underset{\sim}{\pi}') \right\|_1 \tau_1(P^k)$$

from (1.4) ;

$$\leq 2\ \tau_1(P^k)$$

since $\left\| \underset{\sim}{x}'(I - \underset{\sim}{1}\ \underset{\sim}{\pi}') \right\|_1 \leq \left\| \underset{\sim}{x}' \right\|_1 \left\| I - \underset{\sim}{1}\ \underset{\sim}{\pi}' \right\|_1 \leq 2\left\| \underset{\sim}{x}' \right\|_1 = 2$.

Now since $k \geq \gamma$, write $k = m\gamma + r$, $0 \leq r < \gamma$. Then by the
submutiplicative property,

$$\tau_1(P^k) \leq \tau_1^m(P^\gamma)\ \tau_1(P^r) \leq (\tau_1(P^\gamma))^{[k/\gamma]}$$

since $\tau_1(P^r) \leq 1$; $\leq (\tau_1(P^\gamma))^{k/\gamma - 1}$, from which the result follows with

$C = 2\ \tau_1^{-1}(P^\gamma)$, providing $\tau_1(P^\gamma) > 0$. $\square$

This result, more or less known (e.g. Tan, 1983), embodies the spectrum localization property (1.5) in that for any eigenvalue $\lambda \neq 1$ of regular P, $|\lambda| \leq \tau^{1/\gamma}(P^\gamma)$, and illustrates how the easily calculable quantity $\tau_1(P)$ (when $\gamma = 1$) can be used to measure convergence rate.

THEOREM 2.2 . For a regular $(n \times n)$ P with stationary distribution vector $\underset{\sim}{\pi}'$,

$$\|\underset{\sim}{\pi}' - \underset{\sim}{x}'\|_1 \leq \gamma(1 - \tau_1(P^\gamma))^{-1} \|\underset{\sim}{x}'(P - I)\|_1$$

where $\underset{\sim}{x}'$ is an arbitrary $(1 \times n)$ probability vector.

PROOF:

$$\|\underset{\sim}{\pi}' - \underset{\sim}{x}'\|_1 = \|(\underset{\sim}{\pi}' - \underset{\sim}{x}')P^\gamma + \underset{\sim}{x}'(P^\gamma - I)\|_1$$

$$\leq \|(\underset{\sim}{\pi}' - \underset{\sim}{x}')P^\gamma\|_1 + \|\underset{\sim}{x}'(P^\gamma - I)\|_1$$

$$\leq \tau_1(P^\gamma)\|\underset{\sim}{\pi}' - \underset{\sim}{x}'\|_1 + \|\underset{\sim}{x}'(P^\gamma - I)\|_1 ,$$

so that

$$\|\underset{\sim}{\pi}' - \underset{\sim}{x}'\|_1 (1 - \tau_1(P^\gamma)) \leq \|\underset{\sim}{x}'(P - I)(P^{\gamma-1} + P^{\gamma-2} + ... + P^0)\|_1$$

$$\leq \gamma \|\underset{\sim}{x}'(P - I)\|_1 . \qquad \square$$

Note that the bound vanishes at $\underset{\sim}{x}' = \underset{\sim}{\pi}'$; and that if $\underset{\sim}{x}'$ is replaced by $\underset{\sim}{x}'P$ throughout, the bound is non-increasing, since

$$\|\underset{\sim}{x}'P(P - I)\|_1 = \|\underset{\sim}{x}'(P - I)P\|_1 \leq \tau_1(P)\|\underset{\sim}{x}'(P - I)\|_1$$

and $\tau_1(P) \leq 1$. A preliminary version of the above results occur in Schweitzer (1986), who encouraged the present author to simplify them in terms of ergodicity coefficients.

COROLLARY. For P as in Theorem 2.2 and $\bar{P}$ any $(n \times n)$ stochastic matrix and $\bar{\underset{\sim}{\pi}}'$ any stationary distribution corresponding to it

$$\left\| \underset{\sim}{\pi}' - \underset{\sim}{\bar{\pi}}' \right\|_1 \le \gamma (1 - \tau_1(P^\gamma))^{-1} \left\| \bar{P} - P \right\|_1 .$$

<u>PROOF:</u>   Follows since

$$\left\| \underset{\sim}{\bar{\pi}}'(P - I) \right\|_1 = \left\| \underset{\sim}{\bar{\pi}}'(P - \bar{P}) \right\|_1 \le \left\| P - \bar{P} \right\|_1 \left\| \underset{\sim}{\bar{\pi}}' \right\|_1 .$$ □

The Corollary, proved in the case of scrambling P (so $\gamma = 1$) in Seneta (1988) shows that inasmuch as

$$\frac{\left\| \underset{\sim}{\pi}' - \underset{\sim}{\bar{\pi}}' \right\|_1 / \left\| \underset{\sim}{\pi}' \right\|_1}{\left\| \bar{P} - P \right\|_1 / \left\| P \right\|_1} \le \gamma (1 - \tau_1(P^\gamma))^{-1}$$

the right−hand side may be used as a "condition number" measuring relative

stability of the stationary distribution π' when P is "perturbed" to $\bar{P}$.

On this use of ergodicity coefficients see Seneta (1991).

3. Tracking in Markov Chains

The following result occurs as Corollary 1 in Hartfiel (1992), in this degree of generality attributed to the present author. In keeping with our intention of showing the usefulness of ergodicity coefficients as a tool, we give a commensurate proof.

We would need to note the result (e.g. Franklin (1968), p.170) that there are positive numbers α and β such that $\alpha \left\| X \right\|_2 \le \left\| X \right\| \le \beta \left\| X \right\|_2$ for all $(n \times n)$ matrices where $\left\| \cdot \right\|$ is a specific norm, specifically the ℓ_1 or ℓ_∞ norm, to bring Hartfiel's statement, expressed in terms of the ℓ_2 norm, into complete correspondence with ours.

<u>THEOREM 3.1</u> . If P is a finite regular stochastic matrix, and $\varepsilon > 0$ is

arbitrary, there exists a regular stochastic matrix $\bar{P}$ with distinct eigenvalues such that

$$\left\| P^k - \bar{P}^k \right\|_1 < \varepsilon \qquad \underline{\text{for all}} \;\; k \ge 1 .$$

<u>PROOF</u>:   Take $0 < \varepsilon_1 \leq \min(1,\varepsilon)$ and sufficiently small that any stochastic

$\bar{P}$ satisfying $\|P - \bar{P}\|_1 < \varepsilon_1$ has positive elements in at least the same

positions as P , so $\bar{P}$ is also regular.   Choose $k_0 \geq 1$ so that

(3.1) $$2(\tfrac{1}{3} + \tfrac{2}{3}\,\tau_1(P^\gamma))^{k_0} < \varepsilon_1/3 \ .$$

Hartfiel (1992, Lemma 3) shows that for any $(n \times n)$ stochastic A it is possible to find an $(n \times n)$ stochastic B with distinct eigenvalues such that $\|A - B\|_1 < \delta$ for arbitrary δ , $\delta > 0$. Then for any finite fixed positive integer b , we have for $k = 1,...,b$

$$\|A^k - B^k\|_1 = \|(B + (A-B))^k - B^k\|_1 \leq \sum_{s=1}^{k} \binom{k}{s} \|A - B\|_1^s$$

using the properties of norms and the fact that the ℓ_1 norm of a stochastic matrix is unity;

$$\leq (1+\delta)^k - 1$$

$$\leq k(1+\delta)^{k-1}\,\delta$$

$$\leq b(1+\delta)^{b-1}\,\delta \ .$$

In other words, for any stochastic A it is possible to find a stochastic B with distinct eigenvalues such that $\|A^k - B^k\| < h$ for arbitrary $h > 0$ and finite fixed b . Take $A = P$, $b = \gamma k_0$, $h = (\varepsilon_1/3)\,(1 - \tau_1(P^\gamma))/\gamma$, and

write $\bar{P}$ for B . Denote by $\bar{\pi}'$ the unique stationary distribution of $\bar{P}$ (which, by choice of ε_1 , is regular). Then for any $k > k_0\gamma$ it follows from Theorem 2.1 and (3.1) that

(3.2) $\|P^k - \underline{1}\,\underline{\pi}'\|_1 \leq 2\,\tau_1^{k_0}(P^\gamma) \leq 2(1/3 + (2/3)\tau_1(P^\gamma))^{k_0} < \varepsilon_1/3$.

Next

$$\left\| \underset{\sim}{1} \, \bar{\bar{\pi}}{}' - \underset{\sim}{1} \, \underset{\sim}{\pi}{}' \right\|_1 \leq \left\| \bar{\bar{\pi}}{}' - \underset{\sim}{\pi}{}' \right\|_1 \leq \gamma (1 - \tau_1(P^\gamma))^{-1} \left\| \bar{P}{}^\gamma - P^\gamma \right\|_1 \, ,$$

from the Corollary to Theorem 2.2 ;

(3.3) $< \varepsilon_1/3$

from the choice of b and h above. Then, from Theorem 2.1 again, for
$k > k_0 \gamma$

$$\left\| \bar{P}{}^k - \underset{\sim}{1} \, \bar{\bar{\pi}}{}' \right\|_1 \leq 2 \, \tau_1^{k_0}(\bar{P}{}^\gamma) \leq 2 (\tau_1(P^\gamma) + \left\| \bar{P}{}^\gamma - P^\gamma \right\|_1)^{k_0}$$

since $\bar{P}{}^\gamma = P^\gamma + (\bar{P}{}^\gamma - P^\gamma)$, using the triangle inequality and (1.4) ;

$$\leq 2 (\tau_1(P^\gamma) + (\varepsilon_1/3) \, (1 - \tau_1(P^\gamma)))^{k_0}$$

by the choice of b as before; and since $\varepsilon_1 \leq 1$,

(3.4) $\leq 2 (1/3 + (2/3)\tau_1(P^\gamma))^{k_0} < \varepsilon_1/3$

by (3.1) again.

Thus for $k > k_0 \gamma$ the result follows from (3.2) $-$ (3.4) since

$$\left\| P^k - \bar{P}{}^k \right\|_1 \leq \left\| P^k - \underset{\sim}{1} \, \underset{\sim}{\pi}{}' \right\|_1 + \left\| \underset{\sim}{1} \, \underset{\sim}{\pi}{}' - \underset{\sim}{1} \, \bar{\bar{\pi}}{}' \right\|_1 + \left\| \bar{P}{}^k - \underset{\sim}{1} \, \bar{\bar{\pi}}{}' \right\|_1 < \varepsilon_1 \, , \text{ while}$$

for $k \leq k_0 \gamma$, it follows from the choice of $\bar{P}$, h and b above.   □

POSTSCRIPT

It was an honour to be invited to participate in this commemorative conference. My thanks go to all those responsible for the organization, and especially to Heinrich Hering.

Several tributes were paid during the conference to the pioneering work on inhomogeneous Markov chains and ergodicity coefficients of John Hajnal whose participation enhanced the conference. My own work on these topics, beginning in Seneta (1973), was heavily influenced (as that publication shows) also by contributions, occurring at the same time as Hajnal's, of the Uzbek mathematician Tashmukhamed Alievich Sarymsakov (born 10 Sept. 1915), an influence which has continued to the present (see Hartfiel and Seneta (1990)), and I use this historical opportunity to acknowledge this. I learned, in fact as a Masters student at the University of Adelaide in 1964 of Doeblin's contributions in the area from Sarymsakov (1954) which lists three of the earliest, including Doeblin (1937), which motivated me to seek them out. The library of the University of Adelaide had, remarkably, a copy of this rare Russian–language book (4,000 copies printed) of Sarymsakov, who generously sent me, subsequently, his own used personal copy. The book itself deals with inhomogeneous Markov chains only to a limited extent (but does consider the analysis of infinite stochastic matrices by taking finite northwest corner truncations of increasing size, a problem to which I have subsequently devoted a large part of my research).

Sarymsakov (1954) did, however, introduce me to Kolmogorov's (1931) theorem on weak ergodicity of infinite inhomogeneous chains. The generalization of this, using the Birkhoff coefficient, I presented at the conference as a tribute to the great developer of Markov chain theory − along with Doeblin − Kolmogorov on the 60th anniversary of that paper. Space does not permit me to include in this proceedings that generalization, which will appear elsewhere.

REFERENCES

1. Bauer, F.L., Deutsch, E., and Stoer, J. Abschätzungen für die Eigenwerte positiver linearer Operatoren. Linear Algebra and its Applications 2 (1969), 275–301.

2. Dobrushin, R.L. Central limit theorem for non–stationary Markov chains I, II. Theory Prob. Appl. 1 (1956), 65–80, 329–383. [English translation].

3. Doeblin, W. Le cas discontinu des probabilités en chaîne. Publ. Fac. Sci. Univ. Masaryk (Brno), no. 236 (1937).

4. Franklin, J.N. *Matrix Theory.* Prentice–Hall, Englewood Cliffs N.J. (1968).

5. Hartfiel, D.J. Tracking in matrix systems. Linear Algebra and its Applications {to appear} (1992).

6. Hartfiel, D.J. and Seneta, E. A note on semigroups of regular stochastic matrices. Linear Algebra and its Applications 141 (1990), 47–51.

7. Iosifescu, M. *Finite Markov Processes and Their Applications.* Wiley, Chichester; Editura Tehnica, Bucharest (1980).

8. Isaacson, D.L. and Madsen, R.W. *Markov Chain Theory and Applications.* Wiley, New York (1976).

9. Kolmogorov, A.N. Über die analytischen Methoden in der Wahrscheinlichkeitsrechnung. Math. Ann. 104 (1931), 415–458.

10. Markov, A.A. Extension of the law of large numbers to dependent quantities [in Russian]. Izv. Fiz.–Matem. Obsch. Kazan Univ., (2nd ser.) 15 (1906), 135–156 [Also in Markov (1951), pp. 339–361.]

11. Markov, A.A. *Izbrannie Trudy.* A.N.S.S.S.R., Leningrad (1951).

12. Paz, A. *Introduction to Probabilistic Automata* Academic Press, New York (1971).

13. Sarymsakov, T.A. *Osnovi Teorii Protsessov Markova* GITTL, Moscow (1954).

14. Schweitzer, P.J. Posterior bounds on the equilibrium distribution of a finite Markov chain. Communications in Statistics – Stochastic Models 2 (1986), 323–338.

15. Seneta, E. On the historical development of the theory of finite inhomogeneous Markov chains. Proc. Cambridge Phil. Soc., **74** (1973), 507–513.

16. Seneta, E. Coefficients of ergodicity: structure and applications. Adv. Appl. Prob. **11** (1979), 576–590.

17. Seneta, E. *Non–Negative Matrices and Markov Chains.* (2nd Edn.). Springer, New York (1981).

18. Seneta, E. Perturbation of the stationary distribution measured by ergodicity coefficients. Adv. Appl. Prob., **20** (1988), 228–230.

19. Seneta, E. Sensitivity analysis, ergodicity coefficients, and rank–one updates to finite Markov chains. In: W.J. Stewart (ed.) *Numerical Solution of Markov Chains.* Marcel Dekker, New York (1991), pp. 121–129.

20. Tan, C.P. Coefficients of ergodicity with respect to vector norms. J. Appl. Prob. **20** (1983), 277–287.

21. Zarling, R.L. *Numerical Solution of Nearly Decomposable Queueing Networks.* Ph.D. Dissertation, Computer Science, University of North Carolina at Chapel Hill. UMI, Ann Arbor, Michigan (1976).

Current address: School of Mathematics and Statistics, University of Sydney, N.S.W. 2006, Australia
E–mail address: seneta_e@maths.su.oz.au

Contemporary Mathematics
Volume **149**, 1993

APPLICATIONS OF SOME CONSTRUCTIONS OF MARKOV PROCESSES

I. CUCULESCU

1 INTRODUCTION

Since we are commemorating Wolfgang Doeblin, let us begin by mentioning that his work in probability was well known at the Faculty of Mathematics in Bucharest. His contributions were outlined in the book of J. Doob (1953), where references to Doeblin's work were given. His results were also presented in detail at a scientific seminar given by C. Ionescu Tulcea. Moreover, in the book of Sarymsakov (1954) section 18 was entitled "The Kolmogorov-Doeblin method and auxiliary propositions." All this occurred around 1955. However, at least for me who was a student at the time, Doeblin's name seemed to be that of an old man, not one who, if he had still been alive, would have been around 40. This conference is welcome in that it attracts our attention to the personality and the scientific contributions of Doeblin.

In the book by Chung (1960) the following result appears in Chapter 1, correponding to what Sarymsakov referred to as "The method of Doeblin". Let $(x_n)_{n=0,1,...}$ be a homogeneous Markov chain, with transition matrix $Q = (q_{ij})_{i,j \in I}$, on a probability space $(E, \mathcal{K}, P)$, $i \in I$ be a recurrent state, $x_0(w) = i$ for all $w \in E$, let $\{n; \geq 0, x_n(w) = i\}$ be infinite for all $w \in E$, denote it by $\{0 = \tau_0(w) < \tau_1(w) < \ldots < \tau_n(w) < \ldots\}$ and let $y_n^k = x_{\tau_k + n}$ for $n < \tau_{k+1} - \tau_k$, $y_n^k = \delta$ for $n \geq \tau_{k+1} - \tau_k$, where $\delta \notin I$; $k, n = 0, 1, \ldots$. Then $y^k = (y_n^k)_{n=0,1,...}$ is, for every $k = 0, 1, \ldots$, a homogeneous Markov chain, with transition matrix $Q' = (q'_{jk})_{j,k \in I \cup \{\delta\}}$ and with $y_0^k(w) = i$ for

---

AMS 1980 Classification: 60J25; 60J35; 60G40.

This paper is in final form and no version of it will be submitted for publication elsewhere.

all $w \in E$, that is y^k, $k = 0, 1, \ldots$, are identically distributed. Moreover, y^k, $k = 0, 1, \ldots$, are independent.

In this way one can obtain, in such a Markov chain (x_n), many sequences of independent, identically distributed random variables. There is an analogous result for a nonrecurrent i. This paper deals with a generalization of the above stated fact.

Let us remark that the proof of this result does not require the strong Markov property. We know that what happens in one homogeneous Markov chain also happens also in another with the same initial distribution and the same transition matrix. Thus, it is sufficient to take a sequence of independent Markov chains, all starting from i and having transition matrix Q', in order to recover the initial chain from them and to check that the process obtained is a homogeneous Markov chain (x_n) starting from i and with transition matrix Q, calculating $P((x_0 = i_0) \cap \ldots \cap (x_n = i_n))$. This is in fact the initial idea in our constructions.

Since these constructions were given in detail in our papers, Cuculescu (1977, 1979a, 1979b), we shall concentrate on one application, which shows how the "recollement construction" may be included as a particular case of ours.

2 CONSTRUCTIONS

The first construction starts from two transition semigroups, the first, Q_1, on a state space $I_1 \oplus \triangle_1$, and the second, Q_2, on a state space I_2, where I_1, I_2 are separable metric spaces with the Borel fields generated by their topologies. The Borel field on $\triangle_1$ is arbitrary, and there is given a transition probability T from $\triangle_1$ to I_2.

We suppose that, whatever the probability on the first state space, there exists a corresponding (homogeneous) Markov process, having this initial distribution, with the set of visits in $\triangle_1$ of the form $[\zeta_1, \infty)$ (or void), constant on this set, with right continuous sample paths on $[\zeta_1, \infty)$. $\triangle_1$ thus appears as a "cemetery" set. It will be shortly named, a Q_1-process, and the constant will be considered as its value at ∞, while ζ_1 will be named its lifetime.

An analogous hypothesis, not involving a cemetery set, is imposed on Q_2.

The construction leads to a transition semigroup Q_{12} on $I_1 \oplus I_2$ with the same properties, such that, if (x_t), (y_t) are Q_1, Q_2-processes respectively, on the same probability space, constituting, in this order, a Markov family, and if the conditional distribution of y_0 with respect to x_∞ is T on $(\zeta_1 < \infty)$, then, defining $z_t = x_t$ for $t < \zeta_1$ and $z_t = y_{t-\zeta_1}$ for $t \geq \zeta_1$, we obtain a Q_{12}-process (z_t).

We consider that I_2 is invariant to Q_{12} and that the restriction of Q_{12} to I_2 is Q_2.

If Q_2 has a state space $I_2 \oplus \triangle_2$ with cemetery set $\triangle_2$, then Q_2 will have also a cemetery set, namely $\triangle_2$.

3 AN ASSOCIATIVITY PROPERTY

For our purpose, an associativity property will play an important role. Let Q_1, Q_2, Q_3 be as above, on $I_1 \oplus \triangle_1, I_2 \oplus \triangle_2, I_3$ respectively, and let T_{12}, T_{23} be the respective transition probabilities from $\triangle_1$ to $I_2 \oplus \triangle_2$, and $\triangle_2$ to I_3. Applied to Q_1, Q_2, T_{12}, the above construction leads to Q_{12} on $I_1 \oplus I_2 \oplus \triangle_2$; hence this construction may now be applied to Q_{12}, Q_3 and T_{23}, leading to a Q_{123}.

On the other hand, the construction can be applied to Q_2, Q_3 and T_{23}. We get Q_{23} on $I_2 \oplus I_3$. We cannot apply the construction to Q_1, Q_{23} and T_{12} because T_{12} may charge $\triangle_2$ (this means that $T_{12}(\delta, \triangle_2)$ is non null for some $\delta \in \triangle_1$). If it doesn't charge $\triangle_2$, we shall say that it is "good", and the construction may be applied to Q_1, Q_{23} and T_{12} and leads to Q_{123}. If T_{12} is not "good", we have to replace it by $T'_{12}(x, \ ) = \chi_{I_2} T_{12}(x, \ ) + (\chi_{\triangle_2} T_{12}(x, \ )) T_{23}$; the construction then becomes possible with Q_1, Q_{23} and T'_{12} and, moreover, leads to Q_{123}.

4 GROUPING OF THE STATES

Let us recall a construction which we may refer as "grouping of the states". Let Q be as above on $I \oplus \triangle$, let $f : I \oplus \triangle \to J \oplus \Lambda$ be surjective, let $f(I) \subset J, f(\triangle) \subset \Lambda$, f be continuous on I and measurable on $\triangle$, and suppose that, for every measurable $A \subset J \oplus \Lambda$ and $t \geq 0$, $Q_t(x, f^{-1}(A))$ depends not on x but on $f(x)$; denote it by $S_t(f(x), f^{-1}(A))$. Then (S_t) is a transition semigroup as above and, if (x_t) is a Q-process, then $(f \circ x_t)$ will be an S-process.

Let us note two obvious facts. The first relies on applying the construction in Section 2 to Q_1 on $I_1 \oplus \triangle_1, Q_2$ on I_2 and T_{12} to get a Q_{12} on $I_1 \oplus I_2$. If it is possible to group the states in Q_1 with an f which is the identity on $\triangle_1$, leading to a Q_1' on $J_1 \oplus \triangle_1$, then it is possible to group the states in Q_{12} with a g obtained by extending f as the identity on I_2. The result coincides with that of the construction in Section 2 applied to Q_1', Q_2 and T_{12}.

The second is that if, in the above situation, it was possible to group the states in Q_2 with a h and get a Q_2' then it is possible to group the states in Q_{12} with h extended as the identity on I_1 and the result is the same as applying the construction in Section 2 to Q_1, Q_2' and $T_{12}'(x,) = T_{12}(x,) \circ h^{-1}$.

5 SIMILARITY OF TWO CONSTRUCTIONS

Another obvious fact which will be used below is the following. Let Q_1, Q_2, Q_3 be as above, on $I_1 \oplus \triangle_1, I_2 \oplus \triangle_2, I_3$ respectively, and let T_{13}, T_{23} be transition probabilities from $\triangle_1, \triangle_2$ respectively to I_3. Then we may apply the construction in Section 2 to Q_1, Q_3 and T_{13}, as well as to Q_2, Q_3 and T_{23} and get Q_{13}, Q_{23} respectively, on $I_1 \oplus I_3, I_2 \oplus I_3$. Now it is possible to apply the construction in Section 2 to Q_2, Q_{13}, T_{23} on the one hand, and to Q_1, Q_{23}, T_{13} on the other.

The results of these constructions are the same.

6 REPEATING Q WITH T

Let Q_n be transition semigroups as above, on $I_n \oplus \triangle_n$, and let $T_{n,n+1}$ be transition probabilities from $\triangle_n$ to I_{n+1}, $n = 1, 2, \ldots$.

A generalization of the construction in Section 2 leads to a transition semigroup S on $(I_1 \oplus \ldots \oplus I_n \oplus \ldots) \oplus \{\delta\}$, where $\{\delta\}$ is a cemetery set, with the following property. Let $k \geq 1$ and, on the same probability space, $(x_t^k), (x_t^{k+1}), \ldots, (x_t^n), \ldots$ be $Q_k, Q_{k+1}, \ldots, Q_n, \ldots$ processes, respectively, constituting in this order a Markov family. Let ζ_n be their lifetimes, and suppose that, on $(\zeta_k + \ldots + \zeta_n < \infty)$, the conditional distribution of x_0^{n+1} with respect to x_∞^n is $T_{n,n+1}$. Let us define $x_t = x_{t-(\zeta_k+\cdots+\zeta_n)}^{n+1}$ for $\zeta_k + \ldots + \zeta_n \leq t < \zeta_k + \ldots + \zeta_{n+1}$ and $x_t = \delta$ for $t \geq \zeta_k + \ldots + \zeta_n + \ldots$. Then (x_t) is an S-process.

It should be mentioned that one may define a more involved cemetery set for S, that is one allowing many more constructions like that in Section 2.

In the particular case when all Q_n coincide with Q (that is all I_n coincide with the same I and all $\triangle_n$ coincide with the same $\triangle$), and all $T_{n.n+1}$ coincide with the same T (from $\triangle$ to I), it is possible to group the states in S with a function f equal to the identity on every I_k, and also on δ. The result will be named "the repeating of Q with T"; it is a transition semigroup Q_∞ on $I \oplus \{\delta\}$.

The exact statements of the corresponding theorems leading to the above constructions and their proofs, are to be found in the papers by Cuculescu (1977, 1979a, 1979b).

7 "RECOLLEMENT"

We pass now to the main topic of our paper. Let $I = I^1 \cup I^2$ be a separable metric space, with
$I^{12} = I^1 \cap I^2 \neq 0$. Hence $I \backslash I^{12} = (I \backslash I^1) \oplus (I \backslash I^2)$.

Furthermore, let Q^1, Q^2, Q be transition semigroups as above, on I, with cemetery sets
$I \backslash I^1, I \backslash I^2, I \backslash I^{12}$ respectively. Suppose that, if we apply the construction in Section 2 to Q, Q^1 and the transition probability ϵ from $I \backslash I^{12}$ to I, then it is possible to group the states in the resulting transition semigroup denoted by QQ^1 on $(I \backslash I^{12}) \oplus I$, with the I-valued function equal to the identity on both components, and the result is Q^1. We impose an analogous condition on Q, Q^2 requiring that the result should be Q^2.

The transition probability ϵ from A to $B \supset A$ is defined by $\epsilon(x, \ ) = \epsilon_x$ (the unit mass at x).

We consider now the transition semigroup R, obtained by applying the first construction in Section 6 to the sequence $Q^1, Q^2, Q^1, Q^2, \ldots$ and to the transition probabilities all equal to ϵ (since the cemetery sets of Q^1, Q^2 are disjoint, these transition probabilities are "good", as required in Section 6).

Definition. R is a transition semigroup on a state space, which may be described às consisting of δ, of all $(i, 2n + 1)$, with $i \in I^1$, and of all $(i, 2n + 2)$, with $i \in I^2$, n running over $0, 1, 2, \ldots$. Alternatively, the state space consists of δ, of all (i,n) with $i \in I^{12}$, $n \geq 1$, of all $(i, 2n+1)$ with $i \in I^1 \backslash I^2$, $n \geq 0$, and of all $(i, 2n)$ with $i \in I^2 \backslash I^1$, $n \geq 1$.

We now prove in detail, as an illustration of the above techniques, that in R we may group the states, by the function taking every (i,m) into i and δ into δ.

The transition semigroup on $I \oplus \{\delta\}$ obtained will be called "the result of the recollement applied to Q^1, Q^2".

Let us consider the operation of restricting a transition semigroup to an open subset G of its state space H which, in terms of the corresponding processes, consists in reducing their life times to the first visit in $H \backslash G$. Then the "recollement" appears as an operation inverse to that of taking, for a given transition semigroup and a given open covering of its state space, the set of all the restrictions of the semigroup to the subsets of the covering. We considered only the case of a two element covering.

8 THE TRANSITION SEMIGROUPS

We first note the following. Let Q^{12} be the result of the construction of Section 2, applied to Q^1, Q^2 and ϵ, that is a transition semigroup on $I^1 \oplus I^2 \oplus (I \backslash I^2)$, the last term being the cemetery. Then we may first group the states in R with the function g, with values in $I^1 \oplus I^2 \oplus \{\delta\}$, defined by $g(\delta) = \delta$, $g(i, 2n+1) = i \in I^1$, $g(i, 2n) = i \in I^2$. The result is the transition semigroup S on $I^1 \oplus I^2 \oplus \{\delta\}$ which may be also obtained by repeating Q^{12} with ϵ (see Section 6).

The question is, bearing in mind the obvious result concerning a successive grouping of states, to prove that in S we may group the states with a function with values in $I \oplus \{\delta\}$, equal to the identity on I^1, I^2 and δ.

9 ANOTHER TRANSITION SEMIGROUP

In order to use our hypothesis of Section 7, we remark that R may be obtained from a transition semigroup Z, constructed as at the beginning of Section 6 from $Q, Q^1, Q^2, Q^1, Q^2, \ldots$ and the transition probabilities ϵ, by grouping the states $(i, 0)$ and $(i, 1)$, for all $i \in I^{12}$. To be more precise, the transition probability from the cemetery of Q to the state space of the first Q^1 is not "good", since it charges the cemetery set $I \backslash I^1$ of Q^1. But it is the single one in the sequence which is not "good".

Moreover, if we consider Z to be the result of the construction at the beginning of Section 6 applied to $Q, Q^{12}, Q^{12}, \ldots$ and to the transition probabilities ϵ, we see that the transition probability from the cemetery set of Q is now "good". Furthermore, it is possible to group the states in Z, by identifying all the (i, n), with the same i in the state spaces of the Q^{12}'s, and the result is the same as that of applying Section 2 to Q, S and ϵ', where, for $x \in I\backslash(I^1 \cap I^2), \epsilon'(x,)$ is the probability on $I^1 \oplus I^2$ equal to $\epsilon_x \oplus 0$ if $x \in I^1$ and to $0 \oplus \epsilon_x$ if $x \in I^2\backslash I^1$. Denote by QS this resulting transition semigroup.

10 PROPERTIES OF QS

We have $I\backslash(I^1 \cap I^2) = (I\backslash I^1) \oplus (I\backslash I^2)$ so the definition of ϵ' is symmetric in I^1, I^2. Thus QS is also the result of the construction starting with the transition semigroup Z', resulting from the construction at the beginning of Section 6 applied to $Q, Q^1, Q^2, Q^1, Q^2, \ldots$. All the transition probabilities are equal to ϵ, except for the first, which takes and i in the cemetery set of Q into ϵ_i, but on the state space of the first Q^2 rather than the first Q^1 as before. QS is the result of grouping the states in Z', with a function identifying all $(i, 2n+1)$ with $i \in I_1$ and all $(i, 2n+2)$ with $i \in I_2$ (where $n = 0, 1, \ldots$).

We use the obvious remark in Section 5 in order to deduce that Z' also appears as the result of the construction at the beginning of Section 6, applied to $Q^1, Q, Q^2, Q^1, Q^2, \ldots$ where all the transition probabilities are "simple ϵ", except for the first, which takes an $i \in I\backslash I^1$ into ϵ_i on the state space of the first Q^2.

11 GROUPING OF STATES IN Z'

The second hypothesis of the statement of Section 7 now shows that it is possible to group the states in Z' by identifying an $i \in I^1 \cap I^2$ in the state space of Q with the same i in the state space of the first Q^2. The result is nothing other than R introduced in Section 7.

12 Some Equalities

Return now to Section 8. We need to prove that, if $i \in I^1 \cap I^2$ and $A \subset I$ is measurable, then $S_t((i,1), A_1 \oplus A_2) = S_t((i,2), A_1 \oplus A_2)$, where $(i,1)$ belongs to the first component of $I^1 \oplus I^2$, $(i,2)$ to the second, and $A_1 = A \cap I^1 \subset I^1$, $A_2 = A \cap I^2 \subset I^2$ and $t \in [0, \infty)$.

We have, according to Section 8, $S_t((i,1), A_1 \oplus A_2) = R_t((i,1), A_1 \oplus A_2 \oplus A_1 \oplus A_2 \oplus \ldots)$. From section 9, this equals $Z_t((i,0), (A_1 \cap A_2) \oplus A_1 \oplus A_2 \oplus A_1 \oplus A_2 \oplus \ldots)$. If we examine the second part of Section 9, we see that it equals, furthermore, $(QS)_t((i,0), (A_1 \cap A_2) \oplus A_1 \oplus A_2)$.

If we now examine Section 10, we deduce the equality with $Z'_t((i,0), (A_1 \cap A_2) \oplus \emptyset \oplus A_2 \oplus A_1 \oplus A_2 \oplus \ldots) = R_t((i,2), \oplus \emptyset \oplus A_2 \oplus A_1 \oplus A_2 \oplus \ldots) = R_t((i,2), A_1 \oplus A_2 \oplus A_1 \oplus A_2 \oplus \ldots)$. The first equality is due to Section 11, while the second is due to the fact that an R-process starting from $(i,2)$ does not visit the state space of the first Q^1.

We see immediately that the last equals $S_t((i,2), A_1 \oplus A_2)$.

13 A Characterization of Homogeneous Markov Chains

In Cuculescu (1983) and (1989) respectively, we consider a more general construction than that in Section 2. In this, the transition probability T is replaced by a measurable mapping from the cemetery set of Q^1 to the set of entrance laws for Q^2, which correspond to processes having the properties in Section 2, but with time set $(0, \infty)$. The construction in Section 6 also works and leads to the repeating of a transition semigroup Q by using such a T, from Q to Q. If the cemetery set is taken as a product of a sequence of components, all equal to the cemetery set of Q, with the shift invariant Borel field, and if we use a particular case of the generalized Markov chains used by Hunt (1960), we may arrive at a characterization of all the homogeneous Markov chains, with denumerable state space, with no instantaneous state for which, on every sample path, the moments of accumulation of jumps from the right constitute a well ordered set.

REFERENCES

CHUNG, K.L. *Markov Chains with Stationary Transition Probabilities.* Springer-Verlag, Heidelberg (1960).

CUCULESCU, I. Some constructions of Markov processes. I. *Rev. Roumaine Math. Pures Appl.* 22, 10(1977): 1397-1410.

CUCULESCU, I. Some constructions of Markov processes. II. *Rev. Roumaine Math. Pures Appl.* 24, 2(1979a): 205-212.

CUCULESCU, I. Some constructions of Markov processes. III.

Rev. Roumaine Math. Pures Appl. 24, 3(1979b): 357-372.

CUCULESCU, I. Constructions involving Markov processes with time set $(0, \infty)$. I. Contribution to *Studies in Probability and Related Topics.* Papers in Honour of Octav Onicescu on his 90th birthday. Nagard, Rome. (1983): 131-144.

CUCULESCU, I. Constructions involving Markov processes with time set $(0, \infty)$. II. *Stud. Cerc. Mat.* 41, 6(1989): 461-472.

DOOB, J.L. *Stochastic Processes.* John Wiley & Sons, New York (1953).

HUNT, G.A. Markoff chains and Martin boundaries. *Illinois J. of Math.* 4, (1960): 313-340.

FACULTY OF MATHEMATICS, BUCHAREST UNIVERSITY, 14 ACADEMIEI ST, 70109, ROMANIA.

Contemporary Mathematics
Volume **149**, 1993

The Doeblin Decomposition

S.P. Meyn   R.L. Tweedie

Abstract   The Doeblin decomposition is fundamental in the theory of Markov chains for which there are no irreducibility assumptions. It asserts that, under appropriate conditions, the state space is composed of a disjoint collection of "recurrent" sets plus a "transient" set. This paper describes both measure-theoretic conditions, which are in a direct line of descent from work of Doeblin [4], and topological conditions which can be shown to be related to such decompositions. In the latter case we can specify more about the structure of the sets in the decomposition, and some of these consequences are also described.

1   Introduction

This paper gives an overview of the connections between various conditions, each of which implies in some sense that the state space X of a Markov chain $\{\Phi_n\}$ is composed of a (usually countable) collection of disjoint "irreducible recurrent" sets plus a naturally "transient" set. Such a structure is known as a *Doeblin decomposition* in the general state space context, having been introduced by Doeblin in [4].

Based on the well-known behaviour of countable space chains, it is reasonable to hope that such a decomposition might hold. For suppose that X is countable, and the behaviour of the chain is governed by a transition probability matrix $P = P(x, y) = \mathsf{P}_x(\Phi_1 = y)$, $x, y \in X$. The relation of communication, denoted $x \leftrightarrow y$, can be defined (see Chung [2]) as occurring when there exists $n(x, y) \geq 0$ and $m(y, x) \geq 0$ such that $P^n(x, y) > 0$

1991 Mathematics subject classifcation 60J10. This paper is in final form and will not be published elsewhere.

and $P^m(y, x) > 0$. The relation "$\leftrightarrow$" is easily seen to be an equivalence relation, and so the equivalence classes $C(x) = \{y : x \leftrightarrow y\}$ form a disjoint cover of X, with $x \in C(x)$.

Let us denote the probability that $\Phi_n = y$ infinitely often, given $\Phi_0 = x$, by $Q(x, y)$: this notation was introduced in the general state space context by Doeblin (p 67 of [4]). It is an elementary exercise to show that if x and y are in the same communicating class, then $Q(x, y) = 1$ if and only if $Q(y, x) = 1$: in this case the class is called *recurrent*.

When states do not all communicate, then although each state in $C(x)$ communicates with every other state in $C(x)$, there will be states $y \in [C(x)]^c$ such that $x \to y$. For such a class we must have $Q(x, y) = 0$ for each $y \in C(x)$. The class is called *transient* if $Q(x, y) = 0$ for each $y \in C(x)$, and even if $P(y, C) = 1$ for all $y \in C(x)$ (when C is called stochastically closed [2] or *absorbing*) then it can be transient if it is non-finite.

And these modes of behaviour, recurrence and transience, are the only ones available to a communicating class: thus [2] we can classify every class in these ways, leading to the well-known decomposition of the state space

$$\mathsf{X} = \left(\sum_{x \in I} C(x) \right) \cup F$$

where the sum is of disjoint sets corresponding to ordering the equivalence classes so that $C(1), C(2), C(3), \ldots$ correspond to *absorbing* classes C, and F contains those states which are not contained in an absorbing class. We can write this correspondingly as a decomposition of P into block-diagonal form, in the form

$$P = \begin{bmatrix} P(1) & & & & \\ 0 & P(2) & & 0 & \\ 0 & 0 & P(3) & & \\ \vdots & \vdots & \vdots & \ddots & \\ P(F_1) & P(F_2) & P(F_3) & \cdots & P(F) \end{bmatrix}$$

Here $P(F_i)$ will be non-zero if there is some probability of leaving the sets corresponding to F and reaching the set $C(i)$, and then being absorbed in that set forever.

The virtue of such a block decomposition lies largely in the assurance that any chain on $\mathbb{Z}_+$ can be studied assuming irreducibility. The "irreducible absorbing" pieces $C(x)$ can then be put together to deduce most of the properties of a reducible chain: only the behaviour of the remaining states in F must be studied separately.

One goal of Doeblin, in Sections 7-9 of [4], was to find conditions under which this same concept holds without countability of X.

2 More general spaces: the evolution of the Doeblin decomposition

2.1 Doeblin's formulation

The formulation we use for a chain $\{\Phi_n\}$ on a general spaces is little changed from that of Doeblin himself [4]: only the nomenclature has become standardised. We consider a general measure space X, with a σ-field $\mathcal{B}(X)$ of subsets of X, and a time-homogeneous Markov chain whose n-step transition probabilities

$$P^n(x, A) = \mathsf{P}(\Phi_n \in A \mid \Phi_0 = x), \quad x \in X, \ A \in \mathcal{B}(X),$$

are *kernels*, i.e., are assumed to be measures on $\mathcal{B}(X)$ for each $x \in X$ and measurable functions on X for each $A \in \mathcal{B}(X)$. We again follow Doeblin and denote the probability that $\Phi_n \in A$ infinitely often, given $\Phi_0 = x$, by $Q(x, A)$. The probability that Φ_n ever reaches A given $\Phi_0 = x$ is denoted $L(x, A)$: here, for obvious reasons, the notation of Doeblin (which was $\Pr(x, A)$ [4] p 67) has not stood the test of time.

An arbitrary set $A \in \mathcal{B}(X)$ is called *absorbing* if $P(x, A) \equiv 1$ for $x \in A$. Doeblin (p 69 of [4]) called such sets stochastically closed (*"stochastique-ment fermé"*), and this term is still current (see [15]): only the possible confusion when discussing chains on a space with a topology leads us to use the other nomenclature.

In order to discuss the decomposition of the space into "irreducible recurrent" or even "irreducible transient" absorbing subsets, we need a definition of such terms for sets in uncountable spaces, and the development of such concepts has taken a number of decades and a number of directions (see Orey [16], Nummelin [15], Meyn and Tweedie [13]).

The first steps were once again taken by Doeblin. He called an absorbing set *indecomposable* ([4], p 70) if it contains no disjoint pair of absorbing sets. This is a desirable irreducibility property. Defining recurrence or transience is harder. Following Doeblin ([4], p 68), a set $A \in \mathcal{B}(X)$ is called *inessential* if $Q(x, A) \equiv 0$, $x \in X$; and a set which is the union of countably many inessential sets is called improperly essential (*"essentiel impropre"*, [4], p 68). Any other set is called properly or absolutely essential (*"absolument essentiel "* [4], p 68).

In order to develop a general state space decomposition, Doeblin ([4], p 71) first considered the following hypothesis:

($\mathcal{L}$): *There exists in* $\mathcal{B}(\mathsf{X})$ *a family of absorbing sets* L, *indecomposable or improperly essential, with* $\sum L \in \mathcal{B}(\mathsf{X})$ *and such that* $\mathsf{X} - \sum L$ *contains no absorbing set.*

Doeblin himself says that "this hypothesis has no greatly intuitive meaning" and indeed it is not easy to see *a priori* what it might imply. However, in his Théorème I ([4], p 72) Doeblin shows that ($\mathcal{L}$) leads to

(D1) the set $F = \mathsf{X} - \sum L$ is inessential or improperly essential; and

(D2) if F is improperly essential, then there is a collection $F_i \subset F$ of inessential sets, such that from any $x \in F$, either the chain is eventually absorbed in $\sum L$ with probability $L(x, \sum L) > 0$ or the chain remains indefinitely in $\cup F_i$ with probability $1 - L(x, \sum L)$.

The strength of this result lies in the identification of the transient nature of F: nothing is revealed about the structure of the "recurrent" part $\sum L$, and indeed it may clearly contain an absorbing transient (but properly essential) set. Doeblin turns to this in his justly famous Théorème d'Application ([4], p. 74), where he shows that under the condition

($\mathcal{M}$): *There is a finite measure* φ *which gives positive mass to each absorbing subset of* X

the hypothesis ($\mathcal{L}$) is satisfied, and moreover

(D3) the family of absorbing sets L is denumerable.

By extracting out of this family the (countable number of) improperly essential sets, and putting them together with F, one has the basic Doeblin decomposition for $\{\Phi_n\}$: this is said to hold if

($\mathcal{D}$): *There exists a countable disjoint family* $\{D_n\}$ *of absorbing, indecomposable and absolutely essential sets, and an improperly essential set* E, *such that*
$$\mathsf{X} = \sum_n D_n + E.$$

In this representation one loses the property (D2), of absorption from any state in the transient part F with positive probability; but the countability of the representation, and the true "recurrent" nature of the collection $\sum_n D_n$ makes up for this.

2.2   Development since Doeblin

There have been two separate "probabilistic" lines of development of this result since Doeblin, taking place largely in the two decades 1955-1975.

The first of these was somewhat independent of the decomposition idea, and related to identifying different versions of recurrence and transience, then linking these to the Doeblin ideas of essential and inessential sets.

An absorbing set A is called *Harris* if there exists a σ-finite measure φ on $\mathcal{B}(\mathsf{X})$ with $\varphi(A) > 0$ such that $\varphi(B) > 0$ implies $Q(x, B) \equiv 1$, $x \in A$. Obviously a Harris set is both indecomposable and properly essential.

A *transient* set is one for which there is a countable cover with *uniformly transient* sets A such that

$$\sum P^n(x, A) \leq M < \infty, \qquad x \in A.$$

Uniformly transient sets are certainly inessential: not only is the probability of an infinite number of visits to A zero, but the expected number of such visits is actually bounded.

Let us say that a *Harris decomposition* for $\{\Phi_n\}$ holds if

$(\mathcal{H})$: *There exists a countable disjoint family* $\{H_n\}$ *of Harris sets and a transient set E such that*

$$\mathsf{X} = \sum_n H_n + E.$$

We have without any other conditions that $(\mathcal{H}) \Rightarrow (\mathcal{D})$. It is far less clear whether the other implication holds. However, it is in fact true that $(\mathcal{D}) \stackrel{(\sigma)}{\Rightarrow} (\mathcal{H})$, where we write $\stackrel{(\sigma)}{\Rightarrow}$ when a result holds provided $\mathcal{B}(\mathsf{X})$ is countably generated. This beautiful result, linking the Doeblin and the Harris approaches to Markov chains, is due to Jain and Jamison who showed (Theorem 3 of [8]) that indecomposable properly essential sets are (except for at worst an improperly essential φ-null set) Harris sets, and in effect that indecomposable improperly essential sets are transient (Theorem 6 of [8]).

Related work is in [20], and an excellent account of the Doeblin decomposition as at 1970 is given in the still-fresh work of Orey [16]. Nummelin [15], Chapter 3 and Meyn and Tweedie [13] Chapters 8 and 9 also discuss these various concepts of recurrence and transience.

This equivalence of $(\mathcal{D})$ and $(\mathcal{H})$ certainly strengthens the value of the Doeblin decomposition. Simultaneously with the development of recurrence properties in the 1960's, work on conditions leading to such a decomposition was taking place, weakening ($\mathcal{M}$), and the following hypothesis was found to be a valuable one.

($\mathcal{C}$): *There is no uncountable disjoint class of stochastically absorbing subsets on* X.

The implication ($\mathcal{M}$) $\Rightarrow$ ($\mathcal{C}$) is of course trivial. That ($\mathcal{C}$) $\Rightarrow$ ($\mathcal{D}$) was first shown assuming Suslin's Conjecture, by Jamison [10]; and was then shown to hold without Suslin's Conjecture by Winkler [24].

In a sense this is a maximal result, for quite clearly, under the form ($\mathcal{H}$), the recurrent part of the space satisfies ($\mathcal{C}$); and nothing can really be said about the rest of the space. For consider a random walk on $\mathsf{IR} = (-\infty, \infty)$, with increment distribution concentrated on $\mathcal{Q}'$, the set of rationals in (1, 2). Then this chain satisfies ($\mathcal{D}$), since $\mathsf{X} = \cup_n[n, n+1)$, and the number of visits to $[n, n+1)$ is either zero or one with probability one. But of course ($\mathcal{C}$) fails in this case, since there are uncountably many closed sets given by irrational translates of $\mathcal{Q}'$.

3   Topological considerations

3.1   Equivalents of ($\mathcal{D}$)

More recently, for chains on spaces admitting a topology, continuity conditions leading to ($\mathcal{D}$) or ($\mathcal{H}$) have been investigated, and a somewhat surprising cycle of results completed, giving actual equivalences between topological and measure-theoretic results and the Doeblin decompositions.

Following the notation of [13], write the transition law of the discrete time resolvent chain as $K_\theta(x, A) \equiv (1 - \theta) \sum_n P^n(x, A)\theta^n$ for each $A \in \mathcal{B}(\mathsf{X})$, $x \in \mathsf{X}$, and $0 < \theta < 1$. If X admits a topology, and $T(x, A)$ is a kernel, then T is called a *continuous component of K_θ non-trivial at x_0* if

(T1) for each $A \in \mathcal{B}(\mathsf{X})$, $T(x, A)$ is a lower semi continuous (l.s.c.) function of x (i.e., $\liminf\limits_{x_\alpha \to x} T(x_\alpha, A) \geq T(x, A)$);

(T2) for each $A \in \mathcal{B}(\mathsf{X})$ and $x \in \mathsf{X}$, $K_\theta(x, A) \geq T(x, A)$;

(T3) $T(x_0, \mathsf{X}) > 0$.

A point x_0 is called *recurrent* if $Q(x_0, N) = 1$ for every open set N containing x_0, and we let R denote the set of recurrent points. The condition that gives us a Doeblin decomposition is

$(\mathcal{T})$: *There is a countably generated topology with the T_1 seperation property [17], with all open sets in $\mathcal{B}(\mathsf{X})$, and a continuous component T of K_θ for some $0 < \theta < 1$ such that T is non-trivial at $x \in R$.*

The topological result that $(\mathcal{T}) \Rightarrow (\mathcal{H})$ is then proved by Tuominen and Tweedie [21]. In particular, this shows that for any strong Feller chain (where P itself satisfies (T1), and hence trivially (T2) and (T3)), there is a Doeblin decomposition provided only that the topology is T_1 and countably generated.

To complete the cycle of connections, let us introduce the following conditions, one topological and one in the same spirit as $(\mathcal{M})$ or $(\mathcal{C})$.

$(\mathcal{T}')$: *As in $(\mathcal{T})$, but with the continuous component T everywhere non-trivial.*

$(\mathcal{G})$: *There exists no uncountable collection of points (x_α) such that the measures $K_\theta(x_\alpha, \cdot)$ are mutually singular.*

In Tweedie [23] the connections between these conditions and the previously developed conditions $(\mathcal{M})$ and $(\mathcal{C})$ are discussed. In particular, under the assumption that $\mathcal{B}(\mathsf{X})$ is countably generated, then the following simple relationships between the measure-theoretic "non-singularity" conditions, the Doeblin and Harris decompositions, and the topological conditions are shown to hold (where, to avoid trivialities in generating T_1 topologies, we assume that if $P(x, \cdot) \equiv P(y, \cdot)$ then $x = y$).

$$\left[(\mathcal{D} + \mathcal{G}) \overset{(\sigma)}{\Leftrightarrow} (\mathcal{H} + \mathcal{G}) \quad \overset{(\sigma)}{\Leftrightarrow} \quad (\mathcal{T}') \overset{(\sigma)}{\Leftrightarrow} (\mathcal{M}) \overset{(\sigma)}{\Leftrightarrow} (\mathcal{G}) \right]$$
$$\Downarrow$$
$$(\mathcal{C}) \qquad\qquad (1)$$
$$\Downarrow$$
$$\left[(\mathcal{D}) \overset{(\sigma)}{\Leftrightarrow} (\mathcal{H}) \quad \overset{(\sigma)}{\Leftrightarrow} (\mathcal{T}) \right]$$

There is a major open question in this area: what precisely is the position of $(\mathcal{C})$? Certainly $(\mathcal{C})$ is stronger than the bottom set of conditions $(\mathcal{D})$, $(\mathcal{H})$ and $(\mathcal{T})$: these are virtually vacuous when X itself is improperly essential. Following the posing of this question, Maharam [12], Burgess and Mauldin [1] and Gardner [6] have constructed examples of families of probability measures satisfying $(\mathcal{C})$ and not $(\mathcal{G})$, although it is not clear how to extend these examples if one wants the family to satisfy the type of semigroup properties satisfied by the family $K_\theta(x, \cdot)$.

3.2   Consequents for T-chains

A chain satisfying (T') is a particular version of a chain called a *T-chain* (cf [14]): such chains satisfy

$$K_a(x, A) \triangleq \sum P^n(x, A)a(n) \geq T(x, A)$$

for an arbitrary distribution a on $\mathbb{Z}_+$ rather than for the specific choice of the geometric in $a(n) = \theta^n$, and the continuous component T is still assumed to be everywhere non-trivial.

It is straightforward [14] to extend the implication $(T') \Rightarrow (\mathcal{H})$ to show that whenever $\boldsymbol{\Phi}$ is a T-chain, $(\mathcal{H})$ holds.

There are a number of useful consequences for other topological stability conditions which flow from the Doeblin decomposition for T-chains. In particular in [14] the following two recurrence/transience concepts are considered.

Let us say that a sample path of $\boldsymbol{\Phi}$ *converges to infinity* (denoted $\boldsymbol{\Phi} \to \infty$) if the trajectory visits each compact set only finitely often. A Markov chain $\boldsymbol{\Phi}$ will be called *non-evanescent* if $\mathsf{P}_x\{\boldsymbol{\Phi} \to \infty\} = 0$ for each $x \in \mathsf{X}$.

A Harris recurrent set H always admits an invariant measure π_H, and the set is called positive Harris if $\pi_H(H) < \infty$. Introducing a topological condition related to positive recurrence, the chain $\boldsymbol{\Phi}$ is called *bounded in probability* if for each initial $x \in \mathsf{X}$ and each $\varepsilon > 0$, there exists a compact subset $K \subset \mathsf{X}$ with

$$\liminf_{k \to \infty} P^k(x, K) \geq 1 - \varepsilon.$$

In Theorem 2.1 and Theorem 2.2 of [14], we show that for T-chains, compact sets have a simple relationship with the Harris part of the decomposition $(\mathcal{H})$: in particular,

(T1') under (T'), for each compact set $C \subset \mathsf{X}$, $H_i \cap C = \emptyset$ for all but a finite number of $i \in I$;

(T2') under (T'), for each initial $x \in \mathsf{X}$,

$$\mathsf{P}_x\Big\{\{\boldsymbol{\Phi} \to \infty\} \cup \{\boldsymbol{\Phi} \text{ enters } \sum_{i \in I} H_i\}\Big\} = 1,$$

and hence the index set I is non-empty if and only if $P_x\{\boldsymbol{\Phi} \to \infty\} < 1$ for some initial condition $x \in \mathsf{X}$; and thus if $\boldsymbol{\Phi}$ is non-evanescent then $L(x, \sum_{i \in I} H_i) \equiv 1$.

(T3') under (T'), $\boldsymbol{\Phi}$ is bounded in probability if and only if it is non-evanescent, and every Harris set in the Harris decomposition is positive.

Using somewhat strengthened topological conditions we can also identify the topological structure of the sets in the decomposition. Consider

$(\mathcal{T}'')$: *As in* $(\mathcal{T})$, *but for every* x *the continuous component* $T(x, \cdot)$ *is equivalent to the measure* $K_\theta(x, \cdot)$.

This enables us to show, as in Theorem 6.1 of [22]

(**T1"**) under $(\mathcal{T}'')$, in the decomposition $(\mathcal{H})$ each of the sets H_i is topologically closed

(**T2"**) under $(\mathcal{T}'')$, the transient set $E = E' \cup E''$ where E'' is absorbing and topologically closed, and for all $x \in E'$

$$L(x, \sum H_i) > 0.$$

Thus the original Doeblin idea of a transient part E'' directly leading to the recurrent/absorbing part is preserved.

4   A Doeblin decomposition for e-chains

We conclude with a somewhat different continuity condition that leads to a Doeblin-like decomposition. The T-chain condition results from a substantial weakening of the idea of a strong Feller chain: one for which P maps bounded functions into bounded continuous functions.

Alternatively we could strengthen the weak Feller property that P maps bounded continuous functions into bounded continuous functions. We say that $\{\Phi_n\}$ is an *e-chain* if the Markov transition function P is equicontinuous in the sense that for each $f \in \mathsf{C}_c$ the sequence of functions $\{P^k f : k \in \mathbb{Z}_+\}$ is equicontinuous on compact sets, where C_c denotes the class of continuous functions vanishing off compact sets. Now consider the condition

$(\mathcal{E})$: *the space* X *is locally compact separable metric space, and* $\{\Phi_n\}$ *is an e-chain which is bounded in probability on average: that is for each* $x \in \mathsf{X}$ *and each* $\varepsilon > 0$, *there exists a compact subset* $K \subset \mathsf{X}$ *such that*

$$\liminf_{N \to \infty} \bar{P}_N(x, K) \geq 1 - \varepsilon.$$

where $\bar{P}_N(x, K) = N^{-1} \sum_{k=1}^{N} P(x, K)$.

Here we establish an analogue of the Doeblin Decomposition under $(\mathcal{E})$ which extends the work of [11, 9, 19, 18] to Markov chains on a non-compact state space. Indeed, we show that

(E1) if $(\mathcal{E})$ holds, then $\mathsf{X} = \sum S_\alpha + F$, where F is not properly essential and each S_α is topologically closed, absorbing and contains no other closed absorbing proper subsets

(E2) there may be uncountably many of the sets S_α but each admits an invariant probability measure such that as $k \to \infty$

$$\bar{P}_k(y, \, \cdot \,) \overset{\text{vaguely}}{\longrightarrow} \Pi(\alpha, \, \cdot \,) \qquad y \in S_\alpha.$$

Intriguingly, this decomposition returns to the original idea of Doeblin in $(\mathcal{L})$ in the sense that we have no result saying that the recurrent part of the space consists of a countable collection of absorbing sets; and indeed, it may well not do so.

However, in this case we have not made Doeblin's hypothesis that F contains no absorbing sets; and in practice, it may well do so, with the transience coming from trajectories clustering closer and closer to the boundary of the set F , rather than from the motion of the chain leaving this set altogether.

In the remainder of this paper we prove the assertions (E1) and (E2)

Firstly we show that if $\{\Phi_n\}$ is an e-chain, then there exists a kernel Π such that for all $x \in \mathsf{X}$,

$$\bar{P}_k(x, \, \cdot \,) \overset{\text{vaguely}}{\longrightarrow} \Pi(x, \, \cdot \,) \qquad \text{as } k \to \infty$$

where vague convergence is defined as in [13].

Let $\{f_n\} \subset \mathsf{C}_c$ denote a countable collection of continuous functions which is dense, with respect to the uniform norm, as a subset of C_c. By Ascoli's theorem and a diagonal subsequence argument, there exists a subsequence $\{N_i\}$ of $\mathbb{Z}_+$ and functions $\{g_n\} \subset \mathsf{C}$ such that

$$\lim_{i \to \infty} \bar{P}_{N_i} f_n \, (x) = g_n(x)$$

uniformly for x in compact subsets of X for each $n \in \mathbb{Z}_+$. The set of all subprobabilities on $\mathcal{B}(\mathsf{X})$ is sequentially compact with respect to vague convergence, and any vague limit ν of the probabilities $\bar{P}_{N_i}(x, \, \cdot \,)$ must satisfy $\int f_n \, d\nu = g_n(x)$ for all $n \in \mathbb{Z}_+$. Since the functions $\{f_n\}$ are dense in C_c, this shows that for each x there is exactly one vague limit point, and hence a kernel Π exists such that for each $x \in \mathsf{X}$

$$\bar{P}_{N_i}(x, \, \cdot \,) \overset{\text{vaguely}}{\longrightarrow} \Pi(x, \, \cdot \,) \qquad \text{as } i \to \infty.$$

Observe that by equicontinuity, the function Πf is continuous for every function $f \in \mathsf{C}_c$. It follows that Πf is positive and l.s.c. whenever f has these properties.

By the Dominated Convergence Theorem we have for all $k, j \in \mathbb{Z}_+$, $P^j \Pi^k = \Pi$. We now show that $\Pi P = \Pi$, and hence that $\Pi^k P^j = \Pi, k, j \in \mathbb{Z}_+$.

Let $f \in \mathsf{C}_c$ be a continuous, positive function with compact support. It follows from the observation that positive l.s.c. functions on X are the pointwise supremum of a collection of positive, continuous functions with compact support that if $\nu_k \xrightarrow{\text{vaguely}} \nu$ then $\lim \inf_{k \to \infty} \int f \, d\nu_k \geq \int f \, d\nu$ for any positive l.s.c. function f on X. Thus, since the function Pf is positive and continuous,

$$
\begin{aligned}
\Pi(Pf) &\leq \lim_{i \to \infty} \inf \bar{P}_{N_i}(Pf) \\
&= \Pi f
\end{aligned}
$$

By regularity of finite measures on X, this shows that $\Pi P = \Pi$.

Suppose that there exists a different subsequence $\{M_j\}$ of $\mathbb{Z}_+$, and a distinct kernel Π' such that

$$
\bar{P}_{M_j} \xrightarrow{\text{vaguely}} \Pi'(x, \cdot) \qquad \text{as } j \to \infty.
$$

Then for each positive function $f \in \mathsf{C}_c$,

$$
\begin{aligned}
\Pi f &= \lim_{j \to \infty} \Pi \bar{P}_{M_j} f \\
&= \Pi \Pi' f \qquad \text{by the Dominated Convergence Theorem} \\
&\leq \lim_{i \to \infty} \inf \bar{P}_{N_i} \Pi' f \qquad \text{since } \Pi' f \text{ is continuous and positive} \\
&= \Pi' f.
\end{aligned}
$$

Hence by symmetry, $\Pi' = \Pi$, and this completes the proof that $\bar{P}_N$ converges vaguely to Π.

It now follows from the Feller property that for each j, k, $\ell \in \mathbb{Z}_+$ we have

$$
P^j \Pi^k P^\ell = \Pi,
$$

and hence $\Pi(x, \cdot)$ is an invariant subprobability for all $x \in \mathsf{X}$; and moreover if $\{\Phi_n\}$ is bounded in probability on average then $\Pi(x, \mathsf{X}) \equiv 1$ and the kernel Π has the Feller property.

To develop the decomposition define a set $M \subseteq \mathsf{X}$ to be *minimal* if it is closed, absorbing, and does not contain as a proper subset any closed absorbing sets. A minimal set is called an *ergodic kernel* in [7] where strong Feller Markov chains on a compact state space are considered: under those conditions there always exists a non-empty, finite collection of minimal sets.

We will construct minimal sets by considering the behavior of the chain starting from recurrent points. For $x \in R$ we let S_x denote the support of the invariant probability $\Pi(x, \cdot)$, and we let

$$E_x \overset{\Delta}{=} \{y \in \mathsf{X} : Q(y, O) = 1 \quad \text{for every open neighborhood } O \text{ containing } x.\}$$

Since for any measurable set O the set $\{y \in \mathsf{X} : Q(y, O) = 1\}$ is absorbing, and since the intersection of absorbing sets is itself absorbing, it follows that E_x is absorbing for any x.

Under $(\mathcal{E})$ we can now show that

(i) for each $x \in R$, $S_x \subseteq \bar{E}_x$ and $\Pi(y, \cdot) = \Pi(x, \cdot)$ for all $y \in \bar{E}_x$; and

(ii) for each $x \in S$ the set S_x is a minimal set, and every minimal set is of this form.

To see this, let $f \in \mathsf{C}$ and consider the stochastic process $\{\Pi(\Phi_k, f)\}$. By invariance of Π the adapted sequence $(\Pi(\Phi_k, f), \mathcal{F}_k^{\Phi})$ is a bounded martingale, and hence for every initial condition $y \in \mathsf{X}$ there exists a random variable $\tilde{\pi}(f)$ such that

$$\lim_{k \to \infty} \Pi(\Phi_k, f) = \tilde{\pi}(f) \qquad \text{a.s. } [\mathsf{P}_y]$$

with $\mathsf{E}_y[\tilde{\pi}(f)] = \Pi(y, f)$.

If $y \in E_x$ then by continuity of $\Pi(\cdot, f)$ we have

$$\lim_{k \to \infty} \inf |\Pi(\Phi_k, f) - \Pi(x, f)| = 0 \qquad \text{a.s. } [\mathsf{P}_y]$$

which shows that $\tilde{\pi}(f) = \Pi(x, f)$ a.s., and hence that $\Pi(y, f) = \Pi(x, f)$ for $y \in E_x$. By continuity we also have that $\Pi(y, f) = \Pi(x, f)$ for $y \in \bar{E}_x$. Since E_x and hence $\bar{E}_x$ is absorbing, it easily follows that $S_x \subseteq \bar{E}_x$, which completes the proof of (i).

To prove (ii) let $x \in R$, and $C \subset \mathsf{X}$ be closed and absorbing. If $y \in C \cap S_x$ then $P^k(y, C^c) = 0$ for all $k \in \mathbb{Z}_+$, and hence $\Pi(y, C^c) = 0$, showing that $S_y \subset C$. Further, by (i) above $S_x = S_y$, which shows that $S_x \subset C$.

Hence either $S_x \cap C = \emptyset$ or $S_x \subset C$, and this shows that S_x is a minimal set.

Conversely, if S_0 is a minimal set, then from the construction of π and minimality $S_x = S_0$ for all $x \in S_0$.

It now follows that for $f \in \mathsf{C}$,

$$\lim_{N \to \infty} \bar{P}_N f(\Phi_0) = \Pi f(x)$$

for any initial condition $\Phi_0 \in S_x$. The set S_x is absorbing since it is equal to the support of the invariant probability $\Pi(x, \cdot)$, which is always absorbing for a Feller Markov chain [3]. Also, by minimality it follows that for any two states $x, y \in R$, the sets S_x and S_y are either disjoint or identical.

Hence, the sets $\{S_x : x \in R\}$ have the required properties for (E1) and (E2) to hold. Finally, we note that since $x \in \bar{E}_x$ for every recurrent point $x \in R$, $\mathsf{X} - \sum \bar{E}_x = F$ consists entirely of non-recurrent points. It then follows from Proposition 3.3 of [22] that F is not properly essential.

From this construction we find that the set R of recurrent points may be compared to the *center* of X, as defined in [19], and is also equal to the *conservative* part of X under the present conditions [5].

Acknowledgments   This paper was presented at the Blaubeuren conference to mark the 50th anniversary of the death of Doeblin: it seemed an appropriate topic to mark the occasion, even though with the exception of the final section, the material is descriptive rather than original. We are indebted to the organisers for the chance to carry out this development, and hope this overview gives some idea of the current situation, the original ideas of Doeblin, and most importantly, how farsighted Doeblin was in the concepts he originally proposed.

References

[1] J. P. Burgess and R. D. Mauldin. Conditional distributions and orthogonal measures. *Ann. Probab.*, 9:902–906, 1981.

[2] K. L. Chung. *Markov chains with stationary transition probabilities*, volume 104 of *Die Grundlehren der mathematischen Wissenschaften in Einzeldarstellungen mit besonderer Berucksichtigung der Anwendungsgebiete Bd.* Springer-Verlag, Berlin, 2nd edition, 1967.

[3] R. Cogburn. A uniform theory for sums of Markov chain transition probabilities. *Ann. Probab.*, 3:191–214, 1975.

[4] W. Doeblin. Eléments d'une théorie générale des chaînes simples constantes de Markov. *Annales Scientifiques de l'Ecole Normale Supérieure*, 57(III):61–111, 1940. Paris, France.

[5] S. Foguel. The ergodic theory of positive operators on continuous functions. *Annals of Scuola Norm. Sup. Pisa*, 27:19–51, 1973.

[6] R. J. Gardner. A note on conditional distributions and orthogonal measures. *Annals of Probability*, 10:877–878, 1982.

[7] S. Grigorescu. Ergodic decompositions for continuous Markov chains. *Rev. Roumaine Math. Pures et Applied*, 21:683–698, 1976.

[8] N. Jain and B. Jamison. Contributions to Doeblin's theory of Markov processes. *Z. Wahrscheinlichkeitsrechnung.*, 8:19–40, 1967.

[9] B. Jamison. Ergodic decomposition induced by certain Markov operators. *Trans. Amer. Math. Soc.*, 117:451–468, 1965.

[10] B. Jamison. A result in Doeblin's theory of Markov chains implied by Suslin's conjecture. *Z. Wahrscheinlichkeitsrechnung.*, 24:287–293, 1972.

[11] B. Jamison and R. Sine. Sample path convergence of stable Markov processes. *Z. Wahrscheinlichkeitsrechnung.*, 28:173–177, 1974.

[12] D. Maharam. Orthogonal measure: An example. *Ann. Probab.*, pages 879–880, 1982.

[13] S. P. Meyn and R. L. Tweedie. *Markov chains and stochastic stability.* Control and Communication in Engineering. Springer-Verlag, 1992.

[14] S. P. Meyn and R. L. Tweedie. Stability of Markovian processes I: discrete time chains. To appear in the September issue of *Advances Applied Prob.*, 1992.

[15] E. Nummelin. *General Irreducible Markov Chains and Non-Negative Operators.* Cambridge University Press, Cambridge, England, 1984.

[16] S. Orey. *Limit Theorems for Markov Chain Transition Probabilities*, volume 34 of *Van Nostrand Reinhold Mathematical Studies.* Van Nostrand Reinhold, London, 1971.

[17] W. Rudin. *Real and Complex Analysis.* McGraw-Hill, New York, NY, 2nd edition, 1974.

[18] R. Sine. Convergence theorems for weakly almost periodic Markov operators. *Israel Journal of Mathematics*, 19(3):246–255, 1974.

[19] R. Sine. On local uniform mean convergence for Markov operators. *Pacific Journal of Mathematics*, 60(2):247–252, 1975.

[20] P. Tuominen. Notes on 1-recurrent Markov chains. *Z. Wahrscheinlichkeitsrechnung.*, 36:111–118, 1976.

[21] P. Tuominen and R. L. Tweedie. Markov chains with continuous components. *Proceedings London Mathematics Society*, 3(38):89–114, 1979.

[22] P. Tuominen and R. L. Tweedie. The recurrence structure of general Markov processes. *Proceedings London Mathematics Society*, 3(39):554–576, 1979.

[23] R. L. Tweedie. Topological aspects of Doeblin decompositions for Markov chains. *Z. Wahrscheinlichkeitsrechnung.*, 46:299–205, 1979.

[24] W. Winkler. Doeblin's and Harris' theory of Markov processes. *Z. Wahrscheinlichkeitsrechnung.*, 31:79–88, 1975.

S.P. Meyn
University of Illinois
and the Coordinated Science Laboratory
1308 W. Main St., Urbana, IL 61801

R.L. Tweedie
Department of Statistics
Colorado State University,
Fort Collins, CO 80523

Contemporary Mathematics
Volume **149**, 1993

Generalized Resolvents and Harris Recurrence
of Markov Processes

S.P. Meyn   R.L. Tweedie

Abstract   In this paper we consider a φ-irreducible continuous parameter Markov process $\mathbf{\Phi}$ whose state space is a general topological space. The recurrence and Harris recurrence structure of $\mathbf{\Phi}$ is developed in terms of generalized forms of resolvent chains, where we allow state-modulated resolvents and embedded chains with arbitrary sampling distributions. We show that the recurrence behavior of such generalized resolvents classifies the behavior of the continuous time process; from this we prove that hitting times on the small sets of a generalized resolvent chain provide criteria for, successively, (i) Harris recurrence of $\mathbf{\Phi}$ (ii) the existence of an invariant probability measure π (or positive Harris recurrence of $\mathbf{\Phi}$) and (iii) the finiteness of $\pi(f)$ for arbitrary f.

1   Introduction

The stability and ergodic theory of continuous time Markov processes has a large literature which includes many different approaches. One such is through the use of associated discrete time "resolvent chains". The recurrence structure of the process and that of the resolvent chain are essentially equivalent [1] and, since the analysis of discrete time chains is well understood [20], this result simplifies the analysis considerably.

In this paper we develop two generalized forms of the resolvent (1), and isolate a specific type of subset of the state space (a "petite set") such that the behavior of the generalized resolvent chains on such a set provides

---

1991 Mathematics subject classifcation 60J10. This paper is in final form and will not be published elsewhere.

criteria for Harris recurrence and positive Harris recurrence. This extends the approach in [23], where continuity conditions were needed on the transition probabilities of the chain to achieve similar results.

We suppose that $\mathbf{\Phi} = \{\Phi_t : t \in \mathbb{R}_+\}$ is a time homogeneous Markov process with state space $(\mathsf{X}, \mathcal{B})$, and transition semigroup (P^t). For each initial $\Phi_0 = x \in \mathsf{X}$, the process $\mathbf{\Phi}$ evolves on the probability space $(\Omega, \mathcal{F}, \mathsf{P}_x)$, where Ω denotes the sample space. Further details of this framework may be found in [22].

It is assumed that the state space X is a locally compact separable metric space, and that $\mathcal{B}$ is the Borel field on X. We assume that $\mathbf{\Phi}$ is a Borel right process, so that in particular $\mathbf{\Phi}$ is strongly Markovian with right continuous sample paths [22]. When an event $\mathcal{A}$ in sample space holds almost surely for every initial condition we shall write "$\mathcal{A}$ holds a.s. $[\mathsf{P}_*]$".

The operator P^t acts on bounded measurable functions f and σ-finite measures μ on X via

$$P^t f (x) = \int_\mathsf{X} P^t(x, dy) f(y) \qquad \mu P^t (A) = \int_\mathsf{X} \mu(dx) P^t(x, A).$$

The *resolvent* for the process is defined as

$$R(x, A) \triangleq \int_0^\infty e^{-t} P^t(x, A)\, dt, \quad x \in \mathsf{X}, A \in \mathcal{B}. \tag{1}$$

For a measurable set A we let

$$\tau_A = \inf\{t \geq 0 : \Phi_t \in A\}, \qquad \eta_A = \int_0^\infty \mathbf{1}\{\Phi_t \in A\}\, dt.$$

A Markov process is called φ-*irreducible* if for the σ-finite measure φ,

$$\varphi\{B\} > 0 \Longrightarrow \mathsf{E}_x[\eta_B] > 0, \qquad x \in \mathsf{X}.$$

As in the discrete time setting, if $\mathbf{\Phi}$ is φ-irreducible then there exists a *maximal* irreducibility measure ψ such that $\nu \prec \psi$ for any other irreducibility measure ν [20]. We shall reserve the symbol ψ for such a maximal irreducibility measure, and we will let $\mathcal{B}^+$ denote the collection of all measurable subsets $A \subset \mathsf{X}$ such that $\psi(A) > 0$. We will say that A is *full* if $\psi(A^c) = 0$.

Suppose that, for some σ-finite measure φ, the event $\{\eta_A = \infty\}$ holds a.s. $[\mathsf{P}_*]$ whenever $\varphi\{A\} > 0$. Then $\mathbf{\Phi}$ is called *Harris recurrent*: this is the standard definition of Harris recurrence, which is taken from [2]. Clearly a Harris recurrent chain is φ-irreducible.

Using two different forms of generalized resolvent we will show that Harris recurrence is equivalent to a (formally) much weaker and more useful criterion, and we will derive a criterion for Harris recurrence in terms of *petite sets* defined in Section 3. The following summarizes the results proved in Theorem 2.4, Proposition 3.2, and Theorem 3.2:

Theorem 1.1 *The following are equivalent:*

(i) *The Markov chain* Φ *is Harris recurrent;*

(ii) *there exists a* σ-*finite measure* μ *such that* $\mathsf{P}_x\{\tau_A < \infty\} \equiv 1$ *whenever* $\mu\{A\} > 0;$

(iii) *there exists a petite set* C *such that* $\mathsf{P}_x\{\tau_C < \infty\} \equiv 1.$

A σ-finite measure π on $\mathcal{B}$ with the property

$$\pi\{A\} = \pi P^t\{A\} \triangleq \int \pi(dx) P^t(x, A) \qquad A \in \mathcal{B},\ t \geq 0$$

will be called *invariant*. It is shown in [8] that if Φ is a Harris recurrent right process then an essentially unique invariant measure π exists (see also [2]). If the invariant measure is finite, then it may be normalized to a probability measure, and in practice this is the main situation of interest. If Φ is Harris recurrent, and π is finite, then Φ is called *positive Harris recurrent*.

Again through the use of generalized resolvents, we find conditions under which π is finite, and indeed for which $\pi(f) < \infty$ for general functions f. These involve expected hitting times on petite sets, but because of the continuous time parameter of the process we have to be careful in defining such times.

For any timepoint $\delta \geq 0$ and any set $C \in \mathcal{B}$ define $\tau_C(\delta) \triangleq \delta + \theta^\delta \tau_C$ as the first hitting time on C after δ: here θ^δ is the usual backwards shift operator [22]. The kernel $G_C(x, f; \delta)$ is defined for any x and positive measurable function f through

$$G_C(x, f; \delta) \triangleq \mathsf{E}\left[\int_0^{\tau_C(\delta)} f(\Phi_t)\, dt\right], \tag{2}$$

so that in particular for the choice of $f \equiv 1$

$$G_C(x, \mathsf{X}; \delta) = \mathsf{E}_x[\tau_C(\delta)]$$

is (almost) the expected hitting time on C for small δ. The classification we then have is

Theorem 1.2 *If* Φ *is Harris recurrent with invariant measure* π *then*

(a) Φ *is positive Harris recurrent if and only if there exists a closed petite set C such that for some (and then any) $\delta > 0$*

$$\sup_{x \in C} \mathsf{E}_x[\tau_C(\delta)] < \infty; \tag{3}$$

(b) *if $f \geq 1$ is a measurable function on X, then the following are equivalent:*

    (i) *There exists a closed petite set C such that*

$$\sup_{x \in C} G_C(x, f; \delta) < \infty$$

       for some (and then any) $\delta > 0$;

    (ii) Φ *is positive Harris recurrent and $\pi(f) < \infty$.*

Part (a) of this theorem is a special case of part (b), with $f \equiv 1$; and both are proved in Section 4.

The identification of petite sets is therefore important, and in [17] we show that under suitable continuity conditions on the generalized resolvents, all compact sets are petite. These are much weaker than those in the literature, even for special classes of processes, and they certainly hold if the resolvent has the strong Feller property that $R(x, f)$ is continuous if f is bounded. In the special case of diffusion processes on manifolds, necessary and sufficient conditions under which the resolvent of the process possesses the strong Feller property have been obtained in [4] (see also [10,3,11,12]). Strong Feller processes are however a relatively restricted class of processes. Under the condition that the excessive functions for the resolvent-chain are lower semi continuous, characterizations of recurrence are also obtained in terms of hitting probabilities to compact subsets of the state space in [8]: these are similar to those we find for petite sets, but the conditions for petite sets to be compact in [17] appear much weaker than the conditions used in [8].

Thus the results presented here unify and subsume many somewhat diverse existing approaches to recurrence structures for Markov processes.

In [17,18] we use these results in a number of ways, giving not only the characterization of compact sets as petite sets, but also developing a Doeblin decomposition for non-irreducible chains, verifiable characterizations for ergodicity and rates of convergence, and conditions for convergence of the expectations $\mathsf{E}[f(\Phi_t)]$ for unbounded f.

2   State-modulated Resolvents & Harris Recurrence

The central idea of this paper is to consider the Markov process sampled at times $\{T(k) : k \in \mathbb{Z}_+\}$. These times will sometimes form an underlayed renewal process which is independent of the Markov process $\boldsymbol{\Phi}$, or a sequence of randomized stopping times. In either case, the sequence $\{T(k)\}$ will be constructed so that the process $\{\boldsymbol{\Phi}_{T(k)}\}$ is a Markov chain evolving on X, whose recurrence properties under appropriate conditions will be shown to be closely related to those of the original process.

This extends the now-classical form of analysis using the resolvent for the process. The kernel $R(x, A)$ in (1) is clearly such a Markov transition function, with transitions given by sampling the process at points of a Poisson process of unit rate. The Markov chain $\breve{\boldsymbol{\Phi}}$ with transition function R will be called the *R-chain*. If $\breve{\tau}_A$ denotes the first return time to the set A for the R-chain, we shall let $\breve{L}(x, A) = \mathsf{P}_x(\breve{\tau}_A < \infty)$ denote the hitting probability for the R-chain, $\breve{G} = \sum R^n$ its potential kernel, and we set

$$\breve{G}_B(x, A) \triangleq \mathsf{E}_x\Big[\sum_{k=1}^{\breve{\tau}_B} \mathbf{1}_A(\breve{\boldsymbol{\Phi}}_k)\Big]$$

For any fixed constant $\alpha > 0$ let T_α be a random time which is independent of the process $\boldsymbol{\Phi}$ with an exponential distribution having mean α^{-1}. We define

$$R_\alpha(x, A) = \int_0^\infty \alpha e^{-\alpha t} P^t(x, A)\, dt,$$

so that the transition function R_α has the interpretation

$$R_\alpha(x, A) = \mathsf{P}_x\{\boldsymbol{\Phi}_{T_\alpha} \in A\}. \tag{4}$$

If we set

$$U_\alpha(x, A) = \int_0^\infty e^{-\alpha t} P^t(x, A)\, dt,$$

then U_α also has the probabilistic interpretation

$$U_\alpha(x, A) = \mathsf{E}_x\Big[\int_0^{T_\alpha} \mathbf{1}(\boldsymbol{\Phi}_t \in A)\, dt\Big] \tag{5}$$

When $\alpha = 1$, of course both of these expressions coincide and are equal to $R(x, A)$.

We will consider Markov chains derived from $\boldsymbol{\Phi}$, in two different ways which extend the idea of resolvent chains. The definitions of irreducibility, Harris recurrence, and positive Harris recurrence have exact analogues for such discrete parameter chains. See [15,20,21] for these concepts.

In this section, the definition of the chain R_α derived by exponentially sampling at rate α is first generalized so that α, the rate of occurrence of the "sampling time" T_α, may depend upon the value of the state Φ_t. When the rate function is appropriately defined the random time T_α becomes a randomized stopping time for the process. This gives rise to a class of kernels introduced by Neveu in [19] which allow detailed connections between the recurrence structure of a Markov process and the generalized resolvent chains.

Let h be a bounded non-negative measurable function on X, and define the kernels R_h, U_h by

$$R_h(x, A) \triangleq \mathsf{E}_x\Big[\int_0^\infty \exp\Big\{-\int_0^t h(\Phi_s)\,ds\Big\}h(\Phi_t)\mathbf{1}_A(\Phi_t)\,dt\Big], \qquad (6)$$

$$U_h(x, A) \triangleq \mathsf{E}_x\Big[\int_0^\infty \exp\Big\{-\int_0^t h(\Phi_s)\,ds\Big\}\mathbf{1}_A(\Phi_t)\,dt\Big]. \qquad (7)$$

We have $R_h(x, f) = U_h(x, hf)$ and $R_h = R_\alpha$ when $h \equiv \alpha$. A key use of this generalized resolvent occurs when h is taken as the indicator function of a measurable set B: in this case we write $U_B \triangleq U_{\mathbf{1}_B}$, and $R_B \triangleq R_{\mathbf{1}_B}$.

The probabilistic construction of U_h and R_h enables us to write down analogues of (5), (4) for general h. On an enlarged probability space, define a randomized, possibly infinite valued, stopping time T_h by

$$\mathsf{P}_x\{T_h \in [t, t + \Delta] \mid T_h \geq t, \mathcal{F}_\infty^\Phi\} = \Delta h(\Phi_t) + o(\Delta). \qquad (8)$$

In the special case where $h \equiv 1$, T_h is independent of $\mathbf{\Phi}$ and possesses a standard exponential distribution. In general, the distribution of T_h is exponential in nature, but the rate of jump at time t, instead of being constant, is modulated by the value of $h(\Phi_t)$.

With T_h so defined we have

$$R_h(x, f) = \mathsf{E}_x\Big[f(\Phi_{T_h})\mathbf{1}\{T_h < \infty\}\Big] \qquad (9)$$

$$U_h(x, f) = \mathsf{E}_x\Big[\int_0^{T_h} f(\Phi_s)\,ds\Big] \qquad (10)$$

whenever the right hand side is meaningful. This gives a recurrence condition for finiteness of T_h, shown initially in [5].

Theorem 2.1 *The following relation holds:*

$$R_h(x, \mathsf{X}) = U_h(x, h) = 1 - \mathsf{E}_x\Big[\exp\Big\{-\int_0^\infty h(\Phi_s)\,ds\Big\}\Big]. \qquad (11)$$

Hence $R_h(x, \mathsf{X}) = 1$ if and only if $\mathsf{P}_x\{\int_0^\infty h(\Phi_s)\,ds = \infty\} = 1$.

Proof Note that by (9) the kernel R_h is probabilistic if and only if $T_h < \infty$ a.s. $[\mathsf{P}_*]$. We then have

$$1 = R_h(x, \mathsf{X}) = U_h(x, h) = \mathsf{E}_x\left[\int_0^\infty \exp\left(-\int_0^t h(\Phi_s)ds\right)h(\Phi_t)\right]dt.$$

From the change of variables $u = \int_0^t h(\Phi_s)ds$, $du = h(\Phi_t)dt$, we obtain the equality

$$\int_0^\infty \exp\left\{-\int_0^t h(\Phi_s)\,ds\right\}h(\Phi_t)\,dt = 1 - \exp\left\{-\int_0^\infty h(\Phi_s)\,ds\right\},$$

and the equality (11) follows from the definitions of R_h and U_h.   □

There is an analogue of the resolvent equation [19] for the generalized resolvent kernels U_h: for a proof see [13].

Theorem 2.2 *Let $h \geq k \geq 0$. Then U_h and U_k satisfy the generalized resolvent equation:*

$$U_k = U_h + U_h I_{(h-k)} U_k = U_h + U_k I_{(h-k)} U_h. \tag{12}$$

□

We will apply the identities in Theorem 2.3 to connect the probabilistic structure of the R-chain, the kernel U_B and the underlying Markov process.

Theorem 2.3 *For all $x \in \mathsf{X}$, $B, A \in \mathcal{B}$ we have*

(i) $\check{G}_B(x, A) = U_B(x, A);$

(ii) $\check{L}(x, B) = 1 - \mathsf{E}_x[\exp(-\eta_B)];$

(iii) *For all $B \in \mathcal{B}$,*

$$\lim_{t\to\infty} \check{L}(\Phi_t, B) = \lim_{t\to\infty} \mathsf{E}_{\Phi_t}[1 - \exp(-\eta_B)] = \mathbf{1}\{\eta_B = \infty\} \text{ a.s. } [\mathsf{P}_*].$$

Proof   To prove (i) we apply the second form of the identity (12) with $h \equiv 1$ and $k = \mathbf{1}_B$ so that

$$U_B = U_B \mathbf{1}_{B^c} R + R.$$

If we define the n-step taboo probabilities as usual for the R-chain by $_B R^n \triangleq [R \mathbf{1}_{B^c}]^{n-1} R$ (see [20]) then by repeated substitution we see that, for all $n \geq 1$,

$$
\begin{aligned}
U_B &= U_B \mathbf{1}_{B^c} R \mathbf{1}_{B^c} R + R \mathbf{1}_{B^c} R + R \\
&= U_B \mathbf{1}_{B^c} (_B R)^n + \sum_{k=1}^{n} (_B R)^k.
\end{aligned}
$$

Recalling that

$$(_B R)^n (y, f) = \mathsf{E}_y[f(\breve{\Phi}_n) \mathbf{1}\{\breve{\tau}_B \geq n\}],$$

where $\breve{\tau}_B$ is the hitting time for the R-chain, the equality above establishes the result on letting $n \to \infty$.

Result (ii) follows immediately from (i) and Theorem 2.1.

The first equality in (iii) follows from (ii). To see the second, observe that for fixed $s \leq t$, where θ^s, $s \in \mathrm{I\!R}_+$, is the backward shift operator on the sample space [22], we have for any initial condition $x \in \mathsf{X}$,

$$
\begin{aligned}
\mathsf{E}_x[\theta^s (1 - \exp(-\eta_B)) \mid \mathcal{F}_t] &\geq \mathsf{E}_x[\theta^t (1 - \exp(-\eta_B)) \mid \mathcal{F}_t] \\
&\geq \mathsf{P}\{\eta_B = \infty \mid \mathcal{F}_t\}. \quad (13)
\end{aligned}
$$

Applying the martingale convergence theorem gives, as $t \to \infty$,

$$
\begin{aligned}
\mathsf{E}_x[\theta^s (1 - \exp(-\eta_B)) \mid \mathcal{F}_t] &\to \theta^s (1 - \exp(-\eta_B)), \\
\mathsf{P}\{\eta_B = \infty \mid \mathcal{F}_t\} &\to \mathbf{1}\{\eta_B = \infty\}.
\end{aligned}
$$

By the Markov property, $\mathsf{E}_x[\theta^t (1 - \exp(-\eta_B)) \mid \mathcal{F}_t] = \mathsf{E}_{\Phi_t}[(1 - \exp(-\eta_B))]$, so that letting $t \to \infty$ in (13),

$$
\begin{aligned}
\theta^s (1 - \exp(-\eta_B)) &\geq \limsup_{t \to \infty} \mathsf{E}_{\Phi_t}[(1 - \exp(-\eta_B))] \\
&\geq \liminf_{t \to \infty} \mathsf{E}_{\Phi_t}[(1 - \exp(-\eta_B))] \\
&\geq \mathbf{1}\{\eta_B = \infty\}.
\end{aligned}
$$

But as $s \to \infty$, we have

$$\theta^s (1 - \exp(-\eta_B)) \to \mathbf{1}\{\eta_B = \infty\},$$

which establishes the result.

$\square$

Theorem 2.3 (i) is due to Neveu [19], (ii) is due to [5], and (iii) is new. As an immediate application of Theorem 2.3 (ii) and the definitions, we have that Φ is φ-irreducible if and only if the R-chain is φ-irreducible. As a substantially more important use of Theorem 2.3 we now show that the definition of Harris recurrence is equivalent to a (formally) much weaker, and more useful, criterion.

Theorem 2.4 *The Markov chain Φ is Harris recurrent if and only if for some σ-finite measure μ, $\{\tau_A < \infty\}$ a.s. $[\mathsf{P}_*]$ whenever $\mu\{A\} > 0$.*

Proof   The necessity follows from the definition of Harris recurrence. To prove sufficiency set $\varphi = \mu R$. We will prove that if $\mathsf{P}_x\{\tau_A < \infty\} \equiv 1$ for any set A of μ-positive measure, then $\mathsf{P}_x\{\eta_B = \infty\} \equiv 1$ for any set B of φ-positive measure, and hence that Φ is Harris recurrent.

Observe that for any measurable set B we have, by the definitions,

$$\varphi\{B\} > 0 \iff \int \mu(dx)\mathsf{E}_x[1 - \exp(\eta_B)] > 0. \qquad (14)$$

From the assumption of the theorem

$$\int f \, d\mu > 0 \implies \limsup_{t \to \infty} f(\Phi_t) > 0, \qquad \text{a.s. } [\mathsf{P}_*] \qquad (15)$$

for any bounded measurable function $f \colon \mathsf{X} \to \mathbb{R}$. Suppose that $\varphi\{B\} > 0$ so that by (14) and (15),

$$\limsup_{t \to \infty} \mathsf{E}_{\Phi_t}[1 - \exp(\eta_B)] > 0 \qquad \text{a.s. } [\mathsf{P}_*]$$

Then by Theorem 2.3 (iii), $\mathsf{P}_x\{\eta_B = \infty\} = 1$ for all x, which is the desired conclusion.   □

The conditions of Theorem 2.4 are much easier to verify than those of the original definition of Harris recurrence. A simple counter-example shows that the measures μ and φ do not coincide in general. Consider the deterministic uniform motion on the unit circle S^1 in the complex plane described by the equation

$$\Phi_t = e^{2\pi i t}\Phi_0 \qquad t \geq 0. \qquad (16)$$

For this chain, we can take $\mu = \delta_x$ for any x, since the process reaches every point, and in unit time; but since μ is singular with respect to Lebesgue measure, it will not allocate positive mass to sets visited for an infinite amount of time by the chain.

A version of Theorem 2.4 is also given in [9] with a substantially more complicated proof.

3   Sampled Chains, Petite Sets & Harris Recurrence

The second generalization of resovents involves moving from the exponential to a more general, but independent, sampling distribution.

Suppose that a is a general probability on $\mathbb{R}_+$, and define the Markov transition function K_a as

$$K_a \triangleq \int P^t \, a(dt). \tag{17}$$

If a is the increment distribution of the undelayed renewal process $\{T(k)\}$, then K_a is the transition function for the Markov chain $\{\Phi_{T(k)}\}$. In the special case where a is an exponential distribution with unit mean, the transition function K_a is the resolvent R of the process.

We now demonstrate the connections between the recurrence of Φ and the embedded chain $\{\Phi_{T(k)}\}$.

Theorem 3.1 *Suppose that a is a general probability on $\mathbb{R}_+$. If the K_a-chain is Harris recurrent, then so is the process Φ; and then the K_a-chain is positive Harris recurrent if and only if the process Φ is positive Harris recurrent.*

Proof   The first result is a trivial consequence of Theorem 2.4, since the hitting time on a set A by the process occurs no later than the first hitting time in the process sampled with distribution a.

If the process is then positive Harris, it is simple to observe that an invariant measure satisfying $\pi = \pi P^s$ for all s must satisfy $\pi = \pi K_a$ also. To see the converse, we observe first that by the Chapman Kolmogorov equations we have, for any sampling distribution a and any s, the commutative identity

$$\int \left[\int P^t(x, dy) a(dt) \right] P^s(y, A)$$

$$= \int P^{t+s}(x, A) a(dt)$$

$$= \int P^s(x, dy) \left[\int P^t(y, A) a(dt) \right].$$

Let π be the unique invariant probability measure for K_a. Using this identity

$$\int \pi(dy) P^s(y, A)$$

$$= \int \Big[\int \pi(dx) K_a(x, dy) \Big] P^s(y, A)$$

$$= \int \Big[\int \pi(dx) P^s(x, dy) \Big] K_a(y, A).$$

This tells us that πP^s is an invariant probability measure for K_a and since π is unique, we have $\pi = \pi P^s$; thus by definition $\boldsymbol{\Phi}$ is positive Harris. $\qquad \square$

The first of these results extends Theorem 2.2 of [23], which required additional conditions on the process or on a due to the use of the initial form of Harris recurrence in terms of η_A. In the special case of resolvents the second result is proved in [2], but with a rather more difficult proof.

Harris recurrence of the process can under certain conditions also imply that the sampled chain is recurrent: see [23] for details. However, the clock process (16) shows that this is not always true: we merely need a concentrated on $\mathbb{Z}_+$ for K_a to have an uncountable collection of absorbing sets. For more detailed analysis of this situation see [17].

Sampled chains for which a possesses a bounded density will prove to be more tractable than general sampled chains. The following connects such chains to the resolvent.

Proposition 3.1 *Suppose that the distribution a on $(0, \infty)$ possesses a bounded density with respect to Lebesgue measure. Then there exists a constant $0 < M < \infty$ such that*

$$K_a(x, B) \le M \check{L}(x, B) \qquad \forall x \in \mathsf{X}, \ B \in \mathcal{B}.$$

Proof As in the proof of Theorem 3.1 of [23], we may show that if a possesses a bounded density f on $(0, \infty)$ then there exists a constant $M < \infty$ such that

$$\int_0^\infty \mathbf{1}\{\Phi_t \in B\} f(t) \, dt \le M(1 - \exp(-\eta_B)), \qquad B \in \mathcal{B}$$

Taking expectations and applying Theorem 2.3 (ii) we see that

$$K_a(x, B) \le M \mathsf{E}_x[1 - \exp(-\eta_B)] = M \check{L}(x, B) \qquad \forall x \in \mathsf{X}$$

which proves the result. $\qquad \square$

We now introduce the class of *petite sets*. These will be seen to play the same role as the *small sets* of [20]. In particular, we will show that they are test sets or "status sets" [24] for Harris recurrence, as they are in discrete time (see [16,15]).

A non-empty set $A \in \mathcal{B}$ is called φ_a-*petite* if φ_a is a non-trivial measure on $\mathcal{B}$ and a is a probability distribution on $(0, \infty)$ satisfying,

$$K_a(x, \cdot) \geq \varphi_a(\cdot)$$

for all $x \in A$. The distribution a is called the *sampling distribution* for the petite set A.

Harris recurrence may be characterized by the finiteness of the hitting time to a single petite set.

Theorem 3.2 *If C is petite, and $\mathsf{P}_x\{\tau_C < \infty\} = 1$ for all $x \in \mathsf{X}$, then Φ is Harris recurrent.*

Proof We have that C is φ_a-petite. Let b denote a distribution on $\mathbb{R}_+$ with a bounded, continuous density, and define $\varphi = \varphi_a K_b$, so that

$$K_{a*b}(x, B) = K_a K_b(x, B) \geq \varphi_a K_b\{B\} = \varphi\{B\}, \qquad x \in C.$$

To prove the theorem we demonstrate that

$$\varphi\{B\} > 0 \implies \eta_B = \infty \qquad \text{a.s. } [\mathsf{P}_*] \tag{18}$$

Since $a * b$ has a bounded continuous density, by Proposition 3.1, there exists $M < \infty$ for which

$$K_{a*b}(x, B) \leq M\mathsf{E}_x[1 - \exp(\eta_B)] \qquad \forall x \in \mathsf{X}, \ B \in \mathcal{B}.$$

Hence if $\varphi\{B\} > 0$ then,

$$\mathsf{E}_x[1 - \exp(\eta_B)] \geq \frac{1}{M}\varphi\{B\} > 0 \qquad x \in C. \tag{19}$$

The assumption of the theorem is

$$\limsup_{t \to \infty} \mathbf{1}(\Phi_t \in C) = 1 \qquad a.s. \ [\mathsf{P}_*],$$

and hence from (19) and Theorem 2.3 (iii),

$$\mathbf{1}\{\eta_B = \infty\} = \limsup_{t \to \infty} \mathsf{E}_{\Phi_t}[1 - \exp(\eta_B)] \geq \frac{\varphi\{B\}}{M} \qquad a.s. \ [\mathsf{P}_*]$$

which shows that (18) holds.   □

This theorem is not vacuous: a φ-irreducible process admits a large number of petite sets.

Proposition 3.2 *If Φ is φ-irreducible then*

(i) *the state space may be expressed as the union of a countable collection of petite sets, and hence in particular a closed petite set in $\mathcal{B}^+$ always exists.*

(ii) *If C is ν_a-petite then for any $\alpha > 0$, there exists an integer $m \geq 1$ and a maximal irreducibility measure ψ_m such that for any $B \in \mathcal{B}$ and any $x \in C$,*
$$R_\alpha^m(x, B) \geq \psi_m(B).$$

Proof **(i)**   We know that the R-chain and hence every R_α-chain is φ-irreducible for some φ. Applying Theorem 2.1 of [20], it follows that the transition function K_a used in the defining relation for petite sets may be taken to be R_α^n for some $n \in \mathbb{Z}_+$; and there is a countable cover by petite sets for the R_α-chain by Proposition 5.13 of [20].

Hence there is a petite set in $\mathcal{B}^+$, and by the regularity of the measure ψ and the fact that subsets of petite sets are petite, a closed petite set in $\mathcal{B}^+$.

(ii)   Supposing that C is ν_a-petite, by the preceding discussion there exists a set C_n which is ν_n-small for the R_α-chain with $\nu_n\{C_n\} > 0$: that is, in the nomenclature of [20], $R_\alpha^n(x, \cdot) \geq \nu_n(\cdot)$ for $x \in C_n$. The state space for the R_α-chain can be written as the union of small sets, and hence we may assume that $C_n \in \mathcal{B}^+$, from which it follows that ν_n is an irreducibility measure. We now modify ν_n to obtain a maximal irreducibility measure. Choose t_0 so large that
$$\int_0^{t_0} P^t(x, C_n)\, a(dt) = \delta > 0, \qquad x \in C,$$

and hence
$$\int_0^{t_0} P^t R_\alpha^n(x, \cdot)\, a(dt) \geq \delta \nu_n(\cdot), \qquad x \in C \tag{20}$$

We now use the simple estimate $P^t R_\alpha \leq e^{\alpha t} R_\alpha$ to obtain by (20),
$$R_\alpha^n(x, \cdot) \geq e^{-\alpha t_0} \delta \nu_n(\cdot), \qquad x \in C.$$

Applying R_α to both sides of this inequality gives for any $x \in C$
$$R_\alpha^{n+1}(x, \cdot) \geq e^{-\alpha t_0} \delta \nu_n R_\alpha(\cdot)$$

It is easy to see that $\nu_n R_\alpha$ is a maximal irreducibility measure, and hence this completes the proof with $m = n + 1$ and $\psi_m = e^{-\alpha t_0} \delta \nu_n R_\alpha$. □

Even without irreducibility, a countable covering of the state space by petite sets may often be constructed. For example, a diffusion whose generator is hypoelliptic has the property that its resolvent has the strong Feller property. From this it follows that the state space admits a covering by open petite sets [15].

4   Finiteness of $\pi(f)$

We have shown in the previous section that petite sets, like single points for discrete state space processes, may be used to classify a Markov process as Harris recurrent. They have a similar role in the classification of positive Harris chains, and the goal of this section is to prove Theorem 1.2 where we defined that relationship.

We will fix throughout this section a measurable function $f \geq 1$, and assume that $\mathbf{\Phi}$ is Harris recurrent with invariant measure π. The assumption of Theorem 1.2 (b)(i), which we wish to show is equivalent to $\pi(f) < \infty$, is

There exists a closed petite set C such that for some $\delta > 0$

$(\mathcal{R}(\delta))$ $$\sup_{x \in C} G_C(x, f; \delta) < \infty$$

In order to connect this condition with the finiteness of $\pi(f)$ we use the fact that π is also the invariant measure for $\check{\mathbf{\Phi}}$, and hence from Proposition 5.9 of [20] and Theorem 2.3 has the representation

$$\pi(f) = \int_C \pi(dy)\check{G}_C(x, f) = \int_C \pi(dy)U_C(x, f). \tag{21}$$

Proving Theorem 1.2 is thus a matter of obtaining appropriate bounds between the kernels $G_C(x, f; \delta)$ and $U_C(x, f)$ for sets with $\pi(C) < \infty$, such as ([16], Theorem 5.2) petite sets: elucidating these and related results occupies most of this section.

For any x and $f \geq 1$ it is obvious that $G_C(x, f; r)$ is an increasing function of r. The following result gives conditions under which the rate of growth may be bounded.

Lemma 4.1 *Suppose that $G_C(x, f; r)$ is bounded on the closed set C for some $r > 0$. Then there exists $k < \infty$ such that for any $t \geq 0$,*

$$G_C(x, f; t) \leq G_C(x, f; r) + kt, \qquad x \in \mathsf{X}.$$

Hence in particular $(\mathcal{R}(\delta))$ implies $(\mathcal{R}(t))$ for every $t > 0$.

Proof   Let $\tau_C^k(r)$ denote the kth iterate of $\tau_c(r)$ defined inductively by

$$\tau_C^0(r) = 0 \text{ and } \tau_C^{n+1}(r) = \tau_C^n(r) + \theta^{\tau_C^n(r)}\tau_C(r). \qquad (22)$$

We have the simple bound, valid for any positive integer n,

$$G_C(x, f; nr) \le \mathsf{E}_x\Big[\sum_{i=0}^{n-1} \theta^{\tau_C^i(r)} \int_0^{\tau_C(r)} f(\Phi_s)\,ds\Big]$$

and hence by the strong Markov property and the assumption that the set C is closed,

$$G_C(x, f; nr) \le G_C(x, f; r) + (n-1)\sup_{y\in C} G_C(y, f; r)$$

This bound together with the fact that $G_C(x, f; t)$ is an increasing function of t completes the proof.   □

Following discrete time usage [20,15] a non-empty set $C \in \mathcal{B}$ is called f-*regular* if

$$G_B(x, f; \delta) = \mathsf{E}_x\Big[\int_0^{\tau_B(\delta)} f(\Phi_t)\,dt\Big]$$

is bounded on C for any $\delta > 0$ and any $B \in \mathcal{B}^+$.

The next key result shows that the "self regularity" property $\mathcal{R}(\delta)$ for C closed and petite actually implies f-regularity.

Proposition 4.1 *Consider a Harris recurrent Markov process* Φ *such that* $\mathcal{R}(\delta)$ *holds. Then* C *is* f-*regular.*

Proof   Let $B \in \mathcal{B}^+$, and choose $t > 0$ so large that for some $\varepsilon > 0$, $\mathsf{P}_x\{\tau_B \le t\} \ge \varepsilon$ for $x \in C$. This is possible since C is petite. From Lemma 4.1, we have $G_C(x, f; t)$ bounded on C no matter how large we choose t.

Let $\sigma(k) \triangleq \tau_C^k(t)$, the kth iterate of $\tau_C(t)$ defined in (22). Conditioning on $\mathcal{F}_{\sigma(k-1)}$ and using induction gives

$$\mathsf{P}_x\{\tau_B > \sigma(k)\} \le (1-\varepsilon)^k, \qquad k \in \mathbb{Z}_+,\ x \in C. \qquad (23)$$

We will consider the bound

$$\mathsf{E}_x\Big[\int_0^{\tau_B} f(\Phi_t)\cdot dt\Big] \le \lim_{n\to\infty} S_n(x), \qquad (24)$$

where

$$S_n(x) \triangleq \sum_{k=0}^{n} \mathsf{E}_x \left[\int_0^{\sigma(k+1)} f(\Phi_t) \, dt \mathbf{1}\{\sigma(k) < \tau_B \le \sigma(k+1)\} \right]$$

We will exhibit a contractive property for the sequence $\{S_n(x)\}$ which implies that the sequence is uniformly bounded, and hence that $\mathsf{E}_x[\int_0^{\tau_B} f(\Phi_t) \, dt]$ is bounded on C. This together with the estimate

$$\int_0^{\tau_B(t)} f(\Phi_t) \, dt$$

$$\le \int_0^{\tau_C(t)} f(\Phi_t) \, dt + \theta^{\tau_C(t)} \int_0^{\tau_B} f(\Phi_t) \, dt$$

and the strong Markov property will complete the proof of the proposition.

We have for any $n \ge 1$,

$$\begin{aligned}
S_n(x) \;=\; & \mathsf{E}_x \left[\int_0^{\sigma(1)} f(\Phi_t) \, dt \mathbf{1}\{0 < \tau_B \le \sigma(1)\} \right] \\
& + \sum_{k=1}^{n} \mathsf{E}_x \left[\int_0^{\sigma(1)} f(\Phi_t) \, dt \mathbf{1}\{\sigma(1) < \tau_B\} \theta^{\sigma(1)} \mathbf{1}_{\mathcal{B}(k)} \right] \\
& + \sum_{k=1}^{n} \mathsf{E}_x \left[\int_{\sigma(1)}^{\sigma(1)+\theta^{\sigma(1)}\sigma(k)} f(\Phi_t) \, dt \mathbf{1}\{\sigma(1) < \tau_B\} \theta^{\sigma(1)} \mathbf{1}_{\mathcal{B}(k)} \right]
\end{aligned}$$

$$(25)$$

where $\mathcal{B}(k) \triangleq \{\sigma(k-1) < \tau_B \le \sigma(k)\}$.

By conditioning at time $\sigma(1)$ the first two lines on the RHS of this inequality may be bounded by

$$\mathsf{E}_x \left[\int_0^{\sigma(1)} f(\Phi_t) \, dt \right] \left(1 + \sum_{k=1}^{\infty} \sup_{x \in C} \mathsf{P}_x\{\tau_B > \sigma(k-1)\} \right) \le (1+\varepsilon^{-1}) \mathsf{E}_x \left[\int_0^{\sigma(1)} f(\Phi_t) \, dt \right]$$

$$(26)$$

where ε is defined above equation (23).

The last line on the RHS of (25) is bounded in a similar way: by conditioning at time $\sigma(1)$ we may bound this term by

$$\mathsf{P}_x\{\sigma(1) < \tau_B\} \sup_{y \in C} \sum_{k=1}^{n} \mathsf{E}_y \left[\int_0^{\sigma(k)} f(\Phi_t) \, dt \mathbf{1}\{\sigma(k-1) < \tau_B \le \sigma(k)\} \right],$$

which is identical to $\mathsf{P}_x\{\sigma(1) < \tau_B\} \sup_{y \in C} S_{n-1}(y)$. This together with (23), (25), and (26) implies that

$$\sup_{x \in C} S_n(x) \le (1 - \varepsilon) \sup_{x \in C} S_{n-1}(x) + \sup_{x \in C}(1 + \varepsilon^{-1}) \mathsf{E}_x \left[\int_0^{\sigma(1)} f(\Phi_t) \, dt \right].$$

Since by definition $\mathsf{E}_x\left[\int_0^{\sigma(1)} f(\Phi_t)\,dt\right] = G_C(x, f; t)$ this shows that

$$\sup\{S_n(x) : n \in \mathbb{Z}_+,\ x \in C\} < \infty,$$

and hence $\mathsf{E}_x[\int_0^{\tau_B} f(\Phi_s)\,ds]$ is bounded on C, which completes the proof. $\quad\square$

Sets which are f-regular are not hard to locate. We have

Proposition 4.2 **(i)** *If C is f-regular then C is petite.*

(ii) *Suppose that C is closed and f-regular. Then for each n the set*

$$C_n = \{x : G_C(x, f) \le n\}$$

is f-regular whenever it is non-empty, and the union of the $\{C_n\}$ is full. Hence, in particular, there exists a closed petite f-regular set.

Proof **(i)** Let C be f-regular, and let $A \in \mathcal{B}^+$ be any other closed petite set. For any positive measurable function g we have the bound, valid for any $x \in \mathsf{X}$,

$$
\begin{aligned}
R(x, g) &\ge \mathsf{E}_x\left[\int_{\tau_A}^{\infty} g(\Phi_s)e^{-s}\,ds\right] \\
&= \mathsf{E}_x[e^{-\tau_A} R(\Phi_{\tau_A}, g)]
\end{aligned}
$$

Taking $g = \sum 2^{-n} R^{n-1}(\cdot, A)$ any applying Jensen's inequality we see that for any x

$$\sum_{n=1}^{\infty} 2^{-n} R^n(x, A) \ge \exp(\mathsf{E}_x[-\tau_A]) \inf_{x \in A} \sum_{n=1}^{\infty} 2^{-n} R^n(x, A)$$

Since $A \in \mathcal{B}^+$ and A is petite, it follows that the infimum is strictly positive. By regularity the exponential is bounded from below on C, and hence

$$\inf_{x \in C} \sum_{n=1}^{\infty} 2^{-n} R^n(x, A) > 0$$

which shows that C is petite from Lemma 3.1 of [16].

(ii) Since C is f-regular it is ψ_a-petite for a distribution a with finite mean $m(a)$ by Proposition 3.2(ii). By Lemma 4.1 we have for some $k < \infty$, and any $t \ge 0$

$$P^t G_C(x, f; \delta) \le G_C(x, f; \delta + t) \le G_C(x, f; \delta) + (\delta + t)k$$

and hence for any $x \in C$,

$$\int \psi_a(dy) G_C(y, f; \delta) \leq K_a G_C(x, f; \delta)$$
$$= \int_0^\infty P^t G_C(x, f; \delta) \, a(dt)$$
$$\leq G_C(x, f; \delta) + k[m(a) + \delta] < \infty$$

Hence also $\int \psi_a(dy) G_C(y, f; 0) < \infty$, and since ψ_a is maximal this shows that the set of all $x \in \mathsf{X}$ for which $G_C(x, f; 0)$ is finite is full.

That C_n is f-regular follows immediately from the inequality

$$\int_0^{\tau_B(\delta)} f(\Phi_s) \, ds \leq \int_0^{\tau_C} f(\Phi_s) \, ds + \theta^{\tau_C} \int_0^{\tau_B(\delta)} f(\Phi_s) \, ds$$

and the strong Markov property. □

We can now connect f-regularity of a set C with the condition that $\sup_{x \in C} U_C(x, f) < \infty$, which is what we need in order to use (21).

Proposition 4.3 *For a Harris recurrent Markov process Φ the following implications hold for a set $C \in \mathcal{B}$:*

(i) *If C is petite and $\sup_{x \in C} U_C(x, f) < \infty$ then C is f-regular and $C \in \mathcal{B}^+$.*

(ii) *If C is closed and f-regular, and if $C \in \mathcal{B}^+$, then $\sup_{x \in C} U_C(x, f) < \infty$ and C is petite.*

Proof **(i)** It follows from Proposition 3.2 and Theorem 2.3 (i), together with Theorem 11.3.14 of [15] that C is f-regular for the R-chain. That is, $U_B(x, f)$ is bounded on C for any $B \in \mathcal{B}^+$.

Hence for some constant M_B and all $x \in C$,

$$M_B \geq \mathsf{E}_x\left[\int_0^\infty \exp\left(-\int_0^t \mathbf{1}\{\Phi_s \in B\} \, ds\right) f(\Phi_t) \, dt\right]$$
$$\geq \mathsf{E}_x\left[\int_0^{\tau_B(\delta)} \exp\left(-\int_0^t \mathbf{1}\{\Phi_s \in B\} \, ds\right) f(\Phi_t) \, dt\right]$$
$$\geq \mathsf{E}_x\left[\int_0^\delta e^{-\delta} f(\Phi_t) \, dt\right] + \mathsf{E}_x\left[\int_\delta^{\tau_B(\delta)} e^{-\delta} \exp\left(-\int_\delta^t \mathbf{1}\{\Phi_s \in B\} \, ds\right) f(\Phi_t) \, dt\right]$$
$$= \mathsf{E}_x\left[\int_0^\delta e^{-\delta} f(\Phi_t) \, dt\right] + \mathsf{E}_x\left[\int_\delta^{\tau_B(\delta)} e^{-\delta} f(\Phi_t) \, dt\right]$$
$$= e^{-\delta} G_B^\delta(x, f)$$

which shows that (i) holds.

(ii) By Proposition 4.2 the set C is ψ_a-petite with $\psi_a(C) > 0$ so that $P_x\{\eta_C = \infty\} = \delta$ is bounded from below for $x \in C$. Hence, again using the fact that C is petite, there exists $T > 0$ and $\beta < 1$ such that

$$\sup_{x \in C} E_x\left[\exp\left(-\int_0^{\tau_C(T)} \mathbf{1}\{\Phi_s \in C\}\, ds\right)\right] \;\leq\; \sup_{x \in C} E_x\left[\exp\left(-\int_0^T \mathbf{1}\{\Phi_s \in C\}\, ds\right)\right]$$
$$= \;\; \beta, \tag{27}$$

and by f-regularity and ψ-positivity of the set C we have for any T,

$$\sup_{x \in C} E_x\left[\int_0^{\tau_C(T)} f(\Phi_t)\, dt\right] < \infty \tag{28}$$

To avoid dealing with possibly infinite terms, let $h: X \to \mathbb{R}_+$ be a measurable function for which

$$h(x) \geq \mathbf{1}_C(x) \qquad \text{and} \qquad \inf_{x \in X} h(x) > 0,$$

and fix $N \geq 1$. Then we may approximate as follows:

$$E_x\left[\int_0^\infty \exp\left(-\int_0^t h(\Phi_s)\, ds\right) N \wedge f(\Phi_t)\, dt\right]$$
$$\leq \;\; E_x\left[\int_0^{\tau_C(T)} f(\Phi_t)\, dt\right]$$
$$+ E_x\left[\int_{\tau_C(T)}^\infty \exp\left(-\int_0^t h(\Phi_s)\, ds\right) N \wedge f(\Phi_t)\, dt\right] \tag{29}$$

and each of these terms is finite since h is strictly positive. Once we obtain the desired bounds we will let $h \downarrow \mathbf{1}_C$ and $N \uparrow \infty$.

The second term on the right hand side of (29) can be bounded as follows, using the strong Markov property:

$$E_x\left[\int_{\tau_C(T)}^\infty \exp\left(-\int_0^t h(\Phi_s)\, ds\right) N \wedge f(\Phi_t)\, dt\right]$$
$$= \;\; E_x\left[\int_{\tau_C(T)}^\infty \exp\left(-\int_0^{\tau_C(T)} h(\Phi_s)\, ds\right) \exp\left(-\int_{\tau_C(T)}^t h(\Phi_s)\, ds\right) N \wedge f(\Phi_t)\, dt\right]$$
$$= \;\; E_x\left[\exp\left(-\int_0^{\tau_C(T)} h(\Phi_s)\, ds\right) E_{\Phi_{\tau_C(T)}}\left[\int_0^\infty \exp\left(-\int_0^r h(\Phi_s)\, ds\right) N \wedge f(\Phi_r)\, dr\right]\right]$$
$$\leq \;\; E_x\left[\exp\left(-\int_0^{\tau_C(T)} h(\Phi_s)\, ds\right)\right] \sup_{y \in C} E_y\left[\int_0^\infty \exp\left(-\int_0^r h(\Phi_s)\, ds\right) N \wedge f(\Phi_r)\, dr\right]$$

where in the last inequality we have used the assumption that C is closed, so that $\Phi_{\tau_C(T)} \in C$. Hence, using (27), we have from this and (29),

$$\sup_{x \in C} \mathsf{E}_x \left[\int_0^\infty \exp\left(-\int_0^t h(\Phi_s)\,ds \right) N \wedge f(\Phi_t)\,dt \right] \leq \sup_{x \in C} \frac{\mathsf{E}_x \left[\int_0^{\tau_C(T)} f(\Phi_t)\,dt \right]}{1 - \beta} < \infty \tag{30}$$

Letting $h \downarrow \mathbf{1}_C$ and $N \uparrow \infty$ we see that $\sup_{x \in C} U_C(x, f) < \infty$ by monotone convergence, which proves the result.   □

Proof of Theorem 1.2

To see that (i) $\implies$ (ii), observe that if $\mathcal{R}(\delta)$ holds then we have from Proposition 4.2 that there exists an f-regular closed set $C \in \mathcal{B}^+$. Then $U_C(x, f)$ is bounded and C is petite by Proposition 4.3 (ii). Hence $\pi(f)$ is finite from (21).

Conversely, if $\pi(f) < \infty$ we have from (21), Theorem 10.4.6 of [15] and Theorem 2.3 (i) that $U_C(x, f)$ is bounded on some petite set C, and that this implies (i) follows from Proposition 4.3 (i).   □

5   Comments and Extensions

The emphasis of this paper has been on the characterization of Harris recurrence and its refinements. From an applications point of view, it is important to obtain conditions under which bounds such as $(\mathcal{R}(\delta))$ hold.

In discrete time such bounds follow from generalized forms of Foster-Lyapunov criteria [15] applied to the one-step transition laws P. For processes in continuous time, it is more natural to construct bounds by considering the infinitesimal generator for the process. If a drift towards the "center" of the state space can be established using the generator, then an application of Dynkin's formula immediately gives bounds such as $(\mathcal{R}(\delta))$ (see [18]). Even if the generator cannot be easily analyzed, a stability proof often yields as a by-product bounds such as $(\mathcal{R}(\delta))$. One such instance is the stability analysis of generalized Jackson networks given in [14,6]. The bound $(\mathcal{R}(\delta))$ is verified in these papers using a "long time window" stability proof, since it is not possible to establish an infinitesimal drift for the Markov process under consideration in these applications.

We have not discussed here the rich ergodic theory which is a consequence of these results, and which is often the reason for interest in Harris recurrence. Several ergodic theorems which are based upon the results presented here may be found in [7,17,18], and we conclude with two significant applications of the results proved above.

A Markov process $\mathbf{\Phi}$ is called *ergodic* if an invariant probability π exists and

$$\lim_{t \to \infty} \|P^t(x, \cdot) - \pi\| = 0, \qquad x \in \mathsf{X},$$

where $\| \cdot \|$ denotes the total variation norm. Unlike their discrete time counterparts, Harris recurrent processes do not always possess the ergodic property even when π is finite, and additional conditions on the process are required to obtain ergodicity.

If $\mathbf{\Phi}$ is ergodic then it follows that the skeleton chain $\{\Phi_{\Delta n} : n \in \mathbb{Z}_+\}$ is also ergodic for each $\Delta > 0$. Results linking ergodic results of the process and its skeletons are well known for countable space processes. In a general setting, in [23] it is shown that under some continuity conditions (in t) on the semigroup P^t, ergodicity of the process $\mathbf{\Phi}$ follows from the ergodicity of the embedded skeletons or of the resolvent chains. The following extension of those results is taken from [17]: it indicates the value of criteria for positive Harris chains such as those we have presented above.

Theorem 5.1 *If $\mathbf{\Phi}$ is positive Harris recurrent and if some skeleton chain is irreducible, then $\mathbf{\Phi}$ is ergodic.*

By generalizing the kernel U_h further we can also obtain conditions which ensure that $P^t(x, f)$ converges to $\pi(f)$ at an exponential rate, even when the function f is unbounded. Although the kernels U_h and R_h of Section 2 have only been defined with $h \geq 0$, there is no reason why the functions h and k cannot take on negative values in the definitions of U_h and R_h, and in Theorem 2.2. This observation is crucial in the application to analysis of exponentially ergodic chains in [7], from which we take

Theorem 5.2 *For a function $V \geq 1$, suppose that there exists a closed petite set C and $\delta, \varepsilon > 0$ such that*

$$\mathsf{E}_x \Big[\int_0^{\tau_C(\delta)} e^{\varepsilon t} V(\Phi_s) \, ds \Big]$$

is everywhere finite and bounded on C. Then $\mathbf{\Phi}$ is positive Harris recurrent with invariant probability π. If a skeleton chain is irreducible, then *there exists $M < \infty$, $\rho < 1$ such that for any measurable function $|f| \leq V$ and any $x \in \mathsf{X}$,*

$$|P^t(x, f) - \pi(f)| \leq M\mathsf{E}_x\left[\int_0^{\tau_C(\delta)} e^{\varepsilon t}V(\Phi_s)\,ds\right]\rho^t, \qquad t \geq 0$$

Acknowledgments   Part of this work was presented at the Blaubeuren conference to mark the 50th anniversary of the death of Doeblin. Much of it was developed at that conference and in travelling to and from it. We are indebted to H. Hering, H. Cohn, J. Gani and K.L. Chung for providing this stimulating environment.

References

[1] J. Azéma, M. Duflo, and D. Revuz. Propriétés relatives des processus de Markov récurrents. *Z. Wahrscheinlichkeitsrechnung.*, 13:286–314, 1969.

[2] J. Azéma, M. Kaplan-Duflo, and D. Revuz. Measure invariante sur les classes récurrentes des processus de Markov. *Z. Wahrscheinlichkeitsrechnung.*, 8:157–181, 1967.

[3] A. Bonami, N. Karoui, B. Roynette, and H. Reinhard. Processus de diffusion associé un opérateur elliptique dégénéré. *Annals of Inst. Henri Poincaré*, VII:31–80, 1971.

[4] J. M. Bony. Principe du maximum, inégalité de Harnack et unicité du probléme de Cauchy pour les opérateurs elliptique dégénérés. *Annals of Inst. Fourier, Grenoble*, 19:277–304, 1969.

[5] M. Brancovan. Quelques propriétés des résolvantes récurrentes au sens de Harris. *Annals of Inst. Henri Poincaré Sec. B*, 9:1–18, 1973.

[6] D. Down and S. P. Meyn. Stability of multiclass queueing networks. In *Proceedings of the 26th IEEE Conference on Information Sciences and Systems*, Princeton, NJ, March 1992. Princeton University.

[7] D. Down, S. P. Meyn, and R. L. Tweedie. Geometric ergodicity of Markov processes. In preparation, 1992.

[8] R. K. Getoor. *Transience and Recurrence of Markov Processes*, pages 397–409. Springer-Verlag, Berlin, Heidelberg, New York, 1979. J. Azéma and M. Yor, editors.

[9] H. Kaspi and A. Mandelbaum. On Harris recurrence in continuous time. Technical report, Technion–Israel Institute of Technology, 1992. Submitted for publication.

[10] R. Z. Khas'minskii. Ergodic properties of recurrent diffusion processes and stabilization of the solution to the Cauchy problem for parabolic equations. *Theory of Probability Applications*, 5:179–195, 1960.

[11] R. Z. Khas'minskii. *Stochastic stability of differential equations*, volume 7 of *Mechanics – Analysis*. Sijthoff & Noordhoff, Netherlands, Rockville, Md, 1980.

[12] W. Kliemann. Recurrence and invariant measures for degenerate diffusions. *Ann. Probab.*, 15(2):690–707, 1987.

[13] H. Kunita. *Stochastic Flows and Stochastic Differential Equations*, volume 24 of *Cambridge studies in Advanced Mathematics*. Cambridge University Press, Cambridge, MA, 1990.

[14] S. P. Meyn and D. Down. Stability of generalized Jackson networks. To appear in *Annals Applied Prob.*, 1992.

[15] S. P. Meyn and R. L. Tweedie. *Markov chains and stochastic stability*. Control and Communication in Engineering. Springer-Verlag, 1992.

[16] S. P. Meyn and R. L. Tweedie. Stability of Markovian processes I: discrete time chains. To appear in the September issue of *Advances Applied Prob.*, 1992.

[17] S. P. Meyn and R. L. Tweedie. Stability of Markovian processes II: Continuous time processes and sampled chains. To appear in the September issue of *Advances Applied Prob.*, 1993.

[18] S. P. Meyn and R. L. Tweedie. Stability of Markovian processes III: Foster-Lyapunov criteria for continuous time processes, with examples. To appear in the September issue of *Advances Applied Prob.*, 1993.

[19] J. Neveu. Potentiel Markovien récurrent des chains de Harris. *Annals of Inst. Fourier, Grenoble*, 22(2):7–130, 1972.

[20] E. Nummelin. *General Irreducible Markov Chains and Non-Negative Operators*. Cambridge University Press, Cambridge, England, 1984.

[21] S. Orey. *Limit Theorems for Markov Chain Transition Probabilities*, volume 34 of *Van Nostrand Reinhold Mathematical Studies*. Van Nostrand Reinhold, London, 1971.

[22] M. Sharpe. *General Theory of Markov Processes*, volume 133 of *Pure and Applied Mathematics*. Academic Press, Inc., 1988.

[23] P. Tuominen and R. L. Tweedie. The recurrence structure of general Markov processes. *Proceedings London Mathematics Society*, 3(39):554–576, 1979.

[24] R. L. Tweedie. Criteria for classifying general Markov chains. *Adv. in Appl. Probab.*, 8:737–771, 1976.

S.P. Meyn   R.L. Tweedie
University of Illinois   Department of Statistics
and the Coordinated Science Laboratory   Colorado State University,
1308 W. Main St., Urbana, IL 61801   Fort Collins, CO 80523

Contemporary Mathematics
Volume **149**, 1993

Majorization, Monotonicity of Relative Entropy, and Stochastic Matrices

JOEL E. COHEN, YVES DERRIENNIC AND GH. ZBAGANU

ABSTRACT. Schur showed that $x = A y$ implies $\sum_i g(x_i) \le \sum_i g(y_i)$ for any positive probability n-vectors x and y, any doubly stochastic $n \times n$ matrix A, and any convex function $g:(0,1) \to \Re$. We establish a quantitative improvement of Schur's theorem: under the same hypotheses,

$$\sum_i g(x_i) \le \overline{\alpha}(A) \sum_i g(y_i) + \alpha(A) n g(1/n) \le \sum_i g(y_i),$$

where

$$\overline{\alpha}(A) \equiv (1/2) \max_{j,k} \sum_{i=1}^m |a_{ij} - a_{ik}| \text{ and } \alpha(A) = 1 - \overline{\alpha}(A).$$

This improvement follows from a recent quantitative sharpening of the monotonicity theorem of relative entropy. We also establish a converse of the monotonicity theorem of relative entropy (sometimes called the Data Processing Lemma). Specifically, for any positive probability n-vectors x and y and any positive probability m-vectors u and ν, if $H_\phi(u, \nu) \le H_\phi(x, y)$ for every relative ϕ-entropy H_ϕ, then there exists a row-allowable column-stochastic $m \times n$ matrix A such that $u = Ax$ and $\nu = Ay$.

1. Introduction

Let m and n be finite positive integers. Let P_n be the set of positive probability column-vectors with n elements, i.e., $P_n \equiv \{x \in \Re^n : x_i > 0 \, \forall i, \, \sum_i x_i = 1\}$. An $n \times n$ matrix is doubly stochastic if its elements are nonnegative real numbers, every row has sum 1 and every column has sum 1. As usual, a real-valued function h on some convex subset D of a vector space over the reals is called convex if, for all $p \in [0, 1]$ and all $s, t \in D$, $h(ps + [1-p]t) \le ph(s) + (1-p)h(t)$. Recall that a convex function on a convex open subset U of $\Re^n$ is continuous on U (e.g., Roberts and Varberg 1973, p. 93). A fundamental inequality of the theory of majorization (e.g., Marshall and Olkin 1979, p. 108) states:

1991 Mathematics Subject Classification. Primary 15A51; Secondary 15A90, 60J10.

Supported in part by U.S. National Science Foundation grant BSR 87-05047.

This paper is in final form and no version of it will be submitted for publication elsewhere.

THEOREM 1.1. *Let $x, y \in P_n$. The inequality $\sum_i g(x_i) \leq \sum_i g(y_i)$ holds for all convex functions $g:(0, 1) \to \Re$ if and only if there exists a doubly stochastic $n \times n$ matrix A such that $x = Ay$.*

Marshall and Olkin (1979) attribute the equivalent of this result to Hardy, Littlewood and Pólya (1929) and Karamata (1932), and the sufficiency portion (i.e., $x = Ay$ implies $\sum_i g(x_i) \leq \sum_i g(y_i)$) to Schur (1923). See also Hardy, Littlewood and Pólya (1952, ¶3.17) and Alberti and Uhlmann (1982). Csiszár and Körner (1981, p. 58, Exercise 14) use the monotonicity of relative entropy (which they call the Data Processing Lemma (their Lemma 3.11, p. 55)) to establish the sufficiency part of Theorem 1.1 (Schur's theorem) in the special case where $g(s) = s \log s$.

The first purpose of this note is to sharpen the inequality $\sum_i g(x_i) \leq \sum_i g(y_i)$ in Schur's theorem (see Theorem 2.1). The improvement follows from a recent quantitative sharpening of the monotonicity theorem of relative entropy (see Theorem 1.4). An open question is whether, conversely, Schur's theorem (in either its original form or as sharpened in Theorem 2.1) implies the monotonicity of relative entropy (in its original form or as sharpened in Theorem 1.4).

The second purpose of this note is to establish (in Theorem 4.1) a converse of the monotonicity theorem of relative entropy. We shall prove elsewhere that this converse holds in a sharpened, quantitative form.

To state results, some definitions are required. A real-valued function h on some convex cone D of a vector space over the reals is called homogeneous (meaning homogeneous of degree one) if, for all $x \in D$ and all nonnegative λ, $h(\lambda x) = \lambda h(x)$.

DEFINITION 1.2. Let ϕ be a real-valued function on $(0, \infty) \times (0, \infty)$ that is homogeneous and jointly convex in its arguments and satisfies $\phi(1, 1) = 0$. For any two positive n-vectors $x = (x_i)$ and $y = (y_i)$, whether or not x and y are probability vectors, define the relative ϕ-entropy $H_\phi(x, y)$ by $H_\phi(x, y) = \sum_i \phi(x_i, y_i)$.

This generalization of relative entropy has been widely studied under various names and notations (e.g., Liese and Vajda 1987). Any real-valued function g that is convex on $(-1, \infty)$ with $g(0) = 0$ can be used to define ϕ that satisfies Definition 1.2 by putting $\phi(x, y) = xg((y/x) - 1)$. Thus, as examples,

$$g(t) = |t| \Rightarrow H_\phi(x, y) = \sum_i |x_i - y_i|,$$

$$g(t) = t^2 \Rightarrow H_\phi(x, y) = \sum_i \frac{(x_i - y_i)^2}{x_i},$$

$$g(t) = -\log(1 + t) \Rightarrow H_\phi(x, y) = \sum_i x_i \log \frac{x_i}{y_i},$$

$$g(t) = (1 + t)\log(1 + t) \Rightarrow H_\phi(x, y) = \sum_i y_i \log \frac{y_i}{x_i},$$

$$g(t) = t \log(1 + t) \Rightarrow H_\phi(x, y) = \sum_i (y_i - x_i) \log \frac{y_i}{x_i}.$$

Except possibly for constants, the first expression on the right is the l_1 norm, the second is the Pearson χ^2-statistic for goodness of fit, the third is the Kullback-Leibler divergence or relative entropy of information theory (Csiszár and Körner 1981) or the G^2 likelihood ratio statistic in the theory of contingency tables, the fourth is the same with the roles of x and y exchanged, and the last (which is the sum of the preceding two) is the entropy production of statistical physics or the symmetric divergence of information theory. Thus significantly diverse measures are subsumed under the generalization of ϕ-entropy.

As usual, the l_p-norms are defined for a vector x and for $1 \leq p < \infty$ by $\|x\|_p = (\Sigma |x_i|^p)^{1/p}$. A column-stochastic matrix is an $m \times n$ matrix with each element a nonnegative real number and with all column sums 1.

DEFINITION 1.3. For any column-stochastic $m \times n$ matrix A, Dobrushin's (1956) coefficient of ergodicity is

$$\alpha(A) = \min_{j,k} \sum_{i=1}^{m} \min(a_{ij}, a_{ik}) .$$

The complement $1 - \alpha(A)$ will be written

$$\bar{\alpha}(A) \equiv 1 - \alpha(A) = \frac{1}{2} \max_{j,k} \sum_{i=1}^{m} |a_{ij} - a_{ik}|$$

and satisfies (Dobrushin 1956, pp. 69-70)

$$\bar{\alpha}(A) = \sup \left\{ \frac{\|A(x-y)\|_1}{\|x-y\|_1} : x \text{ and } y \text{ are positive } n\text{-vectors}, \ x \neq y, \|x\|_1 = \|y\|_1 \right\}.$$

A matrix is row-allowable if each row contains at least one positive element. Every doubly stochastic matrix is clearly stochastic and row-allowable. A column-stochastic $m \times n$ matrix is called a scrambling matrix (Hajnal 1958, p. 235) if any submatrix consisting of two columns has a row both elements of which are positive; i.e., $A = (a_{ij})$ is scrambling if, for all j and k such that $1 \leq j < k \leq n$, there exists an i such that $1 \leq i \leq m$ and $a_{ij}a_{ik} > 0$. A column-stochastic, row-allowable matrix A is scrambling if and only if $\alpha(A) > 0$.

THEOREM 1.4. *Let A be a column-stochastic, row-allowable $m \times n$ matrix and let $x, y \in P_n$. Then*

$$H_\phi(Ax, Ay) \leq \bar{\alpha}(A)H_\phi(x, y) .$$

The significant feature of this theorem, due to Cohen, Iwasa, Rautu, Ruskai, Seneta and Zbaganu (in press), is that the coefficient $\bar{\alpha}(A)$ is valid regardless of which ϕ-entropy is chosen and for all $x, y \in P_n$. $\bar{\alpha}(A) < 1$ if and only if A is a scrambling matrix.

W. Doeblin developed an early quantitative measure of the contractive action of a stochastic matrix. See Seneta (1973) for a historical perspective.

DEFINITION 1.5. If A is a column-stochastic, row-allowable $m \times n$ matrix, Doeblin's (1937) coefficient of ergodicity δ is

$$\delta(A) = \sum_{i=1}^{m} \min\{a_{ij} : j = 1, ..., n\} .$$

Define $\bar{\delta}(A) = 1 - \delta(A)$.

It is known that $\delta(A) \le \alpha(A)$ with equality if $n = 2$. Thus $\bar{\delta}(A) \ge \bar{\alpha}(A)$ (with equality if $n = 2$). Therefore the inequalities in Theorem 1.4 and Theorem 2.1 hold under the same hypotheses with α replaced by δ and $\bar{\alpha}$ replaced by $\bar{\delta}$.

2. Quantitative majorization in discrete processes

THEOREM 2.1.   *Let $h : \Re \to \Re$ be a convex function, A a doubly stochastic $n \times n$ matrix, $x \in P_n$ (i.e., x is a strictly positive probability vector), and $w = Ax$. Then*

$$\sum_i h(w_i) \le \bar{\alpha}(A) \sum_i h(x_i) + \alpha(A) n h\left(\frac{1}{n}\right) \le \sum_i h(x_i).$$

Proof.   Define the function $\phi : \Re \times \Re \to \Re$ by the requirements that ϕ be homogeneous and that

$$\phi(1, \ 1+s) = h\left(\frac{s+1}{n}\right) - h\left(\frac{1}{n}\right), \qquad\qquad \forall s.$$

Then ϕ is jointly convex in both arguments and $\phi(1, \ 1) = 0$. Therefore ϕ satisfies the hypotheses of Theorem 1.4.

The doubly stochastic matrix A is necessarily column-stochastic and row-allowable and therefore satisfies the hypotheses of Theorem 1.4. Choose y to be the n-vector with each element $1/n$. Then $Ay = y$. Let $\bar{\alpha} = \bar{\alpha}(A)$, $\alpha = 1 - \bar{\alpha}$. Theorem 1.4 implies

$$H_\phi(y, \ w) \le \bar{\alpha} H_\phi(y, \ x).$$

But

$$H_\phi(y, \ x) \ = \ \sum_i \phi\left(\frac{1}{n}, \ x_i\right) \ = \ \sum_i \phi\left(\frac{1}{n}, \ \frac{1}{n} n x_i\right) \ = \ \frac{1}{n} \sum_i \phi(1, \ (nx_i - 1) + 1)$$

$$= \ \frac{1}{n} \sum_i \left[h\left(\frac{(nx_i - 1) + 1}{n}\right) - h\left(\frac{1}{n}\right) \right] \ = \ \left(\frac{1}{n} \sum_i h(x_i)\right) - h\left(\frac{1}{n}\right)$$

and similarly

$$H_\phi(y, \ w) \ = \ \frac{1}{n} \sum_i h(w_i) \ - \ h\left(\frac{1}{n}\right).$$

Therefore

$$\frac{1}{n} \sum_i h(w_i) \ - \ h\left(\frac{1}{n}\right) \le \bar{\alpha}\left(\frac{1}{n} \sum_i h(x_i)\right) \ - \ \bar{\alpha} h\left(\frac{1}{n}\right)$$

or

$$\sum_i h(w_i) \le \bar{\alpha} \sum_i h(x_i) + \alpha n h\left(\frac{1}{n}\right).$$

Because h is convex,

$$\sum_i h(x_i) \ge n h\left(\frac{1}{n}\right)$$

hence

$$\bar{\alpha} \sum_i h(x_i) \ + \ \alpha \, n h\left(\frac{1}{n}\right) \le \sum_i h(x_i). \qquad\qquad \square$$

COROLLARY 2.2. *Under the hypotheses of Theorem 2.1, if in addition $h(1/n) \leq 0$,* then

$$\sum_i h(w_i) \leq \overline{\alpha} \sum_i h(x_i) \, .$$

EXAMPLE 2.3. For $s \in [0, \infty)$, if $h(s) = s \log s$ (with $0 \log 0 = 0$, $x \in P_n$, $w = Ax$ and A any doubly stochastic $n \times n$ matrix,

$$\sum_i w_i \log w_i \leq \overline{\alpha}(A) \sum_i x_i \log x_i - \alpha(A) \log n \leq \overline{\alpha}(A) \sum_i x_i \log x_i \, .$$

EXAMPLE 2.4. If the hypotheses of Theorem 2.1 are weakened only by permitting h to be non-convex, then the conclusion no longer follows. Here is an example of a doubly stochastic A, x and $w = Ax$ such that $\sum h(x_i) \geq \sum h(w_i) > \overline{\alpha}(A) \sum h(x_i)$. For small $\varepsilon > 0$, let $x^T = (1 - \varepsilon, \varepsilon)$ and let

$$A = \begin{pmatrix} 1/4 & 3/4 \\ 3/4 & 1/4 \end{pmatrix}.$$

Then $w = Ax = (1/4 + \varepsilon/2, 3/4 - \varepsilon/2)^T$ and $\overline{\alpha}(A) = 1/2$. Define $h(t) = 0$ for $t \in (0, 1/8)$, $h(t) = 3/8$ for $t \in [1/8, 7/8]$, $h(t) = 1$ for $t \in (7/8, 1)$. Then $\sum h(x_i) = 1$, $\sum h(w_i) = 3/4$, as claimed.

3. Quantitative majorization in continuous processes

Analogous results hold for continuous-time processes. For background, see Alberti and Uhlmann (1982, pp. 30-31). Assume now that all matrices are $n \times n$ and real. A matrix in which all off-diagonal elements are nonnegative and the sum of every column is zero is called an intensity matrix; such matrices have zero or negative elements on the main diagonal. If B is an intensity matrix, it is well known that for all nonnegative real t, e^{Bt} is column-stochastic. A matrix in which all off-diagonal elements are nonnegative and the sum of every column and row is zero is called a double intensity matrix; in this case, for all nonnegative real t, e^{Bt} is doubly stochastic.

For $x(0) \in P_d$, $y(0) \in P_d$, and $t \geq 0$, define $x(t) = e^{Bt}x(0)$ and $y(t) = e^{Bt}y(0)$. The following is an immediate result of combining Theorems 4.1 and 7.1 and Corollary 7.3 of Cohen, Iwasa, Rautu, Ruskai, Seneta and Zbaganu (in press).

THEOREM 3.1. *If B is an intensity matrix, $\omega \geq max_i |b_{ii}|$, and*

$$\alpha \equiv \alpha(\omega^{-1}B + I) \, ,$$

then

$$\frac{d}{dt} \log\{H_\phi(x(t), y(t))\} \leq -\omega\alpha \leq -\beta \, ,$$

where

$$\beta \equiv \min_{j,k:\, j<k} \left\{ \sum_{\substack{i \neq j \\ i \neq k}} [\min(b_{ij}, b_{ik})] + b_{jk} + b_{kj} \right\}.$$

Because $\omega^{-1}B + I$ is column-stochastic, α is meaningful; moreover, α is positive, as previously noted, if $\omega^{-1}B + I$ is scrambling.

THEOREM 3.2. *Let h : $\Re \to \Re$ be a convex function, B a double intensity n × n matrix,* $x(0) \in P_n$, $x(t) = e^{Bt}x(0)$, $t \geq 0$. *Then*

$$\frac{d}{dt} \log\left[\sum_i h(x_i(t)) - nh\left(\frac{1}{n}\right) \right] \leq -\omega\alpha \leq -\beta,$$

where ω, α *and* β *are defined in Theorem 3.1.*

Proof.   Choose $y(0)$ to be the n-vector with each element $1/n$.   Then $y(t) = e^{Bt}y(0) = y(0)$. Define ϕ as in the proof of Theorem 2.1, so that, as in that proof,

$$H_\phi(y(t), x(t)) = \frac{1}{n}\left[\sum_i h(x_i(t)) - nh\left(\frac{1}{n}\right) \right].$$

Taking the logarithm, then the derivative, of both sides of this equation and applying Theorem 3.1 gives the desired result. []

An obvious corollary is

$$\frac{d}{dt} \sum_i h(x_i(t)) \leq 0,$$

which is well-known.   Alberti and Uhlmann (1982, p. 30) state the result and cite earlier sources.

4. A converse for the monotonicity of relative entropy

The nonquantitative monotonicity theorem of relative entropy states that if A is a column-stochastic, row-allowable $m \times n$ matrix and x and y are positive n-vectors, then $H_\phi(Ax, Ay) \leq H_\phi(x, y)$ (e.g., Moran 1961, Csiszár 1963 [p. 90, his Theorem 1], Morimoto 1963).   In this theorem, x and y need not be normalized to be probability n-vectors, whereas in the quantitative monotonicity theorem (Theorem 1.4), it is assumed that x and y are probability vectors.   We now establish a converse of the nonquantitative monotonicity theorem of relative entropy with the additional hypothesis that x and y are probability vectors.   This converse is the analogue, for the monotonicity of relative entropy, of the Hardy-Littlewood-Pólya-Karamata converse of Schur's theorem for majorization.

THEOREM 4.1.   *Let* $u, v \in P_m$, $x, y \in P_n$ *be fixed.   If* $H_\phi(u,v) \leq H_\phi(x,y)$ *for every relative ϕ-entropy H_ϕ that satisfies Definition 1.2, then there exists a row-allowable column-stochastic matrix A such that $u = Ax$ and $v = Ay$.*

The proof of Theorem 4.1 depends on Theorem 4.2, a fundamental result of Choquet theory due to Cartier, Fell and Meyer (1964).   For other statements of Choquet theory and Theorem 4.2, see e.g. Winkler (1985) and Bratteli and Robinson (1987, ¶4.2.1).   Let X be a compact metric space.   Let S be a convex cone of measurable functions $f:X \to \Re$ such that the closure $\overline{S}$ of S (in the uniform topology) is closed under the max operation, i.e., if $f, g \in \overline{S}$, then $\max(f, g) \in \overline{S}$.   Let $B(X)$ be the family of Borel sets of X.   A transition measure $T:X \times B(X) \to \Re$, written as $T(x, dy) = T_x(dy)$, is defined to be a dilation if

$$f(x) \leq T_x(f) \equiv \int_{y \in X} f(y)T_x(dy)   \quad \forall f \in S.$$

THEOREM 4.2. *Under the conditions on X and S just stated, if μ and ν are two positive measures on $(X, B(X))$ such that*

$$\mu(f) \leq \nu(f) \quad \forall f \in S,$$

where e.g. $\mu(f) \equiv \int_X f(x)\mu(dx)$, then there exists a dilation T such that $\nu = \mu T$, i.e., for every bounded measurable f, $\nu(f) = \mu(Tf)$, i.e.,

$$\nu(f) \equiv \int_X f(x)\nu(dx) = \mu(Tf) \equiv \int_X T_x(f)\mu(dx) = \int_X \int_X f(y)T_x(dy)\mu(dx).$$

We also require a simple lemma. Define

$$S = \{f : f{:}(0,\infty) \times (0,\infty) \to \Re, f \text{ is convex and homogeneous}\}$$

and $S_0 = \{f \in S : f(1,1) = 0\}$. S contains all linear functions $f(s,t) = as + bt$.

LEMMA 4.3. *Let $u, v \in P_m$, $x, y \in P_n$. If $H_\phi(u,v) \leq H_\phi(x,y)$ for every $\phi \in S_0$, then $H_f(u,v) \leq H_f(x,y)$ for every $f \in S$.*

Proof. For any $f \in S$, define $\phi(x,y) \equiv f(x,y) - f(x,x) = f(x,y) - xf(1,1)$. Then $\phi \in S_0$ and $H_\phi(u,v) = H_f(u,v) - f(1,1)$, $H_\phi(x,y) = H_f(x,y) - f(1,1)$. Hence $H_f(u,v) \leq H_f(x,y)$. []

PROOF OF THEOREM 4.1. In our case, $X = [0,1] \times [0,1]$. Let S and S_0 be the restrictions to X of the convex cones defined just before Lemma 4.3. Both S and S_0 are closed under the max operation; in fact, they are even closed. Given fixed $u, v \in P_m$, $x, y \in P_n$, we shall suppose that all the points (x_j, y_j) are distinct. (If they are not, a slight modification of the proof is required, which the reader can supply.) Then define

$$\mu = \sum_{i=1}^{m} \varepsilon_{u_i, v_i}, \qquad \nu = \sum_{j=1}^{n} \varepsilon_{x_j, y_j},$$

where ε_{x_1, x_2} denotes the Dirac needle function, i.e., the point measure concentrated on $(x_1, x_2) \in X$. (Thus by definition $\varepsilon_{x_1, x_2}(f) = f(x_1, x_2)$.) Then

$$\mu(f) = \sum_{i=1}^{m} f(u_i, v_i), \quad \nu(f) = \sum_{j=1}^{n} f(x_j, y_j).$$

The hypothesis of Theorem 4.1, $H_\phi(u,v) \leq H_\phi(x,y)$ for every relative ϕ-entropy H_ϕ of Definition 1.2, may be expressed as $\mu(\phi) \leq \nu(\phi)$ for all $\phi \in S_0$. By Lemma 4.3, it follows that $\mu(f) \leq \nu(f)$ for all $f \in S$. By Theorem 4.2, there exists a dilation T such that $\nu(f) = \mu(Tf)$, i.e., for every bounded measurable $f : X \to \Re$,

(1)
$$\sum_{j=1}^{n} f(x_j, y_j) = \sum_{i=1}^{m} T_{u_i, v_i}(f).$$

Define the set $E \equiv \{(x_j, y_j) : 1 \leq j \leq n\}$. When f is $1_{X\text{-}E}$, the indicator of $X - E$, we get from (1)

$$\sum_{i=1}^{m} T_{u_i, v_i}(X - E) = 0.$$

Thus the support of the measures T_{u_i, v_i} is included in the finite set E. Define

$$a_{ij} = T_{u_i, v_i}(\{(x_j, y_j)\}), \quad 1 \leq i \leq m, \quad 1 \leq j \leq n.$$

Then

(2)
$$T_{u_i, v_i} = \sum_{j=1}^{n} a_{ij} \, \varepsilon_{x_j, y_j}.$$

When f is the indicator of the set $\{(x_j, y_j)\}$, (1) becomes

$$1 = \sum_{i=1}^{m} T_{u_i, v_i}(\{(x_j, y_j)\}) = \sum_{i=1}^{m} a_{ij}.$$

Since this holds for every $j = 1, \ldots, n$, the $m \times n$ matrix A with elements a_{ij} is column-stochastic.

Since T is a dilation, we have by definition

(3)
$$f(u_i, v_i) \leq T_{u_i, v_i}(f) \quad \forall \, f \in S.$$

From (2)

(4)
$$f(u_i, v_i) \leq \sum_{j=1}^{n} a_{ij} f(x_j, y_j) \quad \forall \, f \in S.$$

In particular, if $f(s, t) = as + bt$, then $f \in S$ and $-f \in S$ and (4) becomes the equality

(5)
$$au_i + bv_i = \sum_{j=1}^{n} a_{ij}(ax_i + by_i).$$

For $a = 1$, $b = 0$, (5) becomes $u = Ax$. For $a = 0$, $b = 1$, (5) becomes $v = Ay$. In short, if $H_\phi(u, v) \leq H_\phi(x, y)$ for all $\phi \in S_0$, then there must exist a column-stochastic matrix A such that $u = Ax$ and $v = Ay$.

Because u and v are positive, A is necessarily row-allowable. []

ACKNOWLEDGMENTS. We thank an anonymous referee for careful and helpful comments, and Mr. and Mrs. William T. Golden for hospitality to J.E.C. This collaboration originated in part as a result of the exchange program between the U.S. National Academy of Sciences and the Romanian Academy (1985) and in part as a result of the conference in Blaubeuren, Germany (1991) to honor Wolfgang Doeblin.

REFERENCES

Alberti, P. M. and Uhlmann, A. 1982, *Stochasticity and Partial Order: Doubly Stochastic Maps and Unitary Mixing.* Berlin: VEB Deutscher Verlag der Wissenschaften; Dordrecht and Boston: D. Reidel.

Bratteli, O. and Robinson, D. W. 1987 *Operator Algebras and Quantum Statistical Mechanics 1: C*- and W*-Algebras, Symmetry Groups, Decomposition of States,* 2d ed. New York, Berlin: Springer-Verlag.

Cartier, P., Fell, J. M. G., and Meyer, P.-A. 1964. Comparaison des mesures portées par un ensemble convexe compact. Bull. Soc. Math. France 92:435-445.

Cohen, J. E., Iwasa, Y., Rautu, Gh., Ruskai, M. B., Seneta, E., and Zbaganu, Gh. In press, Relative entropy under mappings by stochastic matrices. *Linear Algebra and Its Applications.*

Csiszár, I., 1963, Eine informationstheoretische Ungleichung und ihre Anwendung auf den Beweis der Ergodizität von Markoffschen Ketten, *Magyar Tud. Akad. Mat. Kutató Int. Közl.* 8:85-108. Math. Rev. 29:333, #1671, 1965.

Csiszár, I. and Körner, J. 1981, *Information Theory: Coding Theorems for Discrete Memoryless Systems.* New York: Academic Press; Budapest: Akademiai Kiado.

Dobrushin, R. L., 1956, Central limit theorem for nonstationary Markov chains. I, *Theory of Probability and Its Applications* 1:65-80; II, *ibid.* 1:329-383.

Doeblin, W., 1937, Le cas discontinu des probabilités en chaîne. *Publ. Fac. Sci. Univ. Masaryk (Brno)*, no. 236.

Hajnal, J., 1958, Weak ergodicity in non-homogeneous Markov chains, *Proc. Camb. Phil. Soc.* 54:233-246.

Hardy, G. H., Littlewood, J. E., and Pólya, G. 1929, Some simple inequalities satisfied by convex functions. *Messenger of Mathematics* 58:145-152.

Hardy, G. H., Littlewood, J. E., and Pólya, G., *Inequalities,* 2nd ed., Cambridge University Press, 1952.

Karamata, J. 1932, Sur une inégalité relative aux fonctions convexes. *Publ. math. Univ. Belgrade* 1:145-148.

Liese, F. and Vajda, I. 1987, *Convex Statistical Distances*, Leipzig: Teubner.

Marshall, A. W. and Olkin, I. 1979, *Inequalities: Theory of Majorization and its Applications*, Academic Press, New York.

Moran, P. A. P., 1961, Entropy, Markov processes and Boltzmann's H-theorem, *Proc. Camb. Phil. Soc.* 57:833-842.

Morimoto, T., 1963, Markov processes and the H-theorem. *Journal of the Physical Society of Japan* 18:328-331.

Roberts, A. W. and Varberg, D. E. 1973, *Convex Functions*, New York: Academic Press.

Schur, I. 1923, Über eine Klasse von Mittelbildungen mit Anwendungen auf die Determinanten-Theorie. *Sitzungsber. d. Berlin. Math. Gesellschaft* 22:9-20. *Issai Schur Gesammelte Abhandlungen* A. Brauer and H. Rohrbach, eds. 2:416-427. Springer-Verlag, New York, 1973.

Seneta, E. 1973, On the historical development of the theory of finite inhomogeneous Markov chains. *Proc. Camb. Philos. Soc.* 74:507-513.

Winkler, Gerhard. 1985 *Choquet Order and Simplices, with Applications in Probabilistic Models.* Lecture notes in mathematics, 1145. Berlin: Springer-Verlag.

ROCKEFELLER UNIVERSITY, 1230 YORK AVENUE, BOX 20, NEW YORK, NY 10021-6399, U.S.A.

E-mail address: cohen@rockvax.rockefeller.edu

DEPARTMENT OF MATHEMATICS, FACULTY OF SCIENCES, 6 AV. V. LE GORGEU, 29287 BREST, FRANCE

CENTRE OF MATHEMATICAL STATISTICS, BD. MAGHERU 22, RO-70158, BUCURESTI, ROMANIA

Contemporary Mathematics
Volume **149**, 1993

PRODUCTS OF STOCHASTIC, NONSTOCHASTIC, AND RANDOM MATRICES

HARRY COHN

Abstract

The machinery of Markov chains theory is used to study the limit behaviour of products of stochastic matrices. It is then shown that the behaviour of products of nonnegative nonstochastic matrices may be deduced from the stochastics case by a transformation. This deterministic procedure turns out to be useful in studying products of random matrices as well. The type of limit behaviour is closely related to the tail σ–field structure of a nonhomogeneous Markov chain, notion unknown to Doeblin who nevertheless described accurately a number of interesting cases in [7].

1 INTRODUCTION

Let $P_1, P_2, \ldots$ be a sequence of stochastic matrices, i.e. nonnegative matrices with row sums 1. Assume that all matrices are $p \times p$ with p finite. Write $^mP^n = P_m P_{m+1} \cdots P_n$ and $^mP^n(i,j)$ for the (i,j)th entry of the matrix $^mP^n$. We shall make the blanket assumptions that $\{P_n\}$ are *allowable* i.e., have at least one positive entry in each row and each column, and that for any m there exists n (which may depend on m) such that $^mP^n(i,j) > 0$ for all i and j. It is easy to see that $^mP^{n_0}(i,j) > 0$ for all i and j entails $^mP^n(i,j) > 0$ for $n \geq n_0$ and all i and j, in which case we may consider the limit behaviour of the ratios $^mP^n(i,j)/^mP^n(l,j)$ as $n \to \infty$ for all m, i, j and l. We shall say that $\{P_n\}$ is *weakly ergodic* if for all m, i, j and l

$$\lim_{n \to \infty} \frac{^mP^n(i,j)}{^mP^n(l,j)} = 1. \tag{1}$$

AMS 1980 Classification: Primary 60J80; Secondary 60F25.

This paper is in final form and no version of it will be submitted for publication elsewhere.

There is another notion of weak ergodicity, which we shall refer to as *stochastic ergodicity*:

$$\lim_{n \to \infty} \left({}^{m}P^{n}(i,j) - {}^{m}P^{n}(l,j) \right) = 0 \tag{2}$$

for all m, i, j and l. It is clear that weak ergodicity implies stochastic ergodicity but not vice versa. Indeed, (2) does not impose any restriction on the ${}^{m}P^{n}(i,j)$'s which tend to 0 as $n \to \infty$ whereas (1) requires all the ratios to tend to 1. It is well known that (2) is equivalent to the triviality of the tail σ–field of any Markov chain with transition probability matrices $\{P_n; n \geq m\}$ (see [3] or [4]) which may be established by a number of well–known methods in the Markov chains literature.

Consider now a sequence of allowable nonnegative matrices $\{M_n\}$ with entries $M_n(i,j)$ such that for any m there exists n (which may depend on m) with ${}^{m}M^{n}(i,j) > 0$ for all i and j. Weak ergodicity for such matrices has been subject of intensive study especially in the demographic literature (see [1] and [2]). Using the notion of space-time harmonic functions the study of products of non stochastic matrices may be reduced to that of stochastic ones. Indeed, $h(n,i)$ (or briefly h) is said to be *space–time harmonic* for $\{M_n\}$ if $M_n h_{n+1} = h_n$ for $h_n = h(n,.)$ and $n = 1, 2, \ldots$. It follows that the matrix ${}_{h}P_n$ with entries ${}_{h}P_n(i,j) = h(n+1,j) M_n(i,j)/h(n,i)$ is stochastic for any n. A key property of space-time harmonic functions yields

$$ {}^{m}M^{n}(i,j)/{}^{m}M^{n}(l,j) = h(m,i)/h(m,l) \, {}_{h}{}^{m}P^{n}(i,j)/{}_{h}{}^{m}P^{n}(l,j). \tag{3}$$

where $\{{}_{h}{}^{m}P^{n}(i,j)\}$ are products of stochastic matrices $\{{}_{h}P_n\}$. Thus it will suffice to concern ourselves with stochastic matrices which are related to Markov chains. A nonhomogeneous Markov chain with transition probability matrices ${}_{h}P_n(i,j) = h(n+1,j) M_n(i,j)/h(n,i)$ will be said to be an h–chain. We shall first study the general case to elucidate the limit behaviour of ratios of products of random matrices and then specialize to the case of weak and stochastic ergodicity. The method seems to work for random nonnegative matrices as well, case that we shall outline in the end.

2 TAIL σ–FIELD AND RATIO LIMITS

The main probabilistic tool used to describe the asymptotics of such ratios in general is the tail σ–field of a Markov chain. Write $S = \{1, \ldots p\}$, $\Omega = S \times S \times \cdots$, $X_n(\omega) = \omega_n$ for $\omega = (\omega_1, \ldots, \omega_n, \ldots)$ and $\mathcal{F}_n$ for the σ–field generated by $\{X_k; k \geq n\}$. A positive probability vector $\pi^{(m)} =$

$(\pi_1^{(m)}, \ldots, \pi_p^{(m)})$ and a sequence of $p \times p$ stochastic matrices $\{P_n; n \geq m\}$ uniquely determine a probabilty measure $P^{(m)}$ on $\mathcal{F}_m$ such that $\{X_n; n \geq m\}$ is a nonhomogeneous Markov chain on $(\Omega, \mathcal{F}_m, P^{(m)})$ with $P^{(m)}(X_m = i) = \pi_i^{(m)}$ and $P^{(m)}(X_{n+1} = j | X_n = i) = P_n(i, j)$ for $i, j \in S$ and $n \geq m$. We define $\mathcal{T} = \bigcap_{n=1}^{\infty} \mathcal{F}_n$ to be the tail σ–field of $\{X_n\}$. Let $(\Omega, \mathcal{F}, P)$ be a probability space and Λ a set in $\mathcal{F}$. Let $\mathcal{G}$ be a sub σ–field of $\mathcal{F}$. We shall say that Λ is *atomic* with respect to $\mathcal{G}$ if $P(\Lambda) > 0$ and Λ does not contain any subset Λ' in $\mathcal{G}$ such that $0 < P(\Lambda') < P(\Lambda)$. In what follows we shall refer to an atomic set with respect to $\mathcal{T}$ as atomic. If the only atomic sets are those of probability 1 we say that $\mathcal{T}$ is *trivial*. If $\mathcal{T}$ contains a finite number of disjoint atomic sets, i.e., equivalence classes of atomic sets which are P–a.s. equal, and no other sets of positive probability we say that $\mathcal{T}$ is *finite*. Write now $\{A_n \text{ i.o.}\}$ for $\bigcap_{n=1}^{\infty} \bigcup_{m=n}^{\infty} A_m$ and $\{A_n \text{ ult.}\}$ for $\bigcup_{n=1}^{\infty} \bigcap_{m=n}^{\infty} A_m$. Further $A = B$ a.s. will stand for $P(A \triangle B) = 0$ where $\triangle$ is the symbol of the symmetric difference of two sets. We shall write $\lim_{n \to \infty} A_n = A$ a.s. if $\{A_n \text{ i.o.}\} = \{A_n \text{ ult.}\} = A$ a.s. Consider now the tail σ–field of $\{X_n; n \geq m\}$ defined as $\mathcal{T} = \bigcap_{n=m}^{\infty} \mathcal{F}_n$. It is known (see [3] or [4]) that $\mathcal{T}$ is finite with respect to $P^{(m)}$ and the number t of atomic sets $T_1, \ldots, T_t$ does not exceed p. Under the assumption that the transition probabilities $\{^m P^n\}$ become positive for n large enough, the atomic sets of $\mathcal{T}$ are independent of m. To ease the notation we shall also write P for $P^{(m)}$. It follows from the martingale convergence theorem that $\lim_{n \to \infty} P(X_m = i | X_n) = P(T_k | X_m = i) P(X_m = i) / P(T_k)$ for almost all $\omega \in T_k$. Since $P(X_m = i | X_{n+1} = j) = P(X_m = i) / P(X_{n+1} = j)\, ^m P^n(i, j)$ for $\omega \in \{X_{n+1} = j\}$ we are led to the following result (for details see [3] or [4]).

Proposition 1 *Let $\{X_n; n \geq m\}$ be a Markov chain. Then there exist sequences of eventually disjoint subsets of S $\{E_n^{(k)}; n = 1, 2 \ldots\}$ for $k = 1, \ldots, t$ such that for any i, l, k and m*

(i)
$$\lim_{n \to \infty} \{X_{n+1} \epsilon E_n^{(k)}\} = T_k \quad a.s. \tag{4}$$

(ii)
$$\lim_{n \to \infty} \frac{^m P^n(i, j_n)}{^m P^n(l, j_n)} = \frac{P(T_k | X_m = i)}{P(T_k | X_m = l)} \quad for \quad j_n \in E_n^{(k)}. \tag{5}$$

(iii)   *If $E_n := S \setminus \bigcup_{k=1}^{t} E_n^{(k)}$ are not empty, then $P(\{X_{n+1} \in E_n \text{ i.o.}\}) = 0$.*

3 SOME NONHOMOGENEOUS CHAINS CONSIDERED BY DOEBLIN

In [7] Doeblin considered the class of chains satisfying the following condition.

Condition A *There exists a strictly positive number δ such that for any fixed pair of states (i, j) either $P^n(i, j) > \delta$ or $P^n(i, j) = 0$ for all n.*

Doeblin drew a parallel to the homogeneous case (see e.g. [10]) and asserted that in this case it is possible to decompose the state space S into disjoint *final classes* $G_0, G_1, \ldots, G_v$, and each final class G_u, $1 \le u \le v$ can be firther decomposed into *cyclical subclasses* $C_i(u), i = 1, \ldots, d(u)$. These have the following properties as $n \to \infty$.

Theorem 2 *(i) $P_{i,j}^{(m,n)} \to 0$ for every $i \in S$ and $j \in G_0$;*

(ii) $P_{i,j}^{(m,n)} = 0$ for every $i \in G_\alpha$ and $j \notin G_\alpha$

(iii) if $i, j \in G_\alpha$ with $i \in C_l(\alpha)$, $j \in C_{l'}(\alpha)$ then $P_{i,j}^{(m,n)} = 0$ provided that $n - m \ne (l' - l) \pmod{d}(\alpha)$

(iv) if $i, j \in G_\alpha$ with $i \in C_l(\alpha)$ and $j \in C_{l'}(\alpha)$ then $P_{i,j}^{(m,n)} = P_j^{(n)} + \epsilon_{i,j}^{(m,n)}$ provided that $n - m = (l - l') \pmod{d}(\alpha)$. Here $\epsilon_{i,j}^{(m,n)} \to 0$ exponentially as $n \to \infty$ for any m, i and j and the limit distributions $\{P^{(n)}\}$ satisfy $\sum_{j \in C_{l'}}(\alpha) P_j^{(n)} = 1$

(v) for $i \in \cup_{\alpha=0}^v G_\alpha, j \in C_l(\alpha)$ and some $1 \le l \le d(\alpha), \alpha = 1, \ldots, v$, $P_{i,j}^{(m,n)} = P^{(m)}[i, \bar{C}_{n,l}(\alpha)] P_j^{(m)} + \epsilon_{i,j}^{(m,n)}$ where $P_j^{(n)}, \epsilon_{i,j}^{(m,n)}$ are as in (iv) and $P^{(m)}[i, \bar{C}_{n,l}(\alpha)]$ is the limit as $r \to \infty$ of the probability that $X_{n+rd}(\alpha) \in C_l(\alpha)$.

Doeblin subsequently relaxed his condition to allow positive $p_{i,j}^{(n)}$ to tend to 0 as $n \to \infty$. He distinguished between two cases: when the series of such probabilities converges, and when the series diverges. In the first case the asymptotic behaviour of transition probabilities is similar to the one described above. In the second case it may be completely different as cyclically moving subclasses may be skipped and the trajectory of the chain may be subject to wild fluctuations. We quote here Doeblin's description of such chains:

- *Le mouvement pourra être représenté par des tourbillons et disons un morceau de bois qui est entraîné dans un de ces tourbillons qui y reste assez lontemps, mais qui finit par être expulsé de ce tourbillon et qui sera alors après un certain temps attiré par un autre (ou le même). Alors il se peut que le morceau de bois sera finalement entraîné dans un tourbillon qui ne le lâche plus et qui le fait toujour tourner en rond, ou bien aussi il est éternellement expulsé d'un tourbillon, puis entraîné dans un autre, puis expulsé de nouveau. Mais le durées moyennes de séjour dans un tourbillon tendent vers l'infinité si n augmente indéfinitement.*

The general case is described by Proposition 1. Under Doeblin's Condition A the sets $\{E_n^{(k)}\}$ correspond to the cyclically moving subclasses. Doeblin's description given above corresponds to the case when the tail σ-field is trivial and each cyclically moving set is entered infinitely often with probability 1. For details see [4].

4 Non Stochastic Matrices

Let $\{M_n\}$ be a sequence of finite order $p \times p$ nonnegative matrices, $M(i,j)$ the (i,j) entry of the matrix M and ${}^mM^n = M_m \cdots M_n$ for $m < n$. Assume that for each m all the entries of ${}^mM^n$ become strictly positive for n large enough.

We shall say (see e.g., Hajnal [9]) that $\{M_n\}$ is *weakly ergodic* if for all m, i, j and l

$$\lim_{n \to \infty} \frac{{}^mM^n(i,j)}{{}^mM^n(l,j)} := \gamma_m(i,l). \tag{6}$$

(For stochastic $\{M_n\}$ weak ergodicity necessarily implies $\gamma_m(i,l) = 1$.) Simple manipulations show that for arbitrary k weak ergodicity yields that

$$h(m,i) := \lim_{n \to \infty} \frac{{}^mM^n(i,k)}{{}^1M^n(1,k)} = \left(\sum_{j=1}^{p} {}^1M^{m-1}(1,j)\,\gamma_m(j,i) \right)^{-1} \tag{7}$$

exists for all m and i, $0 < h(m,i) < \infty$, and

$$\sum_{j=1}^{p} M_n(i,j)h(n+1,j) = h(n,i). \tag{8}$$

A function h with the property (8) is said to be *space–time* harmonic for $\{M_n\}$. Up to a multiplicative constant, there is only one space–time harmonic function associated with a weakly ergodic $\{M_n\}$ (see [5]). It follows from (8) that the matrices $\{_hP_n\}$ with $P_n(i,j) = h(n+1,j)M_n(i,j)/h(n,i)$

are stochastic. Write ${}^m_h P^n =_h P_m \cdots_h P_n$. An important property implied by (8) is

$$
{}^m M^n(i,j) = \frac{h(m,i)}{h(n+1,j)} {}^m_h P^n(i,j) \tag{9}
$$

for all m, n, i and j, which allows the study of products of nonnegative matrices to be carried out via stochastic matrices (see [5]).

Making use of the notion of tail σ-field of a nonhomogeneous Markov chain we can use (5) to describe the asymptotic behaviour of ratios of nonnegative matrices. First we give a result that shows that weak ergodicity is related to stochastic ergodicity. The relationship is made clear by the tail σ-field of the associated Markov chains.

Theorem 3 *The following conditions are equivalent:*
(i) The sequence $\{M_n\}$ is weakly ergodic.
(ii) The $_h P$–chain has a trivial tail σ–field for any M–harmonic function h.
(iii) The $_h P$–chain is stochastically ergodic for any M–harmonic function h.

For the proof see [5].
We show now how the Hajnal criterion for (6) follows from (2) via Markov chains theory. Let T be a non–negative matrix with entries $t(i,j)$. Define

$$
\phi(T) = \min_{i,j,k,l} \frac{t(i,k)\, t(j,l)}{t(j,k)\, t(i,l)}
$$

$$
\alpha(T) = 1 - \min_{i,j} \sum_{l=1}^{p} \min((t(i,l), t(j,l))
$$

The functions ϕ and α are called ergodic coefficients. The coefficient ϕ has been used in connection with weak ergodicity and is related to Birkhoff's contraction coefficient, whereas variants of α have proved useful in deriving stochastic ergodicity. Doeblin was the first to consider such a coefficient to ensure (2) (see [7]).

Hajnal [9] proved that $\sum_{n=1}^{\infty} \sqrt{\phi(M_n)} = \infty$ implies weak ergodicity. We show now that this result follows from Theorem 3. Indeed, notice that for any space time harmonic function h

$$
\frac{_h P_n(i,k)\, _h P_n(j,l)}{_h P_n(i,l)\, _h P_n(j,k)} = \frac{M_n(i,k)\, M_n(j,l)}{M_n(i,l)\, M_n(j,k)}
$$

Thus if we write P_n for $_hP_n$ by we only need to show that

$$\sum_{n=1}^{\infty} \sqrt{\min_{i,j,k,l} \frac{P_n(i,k)\, P_n(j,l)}{P_n(i,l)\, P_n(j,k)}} = \infty$$

Fix a matrix P_n and let l_i and k_j be chosen such that $P_n(i,l_i) \geq 1/p$ and $P_n(j,k_j) \geq 1/p$. Then $\min_{i,j,k,l} \dfrac{P_n(i,k)\, P_n(j,l)}{P_n(i,l)\, P_n(j,k)} \leq p^2 \min_{i,j}(P_n(i,k_j), P_n(j,l_i))$
$\leq p^4 \min_{i,j}(\min((P_n(i,k_j), P_n(j,k_j))\, \min((P_n(j,l_i), P_n(i,l_i))))$
$\leq p^4 \left(\min_{i,j}(\sum_{k=1}^{p} \min(P_n(i,k), P_n(j,k)))\right)^2$. Thus we deduce that
$\sum_{n=1}^{\infty}(1 - \alpha(P_n)) = \infty$, which is known to imply stochastic ergodicity for $\{P_n\}$ (see e.g [3]).

Suppose that there are t extremal space time harmonic functions $h_1, \ldots, h_t$. Let $h = h_1 + \cdots + h_t$.

Theorem 4 *(i)* *The tail σ–field $\mathcal{T}$ of the $_hP$–chain has t atomic sets $T_1, \ldots, T_t$ such that*

$$h_k(n,i) = {}_hP(T_k|X_n = i)\, h(n,i) \tag{10}$$

and to each h_k, $k = 1, \ldots, t$, there corresponds a nonempty sequence of eventually disjoint sets $\{E_n^{(k)}\}$ such that $\lim_{n\to\infty} {}_hP(\bigcup_{k=1}^{t} E_n^{(k)}) = 1$, and

$$\lim_{n\to\infty} \frac{{}^mM^n(i,j_n)}{{}^mM^n(l,j_n)} = \frac{h_k(m,i)}{h_k(m,l)} \quad \text{for } j_n \in E_n^{(k)}. \tag{11}$$

for any m, i and l.

(ii) *If $E_n := S \setminus \bigcup_{k=1}^{t} E_n^{(k)}$ are not empty, then there are some non–negative functions $c_k(n,i)$ with $k = 1, \ldots, t$ such that $\sum_{k=1}^{t} c_k(n,j) = 1$ and for any m, i, l and j*

$$\frac{{}^mM^n(i,j)}{{}^mM^n(l,j)} \sim \frac{\sum_{k=1}^{t} c_k(n,j)\, h_k(m,i)}{\sum_{k=1}^{t} c_k(n,j)\, h_k(m,l)} \tag{12}$$

where $\sim$ means that the ratio of the left and right side tends to 1 as $n \to \infty$.

For the proof see [5].

5 PRODUCTS OF RANDOM MATRICES

Consider now a probability space $(\Omega, \mathcal{F}, P)$ with generic element ω and some $p \times p$ random matrices $\{X_n\}$. Denote the corresponding matrix products by $\{{}^m X^n\}$. Assume that almost all sequences $\{X_n\}$ are weakly ergodic and let $H(n, i)$ stand for the space–time harmonic function of $\{X_n\}$ which is also random, being a function of $X_0, X_1, \ldots$. Notice that

$$H(n, i)/H(n+1, j) = \lim_{s \to \infty} {}^n X^s(i, k)/{}^{n+1} X^s(j, k) \text{ a.s} \qquad (13)$$

Write $\{Y_n\}$ and $\{{}^m Y^n\}$ for the attached random stochastic matrices. Simple manipulation of (9) in the random setup yield the representation

$$\log {}^1 X^n(i, j) = \eta_1(i, l) + \sum_{m=0}^{n} \zeta_m(l) + \eta_{n+1}(l, j) + \log {}^1 Y^n(i, j) \qquad (14)$$

where $\eta_m(k, u) = \log(H(m, k)/H(m, u))$ and
$\zeta_m(l) = \log(H(m, l)/H(m+1, l))$ for any l and $m = 1, 2 \ldots$. If we assume that $\{X_n\}$ is stationary then $\{\zeta_n(j)\}$ and $\{\eta_{n+1}(l, j)\}$ are also stationary. Under suitable assumptions, η_n is negligible in limit properties obtained when dividing this representation by constants tending to ∞, and if $\{{}^1 Y^n\}$ does not converge too rapidly to 0 then limit results for suitably normed $\{{}^m X^n\}$ turn out to be deducible from the stationary process $\{\zeta_n(j)\}$. For example, Kingman's subadditive ergodic theorem [11] implies

$$\lim_{n \to \infty} \log \|{}^1 X^n\|/n = \gamma \text{ a.s}$$

where the norm of the matrix $M = (M(i, j))$ is defined as
$\|M\| = \max_{1 \le i \le n} \sum_{j=1}^{p} |M(i, j)|$. See also Furstenberg and Kesten [8]. This may be shown to follow from (14) and Birkhoff's ergodic theorem. Also central limit theorems for weakly dependent stationary random matrices which improves on the variants known in the literature are obtainable in this way.

Consider a sequence $\{\alpha_n\}$ of nonnegative integers such that $\lim_{n \to \infty} \alpha_n = 0$. Write $\mathcal{M}_m^n$ for the σ–field generated by $X_m, \ldots, X_n$, $\mathcal{M}_{-\infty}^n$ for the σ–field generated by $\ldots, X_n$, and $\mathcal{M}_n^\infty$ for the σ–field generated by $X_n, \ldots$. We shall say that the sequence $\{X_n\}$ is strongly mixing if for each k and n, $A \in \mathcal{M}_{-\infty}^k$ and $B \in \mathcal{M}_{n+k}^\infty$

$$|P(A \cap B) - P(A)P(B)| \le \alpha_n.$$

We shall say that the sequence $\{X_n\}$ is ρ–mixing if $\lim_{n \to \infty} \rho_n = 0$

where

$$\rho_n := \sup_{m, f \in L_2(\mathcal{M}_{-\infty}^m), g \in L_2(\mathcal{M}_{m+n}^\infty)} |\mathrm{Corr}(f, g)|.$$

In what follows $\xrightarrow{d}$ denotes convergence in distribution and $N(0, 1)$ is the standard normal law.

Theorem 5 *Suppose that $\{X_n\}$ is a strictly stationary sequence of positive $p \times p$ matrices such that one of the following conditions holds.*
(i) For some $\delta > 0$, $\sum_{n=1}^\infty \alpha_n^{\delta/(2+\delta)} < \infty$ and $E|\log(X_1(i, j))|^{2+\delta} < \infty$ for every i and j.
(ii) $\sum_{n=1}^\infty \rho_n < \infty$ and, $E|\log(X_1(i, j))|^2 < \infty$ for every i and j.
Then there are two constants, $\sigma \geq 0$ and γ such that for all i and j

$$n^{-1/2}\{\log {}^1X^n(i, j) - n\gamma\} \xrightarrow{d} N(0, \sigma).$$

where we interpret $N(0, 0)$ as a point mass at 0.

This theorem improves on the results know in the literature. Indeed, we do away with the condition:

$$1 \leq \max_{i,j} X_n(i, j) / \min_{i,j} X_n(i, j) \leq C$$

for all n and some constant C assumed by all papers dealing with central limit theorems for products of positive matrices. The asymptotic independence conditions are also relaxed. For details see [6].

REFERENCES

[1] Cohen, J.E. Contractive inhomogeneous products of non–negative matrices. *Math. Proc. Cambridge Phil. Soc.*, 86, 351–364, 1979.
[2] Cohen, J.E. Ergodic theorems in demography. *Bull. Amer. Math. Soc.*, 1, 275–295, 1979.
[3] Cohn, H. Finite non–homogeneous Markov chains: asymptotic behaviour. *Adv. Appl. Prob.*, 8, 502–516, 1976.
[4] Cohn, H. Products of stochastic matrices and applications. *Internat. J. Math. & Math. Sci.* 12, 2, 209–233, 1989.
[5] Cohn, H. and Nerman, O. On products of nonnegative matrices, *Ann. Prob.* , 18, 4, 1806-1815, 1990.
[6] Cohn, H., Nerman, O. and Peligrad, M. Weak ergodicity and products of random matrices. To appear in *J. Theor. Prob.*, 1993.
[7] Doeblin, W. Le cas discontinu de probabilités en chaîne. *Pub. Fac.*

Sci. Univ. Massaryk (Brno), 236, 1937.

[8] Furstenberg, H. and Kesten, H. Products of random matrices. *Ann. Math. Statist.*, 31, 2, 457-469, 1960.

[9] Hajnal, J. On products of non–negative matrices. *Math. Proc. Cambridge Phil. Soc.*, 79, 521–530, 1976.

[10] Kemeny, J.G., Snell, J.L. and Knapp, A.W. **Denumerable Markov Chains.** *Springer, New–York*, 1976.

[11] Kingman, J.F.C. Subadditive ergodic theory. *Ann. Prob.*, 1, 883-889, 1973.

DEPARTMENT OF STATISTICS, UNIVERSITY OF MELBOURNE, PARKVILLE, VICTORIA, 3052, AUSTRALIA.

E-mail address: harry@ariel.ucs.unimelb.edu.au

Contemporary Mathematics
Volume **149**, 1993

SHUFFLING WITH TWO MATRICES

J. Hajnal

1. Introduction

This paper discusses an application of the theory of finite non-homogeneous Markov chains - a theory which was pioneered by Doeblin (reference (1)). The theory (for general accounts see (8), (9) and (13)) has had to be adapted for the use to which it is put here; but it is safe to say that everything in this paper would have been easily intelligible to Doeblin.

The problem discussed here derives from P.A.P. Moran, who in 1962 in two publications ((10) and (11)) formulated the need for what he called a general theory of shuffling. This need arose from his researches in genetics in which he used non-homogeneous Markov chain models.

One of Moran's examples, slightly generalized, is as follows.

Example 1. A system can be in one of three states. Two operations are repeatedly applied to it. They can be represented by the following two transition matrices.

$$P = \begin{pmatrix} 1 & 0 & 0 \\ 0 & 0 & 1 \\ 0 & \beta & 1-\beta \end{pmatrix}, \qquad Q = \begin{pmatrix} 0 & 1 & 0 \\ \gamma & 1-\gamma & 0 \\ 0 & 0 & 1 \end{pmatrix} \qquad (1)$$

$$0 < \beta, \gamma < 1.$$

If the system is in the first state application of the operation P will leave it unchanged; but repeated application of P in an obvious sense "shuffles" the other two states. Similarly Q leaves the system in the third state if it is in that state, but repeated application of Q shuffles the other states. What happens in the long run if both P and Q are repeatedly applied?

1991 Mathematics Subject Classification: 60J10, 60J20.
This paper is in final form and no version of it will be submitted for publication elsewhere.

Let $v_j(r) = \Pr$ [System is in state j after r operations have been applied] and

let $\underset{\sim}{v}(r) = [v_1(r), v_2(r), v_3(r)]$. If the two operations are <u>alternately</u> applied, then

$$\text{as } r \to \infty, \ \underset{\sim}{v}(r) \to \underset{\sim}{v}^* = K[\beta\gamma, \beta, 1] \tag{2}$$

$$\text{where } K^{-1} = \beta\gamma + \beta + 1.$$

That this is so may be seen by considering the state of the system after every second transition. This sequence of states is governed by the (homogeneous) Markov chain with transition matrix PQ or QP.

Moran was concerned with the question: what happens to $\underset{\sim}{v}(r)$, as $r \to \infty$, under more general sequences of P's and Q's than strict alternation. To discuss this question we introduce the following notation and definition. The phrase "non-homogeneous Markov chain" will be abbreviated to "non-hom. M.C.".

A non-hom. M.C. is specified by v(0), a vector of initial probabilities, and a sequence $\{H(g)\} = \{H(1), H(2), H(3)...\}$ of transition matrices. (For Example 1 above H(g) is either P or Q for all g.) Let

$$\underset{\sim}{v}(r) = \underset{\sim}{v}(0).H(1).H(2).H(3)...H(r-1).H(r) \tag{3}$$

Def. 1   A non-hom. M.C. <u>shuffles</u> the states if $\lim\limits_{r \to \infty} \underset{\sim}{v}(r) = \underset{\sim}{v}^*$ exists, is strictly positive and is the same for all $\underset{\sim}{v}(0)$.

In non-hom. M.C. theory the term "strong ergodicity" has been used to describe the limiting behaviour of $\underset{\sim}{v}(r)$ that is here termed shuffling, but without the positivity requirement.

We now return to the problem posed by Moran. Suppose there is a system which can be in a finite number of states and that there is a fixed number of operations, each represented by a matrix, which can be applied. Each operation shuffles a proper subset of the states, but the union of these subsets comprises all the states.

What restrictions, Moran asked, on the matrices themselves or on the order in which the matrices are applied will result in shuffling?

In this paper only the case of two matrices (denoted P and Q) is discussed and the only order restriction treated is the least restrictive condition that will produce shuffling, namely the condition that both P and Q occur infinitely often in the sequence $\{H(g)\}$.

A model for the shuffling of cards first formulated by Poincaré and developed by Hadamard is part of the early history of Markov chains (discussed by Doeblin in (2), pp. 24-26). In this model there is a single transition matrix which is doubly stochastic. Doeblin in (1) treated a non-homogeneous generalisation of the Poincaré-Hadamard model in which all transition matrices are doubly stochastic and have the same "pattern" (see Def. 5 in section 3 below). A special case of Doeblin's generalisation was studied, without knowledge of Doeblin's work, in 1948-9 by Horton and Smith ((6) and (7)) in connection with the problem of generating random numbers by physical mechanisms. The same model was treated by Fourier-analytical methods, without reference to Markov chains, by Dvoretzky and Wolfowitz in 1951 ((3)). This type of generalisation of the Poincaré-Hadamard model is of a different character from the subject of the present paper.

The remainder of this paper is arranged as follows. Section 2 sets out, with special reference to Doeblin's work, the parts of non-hom. M.C. theory required for our purposes. Non-hom. M.C. theory is needed for proving our results, but is not necessary for understanding them. The main results are described in section 3, which presupposes only a knowledge of the elementary theory of finite homogeneous Markov chains. Sections 4-6 contain the proofs and additional remarks.

2. Use of Doeblin's theorem

We illustrate the way in which the theory of non-hom. M.C.'s may be applied to Moran's problem by proving Proposition 1.

Proposition 1   A Markov chain {H(g)} where each H(g) is either P or Q of Example 1 above will shuffle provided both P and Q occur infinitely often.

The proof proceeds in two stages which are labelled A and B.
(A) If u,w are positive integers let

$$T(u,w) = H(u+1) . H(u+2)...H(u+w-1) . H(u+w) \quad (4)$$

and let $T(u,w) = \{t_{ij}(u,w)\}$.

Def. 2 The chain {H(g)} satisfies the ergodic principle if

$$\lim_{w\to\infty} |t_{i\ell}(u,w) - t_{j\ell}(u,w)| = 0 \qquad \text{for all u,i,j,}\ell.$$

We here use the terminology employed by Doeblin following Kolmogorov. In the literature on non-hom. M.C.'s a chain which satisfies the ergodic principle is termed weakly ergodic.[1]

Def. 3 For any stochastic matrix $H = \{h_{ij}\}$ let

$$\alpha\,[H] \;=\; \sum_{j} \min_{i}\; h_{ij}\,.$$

Remark on Def 3.: $\alpha[H] > 0$ iff H has at least one strictly positive column. Such a matrix is often called a Markov matrix. The theory of non-hom. M.C.'s is now usually developed in terms of a more general type of matrix called a scrambling matrix.   A stochastic matrix $H = \{h_{ik}\}$ is called a scrambling matrix if for any two states i and j there is at least one k such that both $h_{ik} > 0$ and $h_{jk} > 0$. (Markov matrices are a special case.) There is a more general coefficient corresponding to $\alpha[H]$ in use with scrambling matrices. Coefficients like $\alpha[H]$ or $1-\alpha[H]$ are called ergodicity coefficients. Ergodicity coefficients are now usually denoted by Greek letters, but there is no agreement as to which letter should denote which coefficient. Doeblin had no notation for the coefficient $\alpha[H]$, which he introduced only to state the following theorem.

Doeblin's Theorem A chain $\{H(g)\}$ satisfies the ergodic principle if, and only if, there is a strictly increasing sequence of positive integers g_1, g_2, $g_3 \ldots$ such that (with $w_{k+1} = g_{k+1} - g_k$)

$$\sum_{k=1}^{\infty} \alpha\,[\,T(g_k,\,w_{k+1})\,]\; =\; \infty \tag{5}$$

Def. 4 In the application of Doeblin's theorem the matrices

$$H(g_k+1)\,,\;\; H(g_k+2)\,,\;\; H(g_k+3)\ldots H(g_k+w_{k+1})$$

will be called the (k + 1)-th block of matrices.

---

1   The definition of weak ergodicity in Def.2 is in terms of limiting probabilities for transitions starting at any instant $u = 0,1,2,\ldots$ The corresponding definition of strong ergodicity implied in Def.1 above refers in effect only to $u = 0$. In general non-hom. M.C. theory the definition of strong ergodicity needs to be framed, in analogy to Def.2, with reference to any starting instant $u = 0,1,2,\ldots$ However, under the conditions of this paper (i.e. with a finite number of distinct transition matrices each of which occurs infinitely often) the simpler definition involving only $u = 0$ is equivalent to the more general one.

Note on Doeblin's theorem: This theorem is now usually stated using the ergodicity coefficient associated with scrambling matrices. The proof is exactly the same if Doeblin's coefficient is used. For our purposes Doeblin's coefficient is convenient. That the theory of non-hom. M.C.'s can be developed in term's of Doeblin's coefficient and Markov matrices instead of scrambling matrices was pointed out by Seneta (12).

For the chain of proposition 1, QP can be taken as one block every time P follows Q. Since QP is a Markov matrix Doeblin's theorem implies that the chain satisfies the ergodic principle.

(B). The following theorem, based on work more recent than Doeblin, will also be needed.

Theorem 1   Suppose a chain $\{H(g)\}$ fulfils two conditions

    (i) $\{H(g)\}$ satisfies the ergodic principle and

    (ii)   the matrices $H(g)$ have a common positive stationary stochastic row vector $\underset{\sim}{v}^*$, i.e. $\underset{\sim}{v}^*H(g)=\underset{\sim}{v}^*$ for all g.

           Then

    (a) $\underset{\sim}{v}^*$ is the only stationary stochastic row vector common to all the matrices $H(g)$, and

    (b)   the chain shuffles, i.e.
           $\underset{\sim}{v}(r)\to\underset{\sim}{v}^*$, as $r\to\infty$, for any v(0).

Proof: (a) If $H(g_1)$, and $H(g_2)$ are two matrices of the chain then $\underset{\sim}{v}^*$ is a stationary vector of the product $H(g_1).H(g_2)$ since $\underset{\sim}{v}^*[H(g_1).H(g_2)]=[\underset{\sim}{v}^*H(g_1)].H(g_2)=\underset{\sim}{v}^*H(g_2)=\underset{\sim}{v}^*$.   Thus $\underset{\sim}{v}^*$ is a stationary stochastic vector of T(u,w) for all u and w. But if the ergodic principle is satisfied T(u,w) is a Markov matrix for any u and all sufficiently large w, and a Markov matrix has a unique stationary stochastic vector.

(b) This part of the theorem is widely implicit in the literature. It is a special case of Hajnal (4) Theorem 7, and of Isaacson and Madsen (8) Theorem V.4.3.

Now P and Q of Example 1 have in common the stationary stochastic vector $\underset{\sim}{v}^*$ of formula (2) in section 1 above. Moreover, part A of the present section shows that the chain of Proposition 1 satisfies the ergodic principle. Theorem 1 therefore applies to this chain and the proof of Proposition 1 is complete.

Doeblin's theorem and Theorem 1 are the tools needed in the proof of Theorem 3 below.

3. The main results

We now consider the general situation where two matrices (like those of Example 1 above) are applied to a system with n states ($n \geq 3$) and each matrix shuffles some of the states, but not all.

The following <u>notation</u> and <u>assumptions</u> apply in the remainder of this paper. $P = \{p_{ij}\}$ and $Q = \{q_{ij}\}$ are nxn stochastic matrices. Γ_p and Γ_q are proper subsets of the n states.

$\Gamma_p \cup \Gamma_q = \{1, 2, \ldots n\}$. The numbers of states contained in Γ_p and Γ_q are n_p and n_q respectively ($2 \leq n_p, n_q < n$). P has an $n_p \times n_p$ <u>stochastic</u> submatrix, denoted P*, over the states of Γ_p, while Q has an $n_q \times n_q$ <u>stochastic</u> submatrix, denoted Q*, over the states of Γ_q. Moreover $p_{ii} = 1$ for $i \notin \Gamma_p$ and $q_{ii} = 1$ for $i \notin \Gamma_q$. Thus $p_{ij} = 0$ if <u>either</u> $i \in \Gamma_p$ and $j \notin \Gamma_p$ <u>or</u> $i \notin \Gamma_p$ and $j \neq i$. Similarly $q_{ij} = 0$ if <u>either</u> $i \in \Gamma_q$ and $j \notin \Gamma_q$ <u>or</u> if $i \notin \Gamma_q$ and $j \neq i$.

Theorems 2 and 3 relate to non-hom. M.C.'s defined by a sequence of transition matrices $\{H(g)\}$ where $H(g)$ is either P or Q for all $g = 1, 2, 3, \ldots$

<u>Theorem 2</u> Suppose that

   (a)   the sets Γ_p and Γ_q have just one state in common;
   (b)   both the submatrices P* and Q* are irreducible and aperiodic[2]; and
   (c)   the sequence $\{H(g)\}$ contains both P and Q infinitely often.

Then in a realisation of the chain everyone of the states $1, 2, \ldots n$ is visited infinitely often with probability 1.

Note: The proof of theorem 2 is given in section 5 below.

Condition (b) is necessary if the minimum order restriction (c) is to guarantee that each state is visited i.o. The role of condition (a) is more ambiguous. It is necessary for the theorem to hold that Γ_p and Γ_q have one or more states in common. However, the theorem may fail if $\Gamma_p \cap \Gamma_q$ contains more than one state. This is illustrated by Example 2.

---

   [2]   The terms "irreducible" and "aperiodic" are defined in most expositions of the theory of Markov chains, e.g. in references (8), (9) and (13). An irreducible aperiodic transition matrix H has the property that there exists a positive integer r_0 such that H^r has no zero elements for $r \geq r_0$.

Example 2

$$P = \begin{pmatrix} 0 & 1 & 0 & 0 \\ 0 & 0 & 1 & 0 \\ x & x & 0 & 0 \\ 0 & 0 & 0 & 1 \end{pmatrix}, \quad Q = \begin{pmatrix} 1 & 0 & 0 & 0 \\ 0 & 0 & x & x \\ 0 & 1 & 0 & 0 \\ 0 & 0 & 1 & 0 \end{pmatrix}$$

(x denotes a non-zero element)

Here Γ_p comprises states 1, 2 and 3, while Γ_q comprises states 2, 3 and 4. In the matrix PQ state 2 is absorbing and in QP state 3 is absorbing. Both P* and Q* are irreducible and aperiodic. If $H(g)=P$ for g odd and $H(g)=Q$ for g even when the initial state is 2, then states 1 and 4 will never be reached.

However, if $\Gamma_p \cap \Gamma_q$ contains more than one state the conclusion of the theorem may hold. If we modify a pair of matrices P and Q which satisfy the theorem by changing some of the zeros to positive elements the theorem will still hold since all transitions with positive probability prior to the modification will still be possible. To state this formally we require two definitions.

Def. 5 Two matrices have the same pattern if their strictly positive elements are in the same places, though the magnitudes of these positive elements may differ between the two matrices.

Def. 6 The pattern of matrix A is embedded in matrix B if B has positive elements in at least the same places as A. B may have additional positive elements.

Corollary to Theorem 2 Suppose the patterns of two matrices P and Q which satisfy conditions (a) and (b) of theorem 2 are embedded in matrices U and V respectively. Let {H(g)} be a chain where H(g) is U or V for each g and both U and V occur i.o.

Then in a realisation of {H(g)} each of the n states will be visited i.o. with prob. 1.

The conditions of Theorem 2 are not sufficient to guarantee shuffling. It may happen under the conditions of the theorem that while every state is reached infinitely often any one state is visited only at periodic intervals. This possibility is illustrated by the following example.

<u>Example 3</u> Here n=44. Γ_p comprises the states 1,2,...16; Γ_q comprises the states 16,17,...44.

The positive elements of P are as follows:

$$p_{1,16}=1$$
$$p_{i,i-1}=1 \text{ for } i=2,3,...11 \text{ and for } i=13,14,15 \text{ and } 16.$$
$$p_{12,11}=\beta \text{ and } p_{12,16}=1-\beta \text{ where } 0<\beta<1.$$
$$p_{jj}=1 \text{ for } j=17,18,...44.$$

The positive elements of Q are as follows:

$$q_{jj}=1 \text{ for } j=1,2,...15.$$
$$q_{i,i+1}=1 \text{ for } i=16,17,...21 \text{ and for } i=23,24,...43.$$
$$q_{22,16}=\gamma \text{ and } q_{22,23}=1-\gamma \text{ where } 0<\gamma<1.$$
$$q_{44,16}=1.$$

Both P* and Q* are aperiodic as well as irreducible. Yet PQ and QP have period 11. Thus if P and Q occur in strict alternation each state will be visited only periodically.

To guarantee shuffling the conditions of Theorem 2 must be strengthened. This can be done in two ways. One way is to place restrictions (additional to condition (c)) on the order in which the matrices occur in the chain $\{H(g)\}$. With matrices satisfying conditions (a) and (b) of Theorem 2 appropriate restrictions on the order in which they are applied will ensure that shuffling occurs. This avenue will not be pursued further here. The other way to strengthen the conditions of Theorem 2 is to place additional conditions on the positions of the non-zero elements of P and Q.

If one of the matrices, which may arbitrarily be taken to be P, fulfils either one of two additional conditions then the chain will shuffle. To state these conditions additional notation is needed. Let the single state shared by Γ_p and Γ_q be state s; then p_{ss} will be the corresponding diagonal element of P. Moreover let $\bar{P}$ denote the $(n_p-1)\times(n_p-1)$ submatrix of P* obtained by deleting the s-th row and the s-th column.

We can now formulate Theorem 3.

<u>Theorem 3</u>. Suppose conditions (a) to (c) of Theorem 2 are satisfied and that, in addition,

(d) <u>either</u> (i) $p_{ss}>0$ <u>or</u> (ii) the submatrix $\bar{P}$ is primitive, i.e. $\bar{P}^r$ has no zero elements for some r.

Then shuffling will take place.

The proof of Theorem 3 will be found in section 4 below. Here we illustrate Theorem 3 by a simple example, present a corollary, discuss one of the conditions of the theorem and formulate an extension.

<u>Example 4</u>   For arbitrary n let $\Gamma_p = \{1,2\}$ and $\Gamma_q = \{2,3,\dots n\}$, so that s=2. If P* is any aperiodic and irreducible matrix shuffling will occur.

$$\text{Let } P^* = \begin{pmatrix} \beta & 1-\beta \\ 1-\gamma & \gamma \end{pmatrix} \qquad 0 \le \beta, \gamma < 1 \text{ (but not } \beta = \gamma = 0)$$

If $\beta = 0$, $\gamma > 0$ condition d(i) is satisfied. If $\beta > 0$, $\gamma = 0$ d(ii) is satisfied. If $\beta > 0$ and $\gamma > 0$ both d(i) and d(ii) are satisfied.

<u>Corollary to Theorem 3</u> If two matrices P and Q satisfy conditions (a) and (b) of Theorem 2 and also (d) of Theorem 3 then every product of P's and Q's repeated an arbitrary number of times is both irreducible and aperiodic provided P and Q each occur at least once.

Proof of Corollary: Suppose there is some product of P's and Q's which is either reducible or periodic. Then the chain {H(g)} obtained by repeating this product infinitely often will not shuffle. However, by Theorem 3 there can be no such chain.

An essential condition for shuffling, as is clear from Theorem 1 and indeed implicit in the definition of shuffling, is the existence of a vector $\underset{\sim}{v}^* = \{v^*_1, v^*_2, \dots v^*_n\}$ with $v^*_i > 0$ for all i such that

$$\sum_{i-1}^{n} v_i^* = 1 \tag{6}$$

and $\underset{\sim}{v}^* P = \underset{\sim}{v}^* Q = \underset{\sim}{v}^*$ \hfill (7)

Yet the existence of such a vector is not among the conditions of Theorem 3. Condition (a) of Theorem 2 is sufficient. The significance of this condition for Theorem 2 has already been discussed. Its additional significance for Theorem 3 is set out in Proposition 2.

<u>Proposition 2</u>   If the intersection $\Gamma_p \cap \Gamma_q$ contains just one state there exists a unique positive vector $\underset{\sim}{v}^*$ satisfying (6) and (7).

Proof: The vector $\underset{\sim}{v}^*$ will satisfy (7) if (i) for $i,j \in \Gamma_p$ the ratio v^*_i / v^*_j is equal to the ratio of the corresponding elements of the unique positive stationary vector of n_p components for the irreducible matrix P*, and if (ii) for $i,j \in \Gamma_q$ v^*_i / v^*_j is equal to the ratio of the corresponding elements of the n_q-component stationary vector of Q*. Since there is just one state common to both Γ_p and Γ_q $\underset{\sim}{v}^*$ is determined by these ratios up to a multiplicative constant. $\underset{\sim}{v}^*$ may be normed to satisfy (6). Proposition 2 is thus proved.

Theorem 3 may be used to derive sufficient conditions for shuffling in the case where the inter-section $\Gamma_p \cap \Gamma_q$ contains two or more states. We state these conditions in Proposition 3.

Proposition 3   (a) Suppose that the patterns of two matrices P and Q,
                                   which satisfy conditions (a) and (b) of Theorem 2 and
                                   also condition (d) of Theorem 3, are embedded in
                                   matrices U and V respectively.
                                   (b) Suppose there exists a stochastic row vector $\underset{\sim}{v}^*$ such
                                   that $\underset{\sim}{v}^*U = \underset{\sim}{v}^*V = \underset{\sim}{v}^*$.
                                   (c) Let $\{H(g)\}$ be a chain where $H(g)$ is U or V for each
                                   g and both U and V occur i.o.

                                   Then $\{H(g)\}$ is a shuffling chain.

The proof of Proposition 3 will be found at the end of section 4 below.

As has been said, the existence of a stationary stochastic vector common
to the transition matrices is a necessary condition for shuffling to occur. In
Theorem 3, where $\Gamma_p \cap \Gamma_q$ is assumed to contain only one state, the existence of
a common stationary vector follows from the patterns of P and Q and does not
depend on the magnitudes of the positive elements of P and Q. In Proposition 3 on
the other hand the assumption is made that U and V have elements whose
magnitudes are such that a common stationary vector exists. An obvious example
is the case where both U and V are doubly stochastic, so that the limiting
distribution assigns equal probability to all states. This case is a generalisation of
the classic Poincaré-Hadamard model of card shuffling which was mentioned in the
Introduction.

4. Proof of Theorem 3 and Proposition 3

It is convenient to prove Theorem 3 before Theorem 2. The proof follows
the lines of the proof of Proposition 1 in section 2 above. To apply Doeblin's
theorem it is necessary to show how the sequence $\{H(g)\}$ can be divided into
suitable blocks. (See Def. 4.) Given particular P's and Q's there are often simple
ways to accomplish this, as in the proof of Proposition 1 in section 2 above. Here
we shall describe a way of forming blocks which will in all cases ensure that the
ergodicity coefficient α has a common positive lower bound for all blocks and thus
that Doeblin's theorem can be applied.

Def. 7 A path from state i at step u to state j at step u+w is a sequence of indices
$i = i_0,\ i_1,\ i_2, \ldots i_{w-1}, i_w = j$ such that

$$h_{i_{k-1},\,i_k}(u+k) > 0 \qquad \text{for all } k = 1, 2 \ldots w.$$

Thus $t_{ij}(u,w) > 0$ iff there is at least one path from state i at step u to state j at step
u+w. [$h_{ij}(g)$ is the (i,j)-th element of the matrix $H(g)$. For $t_{ij}(u,w)$ see formula
(4) above.]

The concept of a path (in French "chemin") in the treatment of homogeneous Markov chains was initiated by Doeblin, as he mentions in a footnote on p.5 of reference (2).

Def. 8 A run of P's (or Q's) of length r is an uninterrupted succession of r P's (or Q's) in the sequence $\{H(g)\}$.

The proof of Theorem 3 will be divided into stages labelled A to F.

(A) Since P* and Q* are irreducible and aperiodic there exist integers b and c such that $(P*)^r > 0$ if $r \geq b$ and $(Q*)^r > 0$ if $r \geq c$.

Suppose that condition d(i) holds. Let $H(g_{k-1})$ denote the last matrix of the (k-1)th block (k=2,3,4,...). Each block consists of two sub-blocks. In the first sub-block starting with $H(g_{k-1}+1)$ we include as many matrices as are necessary to include b P's (interspersed with an arbitrary number of Q's) and also include any further P's until the first Q occurs. Let the last P found in this way be denoted $H(g_{k-1}+d_k)$.

The second sub-block includes as many matrices as are necessary to include c Q's (interpersed with an arbitrary number of P's), and any additional Q's until the first P occurs. Let the last Q included be denoted $H(g_k)$; it is the last matrix of the k-th block.

(B) We now show that $\alpha[T(g_{k-1},w_k)] > 0$, where $w_k = g_k - g_{k-1}$, for all k=2,3,4...

First, there is a path from any $i \in \Gamma_p$ at step g_{k-1} to at least one $j \in \Gamma_q$ at step $g_{k-1}+d_k$. For if the first sub-block ends in a run of at least b P's or if $q_{ss} > 0$, then $t_{is}(g_{k-1},d_k) > 0$ for any $i \in \Gamma_p$. However if one or more Q's are interspersed between the P's and $q_{ss}=0$ then $t_{is}(g_{k-1},d_k)$ may be zero for some $i \in \Gamma_p$; but in that case for any such i there must be an integer p $(0 < p < d_k)$ such that both $t_{is}(g_{k-1},p) > 0$ and also $H(g_{k-1}+p+1)=Q$ so that $t_{i\ell}(g_{k-1},p+1) > 0$ for some $\ell \in \Gamma_q$. Since $p_{ss} > 0$ there must be a path wholly within the states of Γ_q from ℓ at step $g_{k-1}+p+1$ to some $j \in \Gamma_q$ at step $g_{k-1}+d_k$.

Secondly from any state $i \in \Gamma_q$ at step $g_{k-1}+d_k$ there are paths to all states $j \in \Gamma_q$ at step g_k since the second sub-block contains at least c Q's and $p_{ss} > 0$. Thus for any $i \in \Gamma_p$ and all $j \in \Gamma_q$ $t_{ij}(g_{k-1},w_k) > 0$.

For similar reasons $t_{ij}(g_{k-1},w_k) > 0$ for all $i \in \Gamma_q$ and all $j \in \Gamma_q$. Thus all those columns of $T(g_{k-1},w_k)$ which correspond to states in Γ_q are wholly positive and so $\alpha[T(g_{k-1},w_k)] > 0$.

(C) We show next that there is a common lower bound greater than 0 for $\alpha[T(g_{k-1},w_k)]$ for all k=2,3,4,...

Let lengths of P runs in the k-th block be denoted $\ell_1^{(k)}, \ell_2^{(k)}, \ell_3^{(k)} \ldots \ell_{t_k}^{(k)}$. The lengths of Q-runs in the k-th block will be denoted by $m_1^{(k)}, m_2^{(k)}, \ldots m_{t_k}^{(k)}$. Then one can

write

$$T(g_{k-1}, w_k) = P^{\ell_1^{(k)}} Q^{m_1^{(k)}} P^{\ell_2^{(k)}} Q^{m_2^{(k)}} \ldots P^{\ell_{t_k}^{(k)}} Q^{m_{t_k}^{(k)}} \tag{8}$$

Now let the smallest non-zero element of the matrix P^r ($r = 1,2,3\ldots$) be denoted $\varepsilon_r > 0$. Then the sequence $\varepsilon_1, \varepsilon_2, \varepsilon_3 \ldots$ has a positive lower bound, say $\varepsilon > 0$. To show that this lower bound exists, consider first those elements of $P^r = \{P_{ij}^{(r)}\}$ where $i, j \in \Gamma_p$. For such i, j $P_{ij}^{(r)}$ tends, as $r \to \infty$, to a positive limit because $P*$ is irreducible and aperiodic. If $i \in \Gamma_q$ and $i \neq s$ then $p_{ii}^{(r)} = 1$ for all r. All other elements of P^r are zero for all r.

Let $\delta > 0$ similarly denote the positive lower bound of all non-zero elements of the sequence $Q, Q^2, Q^3 \ldots$

It now follows from (8) that any path between step g_{k-1} and step g_k has probability greater than $(\varepsilon \delta)^{t_k}$. Thus the smallest positive element of $T(g_{k-1}, w_k)$ exceeds $(\varepsilon \delta)^{t_k}$. Now t_k (the number of runs of P's and of Q's) cannot exceed $b + c - 1$. This maximum value of t_k will occur if the first sub-block has b-1 isolated P's (interspersed with runs of Q's) and the second sub-block has c-1 isolated Q's. Since there are n_q states in Γ_q, $T(g_{k-1}, w_k)$ has at least n_q wholly positive columns. Thus

$$\alpha [T(g_{k-1}, w_k)] > n_q (\varepsilon \delta)^{a+b-1} \tag{9}$$

for all $k = 2, 3, 4 \ldots$

(D) If condition d(ii) of Theorem 3 is satisfied blocks can be formed in a manner similar to that described in (A) above. Each block again consists of two sub-blocks. The first sub-block includes at least c+1 Q's and the second sub-block includes b P's. Arguments along lines similar to (B) and (C) above then show that the ergodicity coefficient α is bounded away from 0 for each block by the same lower bound.

(E) Doeblin's Theorem of section 2 above can be applied to show that the conditions of Theorem 3 guarantee that the chain $\{H(g)\}$ satisfies the ergodic principle.

(F) The use of Proposition 2 (in section 3) in conjunction with Theorem 1(b) (in section 2) completes the proof of Theorem 3.

(G) Proof of Proposition 3.

Let {H(g)} denote a chain which satisfies the conditions of Proposition 3 and let {H*(g)} denote the chain in which P occurs whenever U occurs in {H(g)} and Q occurs whenever V occurs in {H(g)}. [P and Q are the matrices of which the patterns are embedded in U and V.] Then {H*(g)} satisfies the ergodic principle, and so {H(g)} does also, since all transitions with positive probability in products of P and Q also have positive probability in the corresponding products of U and V. Since by assumption U and V have a common stationary vector Theorem 1(b) of section 2 may be invoked to complete the proof.

5. Proof of Theorem 2

We prove that under the conditions of Theorem 2 from any state i eventual transition to any state j ($1 \leq i$, $j \leq n$) is certain. As in the proof of Theorem 3 the argument involves a division of the sequence {H(g)} into suitable blocks. The way this division can be carried out is discussed below.

Given the division into blocks two taboo probabilities may be defined. Let $X(g)$ denote the state occupied after matrix $H(g)$ has been applied. As before $H(g_k)$ is the last matrix of the k-th block. Let

$$\psi(k,m; i,j,f) = Pr[X(g) \neq j \text{ for } g = g_k+1, g_k+2 \ldots g_{k+m}-1 \\ \text{and } X(g_{k+m}) = f \mid X(g_k) = i]$$

for k=1,2,3..., m=1,2,3... and $1 \leq i,j,f \leq n$ with $f \neq j$ (10)
and

$$\phi(k,m; i,j) = \sum_{f \neq j} \psi(k,m; i,j,f)$$ (11)

Thus ψ(k,m; i,j,f) denotes the conditional probability that, if the state occupied at the end of the k-th block is i, state j is not visited in the subsequent m blocks and at the end of the (k+m)-th block the system is in state f ($f \neq j$). ϕ(k,m; i,j) is the probability that, given state i at the end of the k-th block, state j is not visited in the succeeding m blocks.

One may write

$$\phi(k,m+1;i,j) = \sum_{f \neq j} [\psi(k,m;i,j,f) \times \phi(k+m,1;f,j)]$$

$$\leq [\sum_{f \neq j} \psi(k,m;i,j,f)] \times \max_f \phi(k+m,1;f,j)$$

$$= \phi(k,m;i,j) \times \max_f \phi(k+m,1;f,j) \tag{12}$$

As will be shown below, one may write for any k,i,j

$$\phi(k,1;i,j) \leq 1-\theta \text{ with } \theta > 0 \tag{13}$$

Here θ is a lower bound for the probability of a path which visits j within a block given that the state at the beginning of the block was not j. The lower bound θ is common to all blocks and depends only on P and Q and on the method of block formation.

Using (13) and applying induction to (12) one obtains

$$\phi(k,m;i,j) \leq (1-\theta)^m \tag{14}$$

So that, as m→∞,

$$\phi(k,m;i,j) \rightarrow 0 \tag{15}$$

for all k=1,2,3... and all $1 \leq i,j \leq n$.

It remains to discuss the proof of (13). If one of the conditions d(i) and d(ii) of Theorem 3 holds, blocks may be formed in the manner described in part (A) or (D) of the proof of Theorem 3. Arguments on the lines of part (B) then show that the required paths exist within each block and arguments on the lines of part (C) show that formula (13) holds. (Theorem 2 may in any case be deduced directly as a consequence of Theorem 3 when the latter theorem holds.)

Suppose, therefore, that conditions d(i) or d(ii) of Theorem 3 are not satisfied by P and Q, i.e. that $p_{ss}=q_{ss}=0$ and that neither $\bar{P}$ nor $\bar{Q}$ (defined analogously to $\bar{P}$) is primitive. $\bar{P}$ can fail to be primitive into two ways. Either $\bar{P}$ is irreducible, but cyclic or there is at least one pair of states, say i and j both in Γ_p, such that under repeated application of P there is no path from i to j except through state s.

The worst case scenario (i.e. where the most complicated block structure is required) occurs when (i) $p_{ss}=q_{ss}=0$, (ii) there are one or more pairs of states (other than s) in Γ_p such that paths between them under successive applications of P necessarily pass through state s and (iii) there are also pairs of states in Γ_q such that paths between them all pass through s. (Example 3 illustrates this worst case scenario.) To prove (13) for cases of this sort blocks consisting of four sub-blocks are needed, where say the first and third sub-blocks comprise at least b P's while the second and fourth sub-blocks comprise at least c Q's.

In all cases where the conditions of Theorem 2 are satisfied suitable blocks can be constructed for which (13) holds. The conclusion of the theorem then follows straightforwardly from (15).

In its basic idea the proof closely resembles Doeblin's proof (see (2) p.8) that in a realisation of a finite irreducible homogeneous Markov chain every state is visited i.o. with prob. 1. Doeblin notes that in a homogeneous chain, within any block of length equal to the number of states, there is a path from any state i to any state j, and that the probability of any such path exceeds some minimum value, say θ. Doeblin deduces that the probability, that a transition from i to j has not occurred within m blocks, is less then $(1-\theta)^m$ and this quantity goes to zero as $m\to\infty$.

This type of argument does not seem to be found in the more recent literature[3] on Markov chains. Perhaps it does not play a role because, as Doeblin remarks, it does not apply to infinite irreducible chains.

6. Concluding remarks

This paper considers only shuffling with two matrices where each matrix has just one irreducible submatrix. There are other sorts of shuffling two-matrix chains (where each matrix shuffles only some of the states) as the following example shows.

Example 5 Let

$$
P = \begin{pmatrix} x & x & 0 & 0 \\ x & x & 0 & 0 \\ 0 & 0 & x & x \\ 0 & 0 & x & x \end{pmatrix}, \quad Q = \begin{pmatrix} 1 & 0 & 0 & 0 \\ 0 & x & x & 0 \\ 0 & x & x & 0 \\ 0 & 0 & 0 & 1 \end{pmatrix}
$$

(x denotes a non-zero element)

---

[3] Arguments similar to the one under discussion are, however, sometimes used in related contexts (Seneta (13), pp 120-1).

Shuffling will occur if the sequence $\{H(g)\}$ contains both P and Q i.o.

Moran's paper (11) mentions cases of this kind. They do not fit into the context of the present paper, but are more naturally discussed in connection with theorems covering shuffling chains with more than two matrices.

I hope to deal in a subsequent publication with the extension of the results of this paper to chains involving three or more matrices and also discuss other aspects of shuffling chains where each matrix shuffles only a subset of the states. Some other matters are covered in a brief note which has already appeared (reference (5)).

Acknowledgements

This work would not have been carried through to completion without very considerable help and encouragement from Eugene Seneta. I would also like to thank Susan Powell for help with the construction of Example 3 and Mary Auckland for assistance with library search.

References

(1)   DOEBLIN, W. (1937), Le cas discontinu des probabilités en chaîne. Publications de la Faculté des Sciences de l'Université Masaryk (Brno), No.236, pp 3-13.

(2)   DOEBLIN, W. (1938), Exposé de la théorie des chaînes simples constantes de Markoff à un nombre fini d'états. Revue mathématique de l'Union Interbalkanique Vol II, Fasc.1, pp 77-105. (I have used an offprint with pages numbered 1-29.)

(3)   DVORETZKY, A. and WOLFOWITZ, J. (1951), Sums of random integers reduced modulo m. Duke Mathematical Journal, Vol.18, pp 501-507.

(4)   HAJNAL, J. (1958), Weak ergodicity in non-homogeneous Markov chains. Proc. Camb. Phil. Soc. Vol.54, pp 233-46.

(5)   HAJNAL, J. (1987), The theory of shuffling - an application of non-homogeneous Markov chains: In Prohorov, Yu. V. and V.V. Sazonov (eds.) Proceedings of the 1st World Congress of the Bernoulli Society (VNU Science Press - Netherlands) Vol.1, pp 317-320.

(6)   HORTON, H.B. (1948), A method for obtaining random numbers. Ann. Math. Stats. Vol.19, pp 81-85.

(7)   HORTON, H.B. and R.T. SMITH III, (1949), A direct method for producing random digits in any number system. Ann. Math. Stats. Vol.20, pp 82-90.

(8)   IOSIFESCU, M. (1980), Finite Markov processes and their applications. Wiley (Chichester, New York).

(9)   ISAACSON, D.L. and MADSEN, R.W. (1976), Markov chains: theory and applications. Wiley (New York).

(10)   MORAN, P.A.P. (1962a), The statistical processes of evolutionary theory. Clarendon Press (Oxford). See pages 29-37.

(11)   MORAN, P.A.P. (1962b), Polysomic inheritance and the theory of shuffling. Sankhya, Ser. A. Vol.24, pp 63-72.

(12)   SENETA, E. (1973), On the historical development of the theory of finite inhomogeneous Markov chains. Proc. Camb. Phil. Soc. Vol.74, pp 507-513.

(13)   SENETA, E. (1981), Non-negative matrices and Markov chains, 2nd Ed. Springer (New York).

Department of Statistical and Mathematical Sciences, London School of Economics London WC2A 2AE, UK
Address for correspondence: 95 Hodford Road, London NW11 8EH, UK

Contemporary Mathematics
Volume **149**, 1993

Continuous time gambling problems

K. B. ATHREYA

ABSTRACT. A brief survey of problems involving optimization of hitting probabilities and expected hitting times for Ito processes on $[0, \infty)$ is given. These parallel discrete time gambling problems studied by Dubins and Savage.

1. Introduction

Continuous time gambling problems are analogs of discrete time ones studied by L. Dubins and L. J. Savage in their classic *How to gamble if you must* [**4**]. Unlike their discrete counterparts which are usually very hard, the continuous time problems are somewhat easier mainly due to the availability of Ito Calculus. The optimal solutions and values are often computable in explicit fashion.

To describe a typical continuous time problem we start with a simple discrete time problem known as *Red and Black*. Let $\{X_n\}$ be a Markovian sequence of random variables with values in $[0, \infty)$ generated by the transition mechanism in which given $(X_0, X_1, \ldots, X_n)$ the random variables $X_{n+1} - Xn$ takes only two values, $s(X_n)$ or $-s(X_n)$ with probability p and $(1 - p)$ respectively. Here $s(\cdot)$ is a measurable function on $[0, \infty) \to [0, \infty)$ such that $0 \leq s(x) \leq x$. The interpretation of $s(x)$ is that it is the amount the player bets at any step when his present fortunate is x. The problem then is, given p, choose the betting function $s(\cdot)$ such that the probability of reaching a before hitting 0 is maximized for each $a > 0$. More precisely, let $T_a = \inf\{n : n \geq 1, x_n = a\}$, for $0 \leq a < \infty$. Let $F(s, a, x) = P_x(T_a < T_0)$ for $0 < x < a < \infty$ and $s : [0, \infty) \to [0, \infty)$, Borel measurable, $0 \leq s(x) \leq x$. For each $x, a, 0 < x < a < \infty$, choose s_0 such that $F(s_0, a, x) = \max_s F(s, a, x)$. Dubins and Savage [**4**] have solved

1991 *Mathematics Subject Classification.* Primary 93E20; Secondary 60G40.

Key words and phrases. Continuous time, Ito processes, local time, gambling, hitting times.

Research was supported in part by a grant from the National Science Foundation NSF Grant #DMS 9007182.

This paper is in final form and no version of it will be submitted for publication elsewhere.

this problem. Although an optimal strategy is easy to describe the proof of its optimality is not. An optimal strategy is as follows: a)   for $p \leq 1/2$ (called subfair games) *bold play* is optimal, i.e., choose $s(x) = \min\{x, a - x\}$, i.e., bet all one has or enough to reach a. b)   For $p > 1/2$ (superfair game) the *proportional play* is optimal, i.e., choose $s(x) = ax$, where a is positive and small.   For $p \leq 1/2$ the computation of the optimal probability $V(x)$ is not easy.   It can be shown to satisfy a functional equation involving p, x, a. For $p > 1/2$, the proportional strategy yields probability one for reaching a before 0 for all a small and positive. This suggests a new problem, namely, among all strategies which yield probability one for reaching a before 0 find one that minimizes the expected time to reach a. This problem is *unsolved*. L. Breiman (see [4]) has an approximate optimal solution.

We now describe the continuous time analog of this problem.   First let us rewrite the problem as saying

$$X_{n+1} - X_n = \Delta X_n = \lambda s(X_n) + s(X_n)\delta_n$$

where $\lambda = (2p - 1)$ and $\{\delta_n\}$ is a sequence of i.i.d. binary r.v. with mean 0 and variance $p(1 - p)$. This suggests that we look at all processes $\{X(t): \quad t \geq 0\}$ that satisfy

$$dX(T) = \lambda s(X_t)dt + s(X_t)\sigma dw(t)$$

where $W(t)$ is a Brownian motion. The so-called *continuous time red and black problem* then is: Given constants λ and σ choose a strategy function $s(\cdot)$ to maximize $P_x(T_a < T_0)$ where $0 \leq x \leq a$ and $T_a = \inf\{t: \quad t > 0, \ X(t) = a\}$. The solution parallels that of the discrete time case. For $\lambda \leq 0$ bold play i.e., $s(x) = \min\{x, 1 - x\}$ is optimal. For $\lambda > 0$, *timid play* is optimal. Further, in the continuous time case when $\lambda > 0$ the *problem of minimizing expected time to reach a i.e., E_xT_a does have a solution*. We now proceed to give many examples of such problems.

2. Continuous time gambling problems

Let (Ω, B, P) be a probability space, $\{F_t: t \geq 0\}$ a filtration of sub σ-algebras of B, $W(\cdot)$ a Brownian motion adapted to and nonanticipating with respect to $\{F_t\}$.

Consider all processes $\mu(\cdot)$ and $\sigma(\cdot)$ adapted to $\{F_t\}$ and satisfying

(1) $-\infty < \mu(\cdot) < \infty, \ 0 < \sigma(\cdot) < \infty$ and for all $0 < t < \infty$

$$\int_0^t (|\mu(u)| + \sigma^2(u))du < \infty \quad \text{a.s.}$$

Let $X(\cdot)$ be an *Ito process* defined by

(2) $$X(t) \equiv x + \int_0^t \mu(u)du + \int_0^t \sigma^2(u)dW(u).$$

Assume for each $0 < x < a < \infty$ the existence of control sets $C(x)$ in $R \times R^+$ and that the infinitesimal mean $\mu(t)$ and standard deviation $\sigma(t)$ of $X(t)$ are such that $(\mu(t),\ \sigma(t)) \in C(X(t))$. That is, if $X(t) = x$ then $(\mu(t),\ \sigma(t)) \in C(x)$. For each Ito process X as in (2) let

(3) $Q(x,\mu,\sigma) = P_x \left(T_a^X < T_0^X \right)$ where $T_a^X = \inf\{t :\ t \geq 0,\ X(t) = a\}$.

Problem 1. Given a family of sets $\{C(y); 0 \leq y \leq a\}$ and $0 < x < a$ determine $q(x) = \sup Q(x,\mu,\sigma)$ where Q is as in (3). Further, determine if $q(x)$ is attained.

This was solved by Pestien and Sudderth [**6**]. Let

(4) $\rho(x) = \sup\{\mu/\sigma^2 :\ (\mu,\sigma) \in C(x)\}.$

Assume that there exists continuous functions $\mu_0(\cdot)$ and $\sigma_0(\cdot)$ on $[0, a]$ such that

$$\begin{array}{ll}
i) & p(x) = \dfrac{\mu_0(x)}{\sigma_0^2(x)} \text{ for } 0 < x < a, \\[2mm]
(5) \qquad ii) & \displaystyle\inf_{0<x<a} \sigma_0(x) > 0, \\[2mm]
iii) & (\mu_0(x),\ \sigma_0(x)) \in C(x) \text{ for } 0 < x < a.
\end{array}$$

Then an optimal solution is a *diffusion* $Y(\cdot)$ on $(0, a)$ that is absorbed at 0 and a and satisfies

(6) $Y(t) = x + \displaystyle\int_0^t \mu_0(Y(s))ds + \int_0^t \sigma_0(Y(s))dW(s)$

and the optimal probability is

(7) $U(x) = P_x(T_a^Y < T_0^Y) = \dfrac{S(x)}{S(a)}.$

$$\text{where } S(x) = \int_0^x e^{-A(y)}dy,\ 0 < x < \infty \text{ and}$$
(8)
$$A(h) = \int_0^y 2\rho(u)du,\ 0 < y < \infty.$$

REMARK 1. No uniqueness of the optimal solution can be asserted since the optimal probability depends only on x, a and $p(\cdot)$ and so any choice of μ, σ satisfying (5) will do.

PROOF. It is easy to verify that $U(\cdot)$ in (7) satisfies

(9) $U''(x)\dfrac{\sigma_0^2(x)}{2} + U'(x)\,\mu_0(x) = 0,\ 0 < x < a,\ U(0) = 0,\ U(a) = 1.$

Let X be any admissible Ito process, i.e., one for which $(\mu(\cdot), \sigma(\cdot)) \in C(x(\cdot))$. Then, by Ito's formula,

$$
U(X(t)) = U(x) + \int_0^t \frac{du}{dx}(X(s)) \, \sigma(s) dW(s)
$$
(10)
$$
+ \int_0^t (\mu(s)U'(X(s)) + \frac{\sigma^2(s)}{2} U''(X(s))) ds
$$

Since $U(x)$ is monotone increasing in x,

$$
(11) \quad \mu(s)U'(X(s)) + \frac{\sigma^2(s)}{2} U''(x(s)) \leq \frac{\sigma^2(s)}{2} (2\rho(X(s))U'(X(s)) + U''(X(s))
$$

which is zero if $0 < X(s) < a$. Let $\tau_t = \min\{T_a, T_0, t\}$ for $0 < t < \infty$. Doob's optional sampling theorem (see [5]) applied to (10), yields $EU(X(\tau_t)) \leq U(x)$ and on letting $t \to \infty$, we get by bounded convergence theorem

$$
(12) \qquad\qquad Q(x, \mu, \sigma) \leq U(x).
$$

Since Y in (b) is admissible and by (7) $U(x) = Q(x, \mu_0(\cdot), \sigma_0(\cdot))$ we are done.

REMARK 2. It is possible to deduce from the above proof the analogs of Dubins Savage results for the continuous time red and black, i.e., where

$$
(13) \qquad \mu(t) = \lambda s(X_t), \sigma(t) = \sigma, \text{ indep of } X(t), \ 0 \leq s(x) \leq x.
$$

3. Finite time case

Fix $0 < t_0 < \infty$. Consider the setup in section 2, but now maximize $P_x(T_a^X < T_0^X \Lambda t_0)$, where $\alpha \Lambda \beta = \min(\alpha, \beta)$. The discrete time analog of this problem has not been fully solved. The continuous time case was solved by Sudderth and Weerasinghe (1989) [7]. The same Y as in section 2 is the optimal solution. To compute $U(x) = P_x(T_a^Y < T_0^Y \Lambda t_0)$ and prove its optimality, we need to go to the space time process:

$$
(14) \qquad
\begin{cases}
dY_1(t) & = \mu_0(Y_1(t)) dt + \sigma_0(Y_1(t)) dW(t) \\
dY_2(t) & = dt, \\
Y_1(0) & = x_1, Y_2(0) = x_2
\end{cases}
$$

Suppose $Q(s_1, s_2)$ solves the partial differential equation.

$$
(15) \qquad \frac{1}{2} \sigma_0^2(x_1) \frac{\delta^2 q}{\delta x_1^2} + \mu_0(x_1) \frac{\delta Q}{\delta x_1} + \frac{\delta Q}{\delta x_2} = 0
$$

$Q(a, x_2) = 1$, $Q(0, x_2) = 0$, $Q(x_1, t_0) = 0$ for $0 \leq x_1 \leq a$, $0 \leq x_2 \leq t_0$.

Then, using Ito's formula and Doob's optional sampling theorem as in section 2, we get $Q(x_1, x_2) \geq E_{x_1, x_2} Q(X_1(\tau), \ x_2(\tau))$ for any admissible space time process $X(t) = (X_1(t),\ X_2(t))$ and any stopping time τ satisfying $P_{x_1,x_2}(\tau < \infty) = 1$. Pestien (see [6]) has proved a theorem which says that

$$\sup_\tau E_{x_1,x_2} Q(X_1(\tau),\ X_2(\tau)) \geq E_{x_1,x_2} \overline{\lim_t} Q(X(t))$$

Finally, verify that $\overline{\lim}_t Q(X(t)) \geq I(T_a^X < T_0^X \Lambda t_0)$.

These steps are typical in these problems and are of varying difficulty. Very often the associated p.d.e. may not admit a solution and one has to go through a sequence of perturbations of the p.d.e.

REMARK 3. As a corollary to the above result one could obtain a comparison result for Ito processes. Let X be an admissible Ito process and Y the optimal diffusion. Then

$$P_x(T_a^X < T_0^X \Lambda t_0) = P_x(\sup_{0 \leq s \leq t_0} X(s) > a)$$

$$\leq P_x(T_a^Y < T_0^Y \Lambda t_0) = P_x(\sup_{0 \leq s \leq t_0} Y(s) > a)$$

Thus, if $\mu(t) \leq \mu_0(X(t))$, $\alpha(t) \leq \sigma_0(X(t))$ for all t then

$$\sup_{0 \leq s \leq t_0} X(s) \text{ is stochastically dominated by } \sup_{0 \leq s \leq t_0} Y(s)$$

whenever, $X(0) = x \leq Y(0) = y,\ 0 < x \leq y < \infty$.

4. Games with reflection at zero

Let us reconsider the discrete time case. Let $\{X_j\}_0^\infty$ be a Markov chain with state space $\{-1, 0, 1, 2, \ldots, N\}$ and given $X_j = i$, the change $X_{j+1} - X_j$ has a distribution that belongs to a family F_i. Each time the chain hits -1 a fee of one dollar is charged and the chain is moved to 0. The goal is to reach N with a ceiling on the expenditure. That is, let $T = \inf\{n :\ n \geq 1,\ x_n = N\}$ and $A = \sum_{j=0}^{T-1} I(X_j = -1)$. Then the problem is to maximize $P_x(A \leq y)$ where x is the initial fortune and y is the ceiling on expenditure and the minimum is over all possible choices of games at each stage from the given family $\{F_i\ i = 0, 1, 2, \ldots, N - 1\}$ of distribution. Unfortunately, even for very restricted class of $\{F_i\}$ such as coin tossing this problem is not solved yet. But the continuous time formulation does admit a solution. Let $\{Z(t); 0 \leq t < \infty\}$ be an Ito process with controls $\mu(\cdot)$ and $\sigma(\cdot)$ as in (2). Let

$$(16) \qquad L_Z(t) = -\min\{\min_{0 \leq u \leq t} Z(u),\ 0\}$$

Then $L_Z(t)$ is nondecreasing in t, and increases only when Z is zero. Let

$$(17) \qquad X(t) \equiv Z(t) + L_Z(t).$$

This is always nonnegative. Such an X is called an *Ito process with reflection at 0* and the process $L_Z(\cdot)$ is called the *local time at 0 of X*. The r.v. $L_Z(T_a^X)$ should be thought of as the amount of money needed for the process X to reach *a.* A problem analogous to the one discussed in section 2 and the discrete time problem mentioned earlier in this section is this. Given a family of control sets $C(y)$, $0 \leq y \leq a$ consider all Ito processes with reflection at 0 and whose controls $\mu(\cdot)$ and $\sigma(\cdot)$ satisfy $(\mu(\cdot), \sigma(\cdot)) \in C(X(\cdot))$. For this class maximize $P_x(L_Z(T_a^X) \leq y)$. This problem has been solved by Athreya and Weerasinghe [2]. Let $Z_0(\cdot)$ be a diffusion satisfying

$$dZ_0(t) = \mu_0(Z_0(t))dt + \sigma_0(Z_0(t))dw(t)$$

where $\mu_0(\cdot)$ and $\sigma_0(\cdot)$ are continuous functions on $[0, \infty)$ as in (5) and extended to $(-\infty, 0)$ by setting $\mu_0(x) = \mu_0(0)$, $\sigma_0(x) = \sigma_0(0)$ for $x \leq 0$. Let

$$(18) \qquad\qquad X_0(t) = Z_0(t) + L_{Z_0}(t)$$

where $L_{Z_0}(\cdot)$ is as in (16).

The construction of going from Z to X in (16) and (17) is a real variable construction known in the literature as Skorohod's construction, see Chung and Williams [3] and Kartazas and Shreve [4]. It is shown in Athreya and Weerasinghe [2] that $X_0(\cdot)$ as in (18) is an optimal solution. It turns out that one can go from $X_0(\cdot)$ to Z_0 via Tanaka's formula (see [3]) asserting that

$$(19) \qquad \frac{1}{2\varepsilon} \int_0^t I_{[0,t)}(X_0(s, w))ds \to L(t) \text{ as } \varepsilon \to 0 \text{ exists a.s. and } X(t) - L(t)$$

is an unrestricted diffusion with coefficients $\mu_0(\cdot)$ and $\sigma_0(\cdot)$. Also $X_0(\cdot)$ is known as a diffusion reflecting at 0 with coefficients $\mu_0(\cdot)$ and $\sigma_0(\cdot)$.

The proof that $X_0(\cdot)$ is the optimal solution follows the method used in sections 2 and 3.

Step 1. We compute $P_x(L^{X_0}(T_a^{X_0}) \leq y)$. This was computed in Athreya and Weerasinghe [1] where $L^{X_0}(t) = L_{Z_0}(\cdot)$, by definition.

THEOREM 1. *For $0 < x < a$ and X_0 is as in (18).*

$$(20) \qquad Q(x, y) \equiv P_x\left(L^{X_0}\left(T_a^{X_0}\right) \leq y\right) = \frac{S(x)}{S(a)} + \left(1 - \frac{S(x)}{S(a)}\right)\left(1 - e^{-y/s(a)}\right)$$

where $S(\cdot)$ is an (8).

Step 2. Verify that $Q(x, y) \geq EQ(X_1(t), X_2(t))$ where $X_1(t)$ is any admissible reflecting Ito processes and $X_2(t) \equiv y - L^{X_1}(t)$, (the amount of money available at t). This is done using Ito's formula and the fact that $S(\cdot)$ satisfies

$$S''(y) + 2\rho(y)S'(y) = 0, \ 0 < y < \infty$$

and also

$$\frac{\delta Q}{\delta x}(0, y) = \frac{\delta Q}{\delta y}(0, \delta)$$

Step 3. Now use Pestien's Lemma (see [6]) and the fact that

$$E_{x,y} \varlimsup_{t\to\infty} Q(X_1(t),\ X_2(t)) \le E_{x,y} I\left(L^{X_1}\left(T_a^{X_1}\right) \le y\right).$$

It turns out that all these steps are much harder than in the previous cases.

REMARK. If $\phi:\quad R^+ \to R^+$ is nondecreasing then it follows from Theorem 1 that

$$E_x \phi\left(L^{X_0}\left(T_a^{X_0}\right)\right) \le E_x(\phi\left(L^X\left(T_a^X\right)\right)$$

for any admissible Ito process reflecting at 0. In particular, X_0 is optimal for minimizing the expected expenditure. Also by optimality of X_0

$$P_x\left(L^X\left(T_a^X\right) > y\right) \ge P_x\left(L^{X_0}\left(T_a^{X_0}\right) > y\right)$$

and letting $y \downarrow 0$ yields

$$P_x\left(L^X\left(T_a^X\right) > 0\right) \ge P_x\left(L^{X_0}\left(T_a^X\right) > 0\right)$$

which is the same as

$$P_x\left(T_a^X < T_0^X\right) \le P_x\left(T_a^{X_0} < T_0^{X_0}\right)$$

recovering the result of Pestien and Sudderth mentioned in section 2.

5. Conclusion

The examples from continuous time gambling given in sections 2, 3 and 4 illustrates the rich variety of problems that can be formulated and solved using Ito calculus and some partial differential equations theory. References below contain a number of other listings to the growing literature of this interesting area.

REFERENCES

1. Athreya, K.B. and Weerasinghe, A., *Exponentiality of the local time at hitting times for reflecting diffusions and an application, Probability, Statistics and Mathematics: Papers in Honor of Samuel Karlin*, Academic Press, New York, 1989, (T.W. Anderson et al., editors).
2. Athreya, K.B. and Weerasinghe, A., *Reflecting Ito Processes in a Stochastic Control Problem* (1989), To appear in Math. Oper. Res..
3. Chung, K.L. and Williams, R.J., *An Introduction to Stochastic Integration*, Birkhauser, Boston, 1983.
4. Dubins, L.E. and Savage, L.J., *How to Gamble if You Must: Inequalities for Stochastic Processes*, Dover, New York, 1976.
5. Karatzas, I. and Shreve, S.E., *Brownian Motion and Stochastic Calculus, Graduate Texts in Math*, Springer-Verlag, 1988.
6. Pestien, V.C. and Sudderth, W.D., *Continuous-time Red and Black: How to control a diffusion to a goal*, Math. Oper. Res. **10** (1983), 599-611.
7. Sudderth, W.D. and Weerasinghe, A., *Controlling a Process to a Goal in Finite Time*, Math. Oper. Res. **14** (1989), 400-409, No. 3.

DEPARTMENT OF MATHEMATICS AND STATISTICS, IOWA STATE UNIVERSITY, AMES, IA 50011-1210

E-mail address: athreya@pollux.math.iastate.edu

Contemporary Mathematics
Volume **149**, 1993

Stochastic processes with long range interactions of the paths

Erwin Bolthausen, Universität Zürich

1. Introduction

The problems discussed here are variations of the following trivial example:

Example 1.1 Let X_1, X_2, ... be a sequence of fair coin tosses, i.e. $X_i = 0$ or 1 w.p. $\frac{1}{2}$. If $S_n = \sum_{i=1}^{n} X_i$, then for $\alpha > \frac{1}{2}$, the law of the sequence X_i, conditioned on $S_n/n \geq \alpha$, obviously converges to coin tossing with success probability α.

In the above example, the conditional law of the X_i, i=1,...,n, is symmetric with respect to permutations of the variables. If this symmetry is only slightly broken, then it becomes much less obvious what happens.

Example 1.2 Let $T_n = \sum_{i=1}^{n-1} X_i X_{i+1}$. If $\alpha > \frac{1}{4}$, what is the law of X_1 conditioned on $T_n/n > \alpha$?

The example is mentioned as a partly open problem in a paper by Choi, Cover and Csiszar [CCC]. We will discuss it later on. It should, however, be remarked that if T_n is replaced by the slightly different statistic $T'_n = T_n + X_n X_1$, which makes it symmetric with respect to cyclic permutations, then the answer is fairly easy. We will see this later on.

1991 Mathematics Subject Classification. Primary 60Jxx, 60F10

Supported by the Swiss National Science Foundation Grant Nr. 21-298333.90

This is a survey paper. Details of the proofs of the results are published elsewhere

The examples above are (at least formally) very special cases of a class of problems where a "simple" stochastic process, like coin tossing, is transformed by introducing an interaction which is long range in time. This interaction enters by a density which is a functional of the empirical measure.

Let us fix some notations:
Let X_1, X_2, ... or X_s, $s \geq 0$, be a discrete or continuous time stochastic process defined on some probability space $(\Omega, \mathcal{F}, \mathbb{P})$ with values in some Polish space E, which is equipped with the Borel σ-field $\mathcal{E}$. In the continuous time case we tacitly always assume that the process has caldlag paths. We always assume that Ω is an appropriate path space : $E^{\mathbb{N}}$ in the discrete time case, and the Skorhod space $D([0,\infty) \to E)$ or $C([0,\infty) \to E)$ in the continuous time case. If $T > 0$, we denote by L_T the empirical process, i.e.

$$L_T = \frac{1}{T} \sum_{i=1}^{T} \delta_{X_i} \ , \ L_T = \frac{1}{T} \int_0^T \delta_{X_s} \, ds \ , \text{ respectively.}$$

Let $\mathcal{M}_1^+(E)$ be the set of probability measures on $(E, \mathcal{E})$. We equip $\mathcal{M}_1^+(E)$ with the weak topology and its Borel field. L_T is a $\mathcal{M}_1^+(E)$-valued random variable. If $F : \mathcal{M}_1^+(E) \to [-\infty, \infty)$ is measurable with respect to the Borel field of the weak topology, a_T are real numbers and

(1.3) $\mathbb{P}(F(L_T) \neq -\infty) \neq 0$

then we can define the transformed measures

(1.4) $d\widehat{\mathbb{P}}_T = \exp(a_T F(L_T)) d\mathbb{P} / z_T \ ,$

where $z_T = \mathbb{E}(\exp(a_T F(L_T)))$.

Taking conditional distributions is a special case : If $A \subset \mathcal{M}_1^+(E)$ is a measurable subset which satisfies $\mathbb{P}(L_T \in A) > 0$ for all T, then $\widehat{\mathbb{P}}_T$ with a_T arbitrary, $F = -\infty 1_{A^c}$, is just $\mathbb{P}(\ . \ | L_T \in A)$. In the Example 1.2 above, one has to take the empirical measure of the bivariate chain (X_1, X_2), $(X_2, X_3), \ldots$.

Perhaps the most famous examples of this type are the self-avoiding and the self-repellent random walk on the d-dimensional lattice $\mathbb{Z}^d$. Here $X_0 = 0$, X_1, X_2, ... is an ordinary symmetric, nearest neighbour random walk on the lattice. The self-avoiding walk of length T is given by condition the walk on the event that there are no self-intersections up to time T, i.e. that the empirical measure TL_T gives weight 0 or 1 to every point on the lattice. The self-repellent walk is defined by

$$(1.5) \qquad d\widehat{\mathbb{P}}_T = \exp\!\Big(- \beta \sum_{x \in \mathbb{Z}^d} (TL_T(x))^2 \Big)\, d\mathbb{P}/z_T \,,$$

with a coupling parameter $\beta > 0$. $\widehat{\mathbb{P}}_T$ obviously is of the above form (1.4). The behaviour of these measures for $T \to \infty$ is an unsolved problem for $d = 2,3,4$. (See [HS] for $d \geq 5$ and [GH2] for $d = 1$, or [B4] for the not nearest neighbour case.) Unfortunately, the techniques presented here are not of much use (except for $d = 1$) for this problem.

We will discuss here transformations which are strongly related to the large deviation theory of Donsker and Varadhan (see e.g.[DS]). The large deviation theory yields information on the "partition function" z_T.

If $a_T \uparrow \infty$ for $T \uparrow \infty$, we say that L_T satisfies a strong a_T – large deviation principle (a_T-LDP for short) with a rate function $I : \mathcal{M}_1^+(E) \to [0,\infty]$, if for any measurable $A \subset \mathcal{M}_1^+(E)$, we have

$$(1.6u) \qquad \limsup_{T \to \infty}\ a_T^{-1} \log \mathbb{P}(L_T \in A) \ \leq\ -\inf\{\, I(\mu) : \mu \in cl(A)\,\}\,,$$

$$(1.6l) \qquad \liminf_{T \to \infty}\ a_T^{-1} \log \mathbb{P}(L_T \in A) \ \geq\ -\inf\{\, I(\mu) : \mu \in int(A)\,\}\,, \text{ and}$$

$$(1.7) \qquad \{\, \mu : I(\mu) \leq c \,\} \text{ is compact for all } c \in (0,\infty)\,.$$

If (1.6u) is satisfied only for sets with compact closure and I is only assumed to be lower semicontinuous, then one speaks of a weak LDP.

It is well known that under (1.7), (1.6) is equivalent to a somewhat stronger statement: Let F be as defined above with the additional requirement to be bounded above. $\overline{F}$ and $\underline{F}$ denote the upper and lower semicontinuous regularizations of F. Then (1.5) is equivalent to

$$(1.8u) \qquad \limsup_{T \to \infty}\ a_T^{-1} \log E\!\big(\exp(a_T\, F(L_T))\big) = b(\overline{F})\,,$$

$$(1.8l) \qquad \liminf_{T \to \infty}\ a_T^{-1} \log E\!\big(\exp(a_T\, F(L_T))\big) = b(\underline{F})\,,$$

where $b(G) = \sup\{G(\nu) - I(\nu) : \nu \in \mathcal{M}_1^+(E)\}$. This is Varadhans Lemma.

The requirement that F is bounded above can often be relaxed, but this needs some additional estimates.

If $b(\overline{F}) = b(\underline{F})$, then we have a limit result in (1.8) giving an evaluation of $\lim_T a_T^{-1} \log z_T$. This has found many application to models which, in contrast to (1.5), are more of a self attracting type, e.g. in [DV1] and [DV2]. We will come back to this in the Sections 5 and 6. It is, however, not quite clear how to come from a rough estimation of z_T to a discussion of the path measures. In Section 2, we discuss some points in the rather trivial i.i.d. case and allude at some possible extensions. In Section 3, we present a result for continuous time Markov process. In Section 4, we give some indications, why a more precise asymptotics is necessary in some cases. The Sections 5 and 6 give applications to special classes of transformations, in Section 5 to transformations by the number of points visited by a random walk and Section 6 treats polaron type models.

2. The role of symmetry: The i.i.d. case and some extensions

It was remarked in Section 1 that symmetry considerations play an important role in some cases. This particularly applies to i.i.d. sequences $X_1 , X_2 , \ldots$. The discussion of the law of the sequence under $\widehat{\mathbb{P}}_T$ introduced in (1.4), $\mathcal{L}_{\widehat{\mathbb{P}}_T} (X_1 , X_2 , \ldots)$, splits into two parts:

(I) A discussion of the law Γ_T of L_T under $\widehat{\mathbb{P}}_T$, i.e. of $\Gamma_T = \widehat{\mathbb{P}}_T(L_T)^{-1} \in \mathcal{M}_1^+(\mathcal{M}_1^+(E))$

(II) Transfer of the information obtained in (I) to discuss $\mathcal{L}_{\widehat{\mathbb{P}}_T} (X_1 , X_2 , \ldots)$

For (I) partial information can been obtained from a LDP. In the i.i.d. case, this is Sanov's Theorem (see [DS]) which is a strong T-LDP with rate function $I(\nu)$, the Kullback-Leibler information $k(\nu|\mu) = \int \log(d\nu/d\mu)d\nu$, where μ is the law of the X_i. Let us assume that F is bounded above and $b(\overline{F}) \neq -\infty$. Then

$$K_F = \{\gamma \in \mathcal{M}_1^+(E) : \overline{F}(\gamma) - I(\gamma) = b(\overline{F})\}$$

is a nonvoid and weakly compact subset of $\mathcal{M}_1^+(E)$. If

(2.1)
$$b(\overline{F}) = b(\underline{F}) \ ,$$

is satisfied, then for any open neighbourhood U of K

$$\widehat{\mathbb{P}}_T(L_T \notin U) = \mathbb{E}(\exp(TF_U(L_T)))\Big/ \mathbb{E}(\exp(TF(L_T))) \ ,$$

where $F_U = -\infty 1_{U^c} + F$. From (1.8) we get

$$\underset{T\to\infty}{\limsup} \tfrac{1}{T} \log \mathbb{E}(\exp(TF_U(L_T))) \ \leq \ \overline{F}_U < \overline{F} \ ,$$

and therefore

(2.2)
$$\lim_{T\to\infty} \Gamma_T(U^c) = \lim_{T\to\infty} \widehat{\mathbb{P}}_T(L_T \notin U) = 0 \ .$$

This implies that Γ_T is relatively compact in $\mathcal{M}_1^+(\mathcal{M}_1^+(E))$ and any limit point is supported by K_F. If, by good luck, K_F just contains one element

(2.3)
$$K_F = \{\nu_o\} \ ,$$

then we get

$$\lim_{T\to\infty} \Gamma_T = \delta_{\nu_o} \ .$$

Using the strong convexity of I one sees that (2.2) is true if F is concave, but in general, this is a delicate point to decide. If K_F does contain more than one point, the rough large deviation principle is not sufficient to determine if Γ_T converges or not, and if it does, to which limiting measure. We will give a short discussion of this point in Section 4 and **do** assume for the moment that

(2.4)
$$\Gamma = \lim_{T\to\infty} \Gamma_T$$

exists. By (2.2) we then automatically have that Γ is supported by K_F.

Remark The above discussion does use only the strong LDP and is therefore true whenever this holds. This especially applies to empirical measures of uniformly ergodic Markov chains and processes, or to Gibbs fields (see [DS]).

We discuss now the step (II) above. For general Markov processes, this is delicate. We will come back to this later on. However, in the i.i.d. situation, this is a rather trivial point due to a simple symmetry argument: Let $\phi : E^k \to \mathbb{R}$ be a bounded and smooth function. Then

$$\widehat{\mathbb{E}}_T(\phi(X_1, X_2, \ldots, X_k)) = \int \mathbb{E}(\phi(X_1, X_2, \ldots, X_k) \mid L_T = \nu) \, \Gamma_T(d\nu) .$$

Conditioned on $L_T = (1/T) \sum_{i=1}^T \delta_{\xi_i}$, $(T \geq k)$, $X_1, X_2, \ldots, X_k$ is just drawing without replacement from the set $\{\xi_1, \ldots, \xi_T\}$. If T is large (k fixed), this is close to drawing with replacement, i.e.

$$\mathbb{E}(\phi(X_1, X_2, \ldots, X_k) \mid L_T = \nu) \approx \int \phi \, d\nu^{\otimes k}$$

From this, one easily sees that if (2.4) holds, then

$$\lim_{T \to \infty} \widehat{\mathbb{E}}_T(\phi(X_1, X_2, \ldots, X_k)) = \int_K \left(\int_{E^k} \phi \, d\nu^{\otimes k} \right) \Gamma(d\nu) , \quad \text{i.e.}$$

$$\lim_{T \to \infty} \mathcal{L}_{\widehat{\mathbb{P}}_T}(X_1, X_2, \ldots) = \int_K \nu^{\otimes \mathbb{N}} \Gamma(d\nu) .$$

The argument essentially carries over to certain discrete time Markov chains, but an easy symmetry argument as that given above works only if the transition kernel of the chain is bounded from below (see [BS]). I don't discuss this here in details as a "general" result in the more delicate continuous time case is presented in the next Section.

It may be instructive to look at the Example 1.2 , which belongs to a class discussed in [CCC]. The large deviation behaviour of T_n and T'_n can be obtained from the large deviation behaviour of the bivariate empirical measures

$$L_n^{(2)} = \frac{1}{n} \sum_{i=1}^{n-1} \delta_{(X_i, X_{i+1})} \quad , \quad L_n'^{(2)} = L_n^{(2)} + \frac{1}{n} \delta_{(X_n, X_1)} .$$

Both satisfy a n-LDP with rate function $I(\mu) = k(\mu|\mu_1 \otimes \pi)$, for $\mu \in \mathcal{M}_1^{0,+}(E \times E) = \{ \mu \in \mathcal{M}_1^+(E \times E) : \mu_1 = \mu_2 \}$, where μ_1 , μ_2 are the two marginals of μ , and π is the distribution of the X_i . $I(\mu) = \infty$ if $\mu \notin \mathcal{M}_1^{0,+}(E \times E)$ (see [CCC] or [B3]). Our functional $F : \mathcal{M}_1^+(E \times E) \to [-\infty, \infty)$ is just $-\infty 1_{\{\mu(1,1) < \alpha\}}$. One easily checks that $b(\overline{F}) = b(\underline{F})$ and that K_F contains just one point, namely the measure

$$\mu_\alpha = (\mu_\alpha(i,j))_{i,j=1,2} = \begin{bmatrix} \alpha & \beta(\alpha) \\ \beta(\alpha) & 1\text{-}\alpha\text{-}2\beta(\alpha) \end{bmatrix},$$

where $\beta(\alpha)$ is the unique solution of the cubic equation $x^2(1\text{-}\alpha\text{-}x) = (\alpha+x)(1\text{-}\alpha\text{-}2x)^2$ in the interval $(0,(1\text{-}\alpha)/2)$. The discussion of the point (I) given above yields

$$\lim_{n \to \infty} \widehat{P}_n(L_n^{(2)})^{-1} = \delta_{\mu_\alpha} \; ,$$

and the same for $L_n^{'(2)}$.

How about (II)? A symmetry argument works for $L_n^{'(2)}$, just because everything is then symmetric with respect to cyclic permutations. Using such an argument one can easily get:

$$\mathbb{P}\Big(X_1 = i_1 , \ldots X_k = i_k \mid L_n^{'(2)} = \nu\Big) \approx \nu_1(i_1)P_\nu(i_1,i_2) \ldots P_\nu(i_{k-1},i_k) \; ,$$

for $n \gg k$, with $P_\nu(i,j) = \nu(i,j)/\nu_1(i)$. This yields

(2.5) $$\lim_{n \to \infty} \widehat{P}_n = \mu_{\alpha,1} \otimes P_{\mu_\alpha} \otimes P_{\mu_\alpha} \otimes \ldots$$

This symmetry argument does not work for $L_n^{(2)}$ and it seems to be unavoidable to use precise large deviation asymptotics. We will give some indications of this in Section 4 (Although at first it may seem surprising, the answer is different from (2.5)).

The arguments given in this Section extend to Gibbs measures, but again only with periodic boundary conditions. This has been investigated under somewhat different viewpoints (including a new proof of the LDP) in a recent paper by Georgii [G]. For a discusion of related question, see also [DSZ].

3. A result for continuous time processes

We state here a result for continuous time Markov processes, which satisfy a quite restrictive recurrence condition and where the functional F satisfies an additional differentiability property. We give a sketch of the main point of the argument, which reveals some interesting points, which will also be important in the examples in the Sections 5 and 6. Details of the resuts are in [BDS], Section 2.

Assumption 3.1

a) X_t , $t \geq 0$, is a Markov process on a compact state space E , having cadlag sample paths and transition densities $p_t(x,y)$ with respect to a stationary measure π which are continuous, bounded and bounded away from 0 for each fixed t . The starting point is (for the sake of simplicity) fixed and denoted by x_o .

b) $F : \mathcal{M}_1^+(E) \to \mathbb{R}$ is bounded, continuous and differentiable in the following sense: There exists a continuous mapping $DF : \mathcal{M}_1^+(E) \to C(E)$, such that for $\lambda \in [0,1]$, μ , $\nu \in \mathcal{M}_1^+(E)$

$$F((1-\lambda)\mu+\lambda\nu) = F(\mu) + \lambda\langle DF(\mu),\nu-\mu\rangle + R(\lambda;\mu,\nu) ,$$

where $\langle . , . \rangle$ denotes the pairing between $\mathcal{M}_1^+(E)$ and C(E) , and where the rest term R converges to 0 as $\lambda\to0$ uniformly in μ for each ν .

Example 3.2 $F(\mu) = \int \int V(x,y)\mu(dx)\mu(dy)$, where   V   is a continuous function on $E \times E$ satisfies the Assumption 3.1.

Under the above assumption a), L_T satisfies a strong T-LDP with rate function

$$I(\mu) = -\inf\left\{ \int \frac{Lu}{u} : u \in \mathcal{D}(L) , u > 0 , \log(u) \text{ bounded}\right\},$$

where   L   is the infinitesimal generator, and $\mathcal{D}(L)$ its domain (see [DS]). At least if   F   is bounded and continuous, the analysis of step (I) works well, so we know that Γ_T is relatively compact for $T \to \infty$ and every limit point is supported by the set K_F . However, the transfer of this information to the path measure is in general delicate. It is this point where we need the differentiability assumption.

Theorem 3.3 $d\widehat{P}_T = \exp(TF(L_T))dP/z_T$ is relatively compact as $T \to \infty$ in $\mathcal{M}_1^+(D(E))$, where $D(E)$ is the set of cadlag paths $[0,\infty) \to E$ with the Skorohod topology. Each limit point $\widehat{P}$ of $\widehat{P}_T$ has a representation

$$\widehat{P} = \int_{K_T} Q^\mu \, \Gamma(d\mu) \, ,$$

where Q^μ is a special h-path transformation of the original measure P which is described in the following way: For $\phi \in C(E)$, the kernel P_t^ϕ is defined by

$$P_t^\phi(x,A) = \mathbb{E}_x \left(e^{\int_0^T \phi(X_u)du} \, 1_A(X_t) \right).$$

The logarithmic spectral radius of this kernel is $\Lambda_\phi t$ where $\Lambda_\phi = \sup\{\langle \phi,\mu \rangle - I(\mu) : \mu \in \mathcal{M}_1^+(E)\}$. By the Krein-Rutman theory of positive operators (see [S]), we find a function $h^\phi \in C^+(E)$, which is unique up to multiplication by a constant, which satisfies $P_t^\phi h^\phi = \exp(\Lambda_\phi t)h^\phi$. Put $Q_t^\phi(x,A) = \exp(-\Lambda_\phi t)(1/h^\phi(x))P_t^\phi(x,dy)h^\phi(y)$ and use this family of kernels to construct the path measure Q^ϕ .

Our path measures Q^μ are just $Q^{DF(\mu)}$, and we write h^μ , l^μ , Λ_μ instead of $h^{DF(\mu)}$ etc. (We us P , $\mathbb{E}$ without lower index if the starting point is x_o).

Remarks 3.4 a) Q_t^ϕ has a stationary measure $\pi^\phi(dx) = h^\phi(x)l^\phi(x)\pi(dx)$, where l^ϕ corresponds to the time reversed process.

b) If $\mu \in K_F$ then $\pi^{DF(\mu)} = \mu$.

c) In the time reversible case, we have $h^\phi = l^\phi = \sqrt{\dfrac{d\pi^\phi}{d\pi}}$.

We give a **sketch of the of the proof** of the above Theorem.

Let $0 < s \ll t \ll T$, and f be a real valued bounded continuous function on $D(E)$ which depends only on the restrictions of the paths on the time slot $[0,s]$. We want to discuss how $\widehat{\mathbb{E}}_T(f)$ behaves for large T . Using the differentiability of F we get

$$TF(L_T) = TF(L_{t,T} + \tfrac{t}{T}(L_t - L_{t,T})) \approx TF(L_{t,T}) + t\langle DF(L_{t,T}), L_t - L_{t,T}\rangle \,,$$

where $L_{t,T} = \frac{1}{T-t} \int_t^T \delta_{X_s} ds$. Using formally δ as a Dirac function with respect to the stationary measure π, we get

$$\mathbb{E}\big(f \exp(TF(L_T))\big) = \int \pi(dx) \int \pi(dy) \, \mathbb{E}\Big\{ f \exp\Big(\int_0^s DF(L_{t,T})(X_u)du\Big) \delta(X_s - x)$$

$$\times \exp\Big(\int_s^t DF(L_{t,T})(X_u)du\Big)\delta(X_t - y) \exp\Big(TF(L_{t,T}) - t\langle DF(L_{t,T}), L_{t,T}\rangle\Big)\Big\}$$

$$= \int \pi(dy) \int \mathbb{P}_y(L_{t,T} \in d\mu) \exp\Big(TF(\mu) - t\langle DF(L_{t,T}), L_{t,T}\rangle\Big)$$

$$\times \underbrace{\int \pi(dx) \, \mathbb{E}\Big\{ f \exp\Big(\int_0^s DF(\mu)(X_u)du\Big) \delta(X_s - x)\Big\}}_{\dfrac{h^\mu(x_o)}{h^\mu(x)} e^{s\Lambda_\mu} \, \mathbb{Q}^\mu(f\,\delta(x - X_s))} \underbrace{\mathbb{E}_x\Big\{\exp\Big(\int_s^t DF(\mu)(X_u)du\Big) \delta(X_t - y)\Big\}}_{\substack{\dfrac{h^\mu(x)}{h^\mu(y)} e^{(t-s)\Lambda_\mu} \underbrace{q_{t-s}^{DF(\mu)}(x,y)}_{\approx h^\mu(y)l^\mu(y)}}}$$

The last approximation holds if $t-s$ is large. We see that $h^\mu(x)$ and $h^\mu(y)$ cancel and we can perform the $\pi(dx)$ integration which results in just dropping the $\delta(x - X_s)$ term. Introducing now the following measure on $\mathcal{M}_1^+(E)$:

$$\Gamma_{t,T}(d\mu) = \int \pi(dy) \, l^\mu(y) \, \mathbb{P}_y(L_{t,T} \in d\mu) \, e^{t\Lambda_\mu} \exp\Big(TF(\mu) - t\langle DF(\mu), \mu\rangle\Big)\Big/ z_{t,T},$$

$z_{t,T}$ being the appropriate norming, we get for $s \ll t$ and $t \ll T$:

$$(3.5) \qquad \hat{\mathbb{E}}_T(f) \approx \int \mathbb{Q}^\mu(f) \, h^\mu(x_o) \, \Gamma_{t,T}(d\mu) \Big/ \int h^\mu(x_o) \, \Gamma_{t,T}(d\mu) \,.$$

It is not difficult to see that for $T \gg t$, $\Gamma_{t,T}$ is concentrated near K_F. From this, the above Theorem 3.3 follows. Of course, a justification of the different $\approx$-arguments needs quite some care (see [BDS]), but I hope to have convinced the reader that the argument is essentially simple. However, the differentiability assumption is quite a heavy restriction. Also the assumption of uniform ergodicity excludes applications e.g. directly to Brownian motion. At least, results for Brownian motion need a further justification. Nevertheless, I am convinced, that the above picture is correct under quite general conditions, even when the assumptions are not satisfied. We will look at such situations in the Sections 5 and 6.

As in the i.i.d. case, and even more so here, the determination of the limiting law of $\Gamma_{t,T}$ (first $T \to \infty$, then $t \to \infty$) if it exists at all , is difficult to determine except, of course in the case where K_F just contains one point. However, there is an important special case, where this can be discussed easily with the help of symmetry considerations.

Let G be a compact and metrizable topological group which acts continuously on E . If $(g,x) \in G \times E$, we write $gx \in E$. G then acts also in an obvious way on $\mathcal{M}_1^+(E)$. We assume that F and the transition kernels are G-invariant. Obviously, we then have that the set K_F is G-invariant, but in general the elements $\mu \in K_F$ are not G-invariant. We make the restrictive assumption that the solution of our variational problem is unique up to the action of G :

Assumption 3.6 There exists $\mu \in K_F$ such that $K_F = \{g\mu : g \in G\}$.

To check if this assumption is satisfied is usually a very heavy task. Both in the Wiener sausage case and in the polaron case discussed later on, this is true, but are difficult analytic result.

Theorem 3.7 Under this assumption, we have

$$\lim_{T \to \infty} \widehat{\mathbb{P}}_T = \int_G \mathbb{Q}^{g\mu} \, h^{g\mu}(x_o)\sigma(dg) \Big/ \int h^{g\mu}(x_o)\sigma(dg) ,$$

where σ is the Haar measure on G .

The Theorem follows fairly easily from (3.5) and the fact that our measures $\Gamma_{t,T}$ must, by our assumptions, for $0 \ll t \ll T$, be close to the uniform distribution on K_F , i.e. that induced by the Haar measure.

Remark In the case where the transition kernels P_t are reversible with respect to π , i.e. $\pi \otimes P_t$ is a symmetric measure on $E \times E$, we have $h^\phi = 1^\phi$ and therefore

$$h^{g\mu}(x_o) = \sqrt{\frac{d(g\mu)}{d\pi}(x_o)} = \sqrt{\frac{d\mu}{d\pi}(g(x_o))} .$$

4. The role of precise asymptotics

If the set K_F of solutions of the variational problem contains more than one point, and symmetry considerations as in Theorem 3.7 are not applicable, then it is unavoidable to look at an asymptotic evaluation of the partition function z_T (and of related expressions) which go beyond (1.7). For a discussion of the limit of $\Gamma_T = \widehat{P}_T(L_T)^{-1}$, one needs an evaluation of

$$\int \varphi \, d\Gamma_T = \mathbb{E}\big(\exp(a_T F(L_T))\, \varphi(L_T)\big)\big/ z_T$$

for suitable test functions $\varphi : \mathcal{M}_1^+(E) \to \mathbb{R}$. In order to obtain this, we need an evaluation of these expectations up to a factor $(1 + o(1))$ and not $\exp(a_T o(1))$, which is provided by (1.8). There are still not many results of this type available and there are a lot of open problems. In the i.i.d. infinite dimensional case, this has been discussed in [B1] and [B2] and for symmetric Markov processes in [KT]. Symmetric finite state Markov processes are also treated in [BM] with the help of noncommutative Grassmanian ("Fermionic") integration. Nonsymmetric Markov processes are treated in a forthcoming paper [BDT]. I give here a rough outline of this.

An essential role is played by the notion of a nondegenerated solution of the variational problem. Roughly, this is a point $\mu \in K_F$ where the second derivative of $F - I$ is strictly negative definite. As I, in general is not a smooth function, this notion needs some explanation.

It is convenient to work in a Banach space or even a Hilbert space instead the space of measures $\mathcal{M}_1^+(E)$. There are several ways to do this, one possibility is described in the paper by Kusuoka and Tamura [KT]. Essentially, one is embedding $\mathcal{M}_1^+(E)$ in a Hilbert space H and assumes that F is extensible to a function defined on H. One of the advantages to work in a Hilbert space is that central limit theorems are readily available, but also the derivation of some very crucial exponential inequalities is simpler in such a setting.

However, I don't want to go into technicalities of this type here, and for this reason, I assume that the state space E is finite, and F is a smooth bounded function $\mathbb{R}^E \to \mathbb{R}$. It should however be remarked that one of the technical difficulties, namely the nonsmoothnes of I is absent it this setting.

Let us look first at the case where K_F just contains one point, i.e. $K_F = \{\mu\}$. As in Section 3, we consider the measure Q^μ. We can describe the derivative of $\mathbb{P}$ with respect to Q^μ on $\mathcal{F}_T = \sigma(X_s : s \leq T)$ by

$$\frac{d\mathbb{P}}{dQ^{\mu}}\bigg|_{\mathcal{F}_T} = h^{\mu}(x_o)\, e^{T\Lambda_{\mu}} \exp\left(-\int_0^T DF(\mu)(X_u)\, du \right) \frac{1}{h^{\mu}(X_T)} \ .$$

(x_o is the starting point) . Therefore

$$\mathbb{E}\left(e^{TF(L_T)}\right) = h^{\mu}(x_o)\, e^{T\Lambda_{\mu}}\, \mathbb{E}_{Q^{\mu}}\left\{ \exp[TF(L_T) - \langle DF(\mu), L_T\rangle] \frac{1}{h^{\mu}(X_T)} \right\} \ .$$

It is easy to see that (for $\mu \in K_F$!), $\Lambda_{\mu} = \langle DF(\mu), \mu\rangle - I(\mu)$, so

$$\mathbb{E}\left(e^{TF(L_T)}\right) = h^{\mu}(x_o)\, e^{Tb(F)}\, \mathbb{E}_{Q^{\mu}}\left\{ \exp[TF(L_T) - \langle DF(\mu), L_T - \mu\rangle] \frac{1}{h^{\mu}(X_T)} \right\} \ .$$

Let now $A = (a_{ij})_{i,j\,\in\,E}$ be the matrix of the second derivatives of F at μ : $a_{ij} = \partial^2 F(\mu)/\partial x_i \partial x_j$, and set $l_T(i) = \sqrt{T}(L_T(i) - \mu(i))$. Then we approximate the exponent on the right hand side above :

$$T[F(L_T) - \langle DF(\mu), L_T - \mu\rangle] \approx \tfrac{1}{2} \sum_{i,j} a_{ij}\, l_T(i) l_T(j) \ ,$$

and we apply now a central limit theorem for l_T . Q^{μ} is indeed uniformly ergodic as well as $\mathbb{P}$ and l_T satisfies a central limit theorem. So

$$\mathcal{L}_{Q^{\mu}}(\, l_T(\,\cdot\,)) \to \gamma \ ,$$

where γ is a centred Gaussian measure on $\mathbb{R}^E$. The covariance matrix is well known to be the recurrence potential matrix

$$g(i,j) = \int x_i x_j \gamma(dx) = \mu(i)\int_0^{\infty}(q_t^{\mu}(i,j) - \mu(j))dt + \mu(j)\int_0^{\infty}(q_t^{\mu}(j,i) - \mu(i))dt \ .$$

One is tempted to argue that

(4.1)

$$\mathbb{E}_{Q^{\mu}}\left\{ \exp[T(F(L_T) - \langle DF(\mu), L_T - \mu\rangle)] \frac{1}{h^{\mu}(X_T)} \right\}$$

$$= (1 + o(1))\int_{\mathbb{R}^E} \exp\left[\tfrac{1}{2}\sum a_{ij}\, x_i x_j \right] \gamma(dx) \int_{\mathbb{R}} \frac{1}{h^{\mu}(x)}\, \mu(dx) \ ,$$

but a justification of this needs some care as the exponents on both sides are not bounded. The left hand side is finite if and only if $I - AG$ is positive definite, and in this case

$$\int_{\mathbb{R}^E} \exp\left[\tfrac{1}{2} \sum a_{ij}\, x_i x_j \right] \gamma(dx) = (\det(I - AG))^{-1/2} \,.$$

It can be proved that $I - AG$ is always positive semidefinite. This just comes from the fact that $F - I$ is maximal at μ. It turns out that (4.1) is correct if and only if $I - AG$ is positive definite. In this case, we call μ a **nondegenerated** maximum.

Assume now that K_F contains a finite number of points : $K_F = \{\mu_1, \dots, \mu_k\}$. This leads to matrices $A_1, \dots, A_k$; $G_1, \dots, G_k$. If we assume that each μ_i is nondegenerated, the numbers

$$\beta_i = h^{\mu_i}(x_o)\,(\det(I - A_i G_i))^{-1/2} \int_{\mathbb{R}} \frac{1}{h^{\mu_i}(x)}\, \mu_i(dx)$$

will be the relative weights of $\widehat{P}_T(L_T)^{-1}$ on μ_i , i.e. $\Gamma = \lim_{T\to\infty} \widehat{P}_T(L_T)^{-1}$ exists and $\Gamma(\mu_i) = \beta_i / \sum_i \beta_i$. Together with the analysis in Section 3, this leads to

Theorem 4.2 Under the above conditions, we have

$$\lim_{T\to\infty} \widehat{P}_T = \sum_{i=1}^k \beta_i\, Q^{\mu_i} \Big/ \sum_{i=1}^k \beta_i \,.$$

As remarked above, the analysis is not restricted to a finite state space. More challenging is the treatment of degeneracies and manifolds of maxima. See [B2] for the i.i.d. case and [KT] for the symmetric Markov case.

The delicacy in such kind of problems is already revealed in the Example 1.2 when one takes $T_n = \sum_{i=1}^{n-1} X_i X_{i+1}$. Then

$$\widehat{P}_n = \frac{P(X_1 = 1\,,\, T_n \geq \alpha n)}{P(T_n \geq \alpha n)} = 1 - \frac{1}{2}\frac{P(T_n \geq \alpha n \mid X_1 = 0)}{P(T_n \geq \alpha n)} = 1 - \frac{1}{2}\frac{P(T_{n-1} \geq \alpha n)}{P(T_n \geq \alpha n)} \,.$$

One is now tempted to do the following formal calculation :

$$P(T_{n-1} \geq \alpha n) = P\big(\tfrac{T_{n-1}}{n-1} \geq \alpha(1 + \tfrac{1}{n-1})\big) \approx \exp\big(-(n-1)j(\alpha + \tfrac{\alpha}{n-1})\big)$$

$$\approx \exp(-nj(\alpha)) \exp(j(\alpha) - \alpha \ j'(\alpha)) \approx \mathbb{P}(T_n \geq \alpha n) \exp(j(\alpha) - \alpha \ j'(\alpha)) \ ,$$

where $j(\alpha) = k(\mu_\alpha | \mu_{\alpha,1} \otimes \pi)$ and therefore

$$(4.3) \qquad \lim_{n \to \infty} \widehat{P}_n(X_1 = 1) = 1 - \tfrac{1}{2} \exp(j(\alpha) - \alpha \ j'(\alpha)) \ .$$

Of course, the above calculation contains many inaccuracies. E.g. $\mathbb{P}(T_n \geq \alpha n)$ is certainly not $c(1 + o(1)) \exp(-nj(\alpha))$. The reader is invited to provide a justification of (4.3) as an exercise.

5. Transformation by the number of points visited by a random walk

Let X_t , $t \geq 0$, be a symmetric nearest neighbour random walk on $\mathbb{Z}^d$ starting at 0 , and let N_T be the number of points visited by this walk. We are interested in the behaviour of the path measure

$$d\widehat{P}_T = \exp(-N_T) d\mathbb{P}/z_T$$

for large T . In order to turn it into a problem which looks similar to the ones discussed in Section 3, we do a Brownian rescaling: Let $\xi_t = T^{-1/(2+d)} X_{tT^{2/(2+d)}}$, for $t \leq \tau = T^{d/(2+d)}$. Then $N_T = \tau |\text{supp}(l_\tau)|$, where the empirical density $l_\tau(x) = \int_0^\tau 1_x(\xi_s) ds$, for $x \in T^{-1/(2+d)} \mathbb{Z}^d$. It is convenient to extend l_τ to a (random) mapping $\mathbb{R}^d \to \mathbb{R}$ by keeping it constant on the cubes of $T^{-1/(2+d)} \mathbb{Z}^d$. Remark that then $\int l_\tau(x) dx = 1$, so it is a random probability density. Donsker and Varadhan have proved

Theorem 5.1 (Donsker&Varadhan, [DV1])

$$\lim_{T \to \infty} \tfrac{1}{\tau} \log z_T = -\inf\{\lambda(\text{supp}(\mu)) + I(\mu) : \mu \in \mathcal{M}_1^+(\mathbb{R})\} \ ,$$

where I is the rate function of Brownian motion, i.e. $I(\mu) = \tfrac{1}{2} \int \left| \nabla \sqrt{\dfrac{d\mu}{d\lambda}} \right|^2 d\lambda$, where λ is Lebesgue measure and $\mathrm{supp}(\mu)$ the topological support of μ .

It is well known that the variational problem has a unique solution up to translations. This is the famous Haber-Krahn Theorem. It states that K_F , the set of measures μ for which $\lambda(\mathrm{supp}(\mu))+I(\mu)$ is minimal, is a family $\{\mu_x : x \in \mathbb{R}^d\}$, where $d\mu_x/d\lambda = \psi_x^2$, and where ψ_x is the L_2-normalised eigenfunction of the Laplacian with Dirichlet boundary condition in a ball $B_\rho(x)$ with a fixed radius $\rho(d)$ and centre x. ψ_x is extended to be 0 outside B(x). Actually, the Haber-Krahn theorem states that for bounded open sets G of a given size $\lambda(G)$, $I(\mu)$ with support in G is minimal if and only if G is a ball and $d\mu/d\lambda$ the square of an eigenfunction as described above. There is then a further trivial minimisation problem to determine the optimal radius $\rho(d)$. Although the assumptions 3.1 are not satisfied - our F here is not smooth and Brownian motion is not uniformly ergodic - one is tempted to guess that the answer given in Section 3 is correct and we have:

$$(5.2) \qquad \lim_{T \to \infty} \mathcal{L}_{\widehat{\mathbb{P}}_T} (\xi_s , s \geq 0) = \int_{B(0)} dx \; \psi_o(x) \, \mathbb{Q}^x ,$$

where $\mathbb{Q}^x$ is the Brownian taboo process for the ball B(x) , i.e. the process conditioned to stay in this ball forever (by an appropriate limiting procedure). $\mathbb{Q}^x$ is the diffusion process given by the generator $\frac{1}{2}\Delta + \nabla(\log \psi_x)\nabla$. (Although our F is not differentiable, this is formally the h-path transformation described in Section 3). (5.2) would mean that the unscaled process X_t , $0 \leq t \leq T$, has a typical displacement of order $T^{1/(2+d)}$. Actually, although it is not implied by (5.2) , there should also hold a strong form of this confinement in the sense that for any $\epsilon > 0$

$$(5.3) \qquad \widehat{\mathbb{P}}_T\Big(\bigcup_{x \in B(0)} \{ d_H(\mathrm{supp}(l_T),B(x)) \leq \epsilon \} \Big) \to 0 , \text{ for } T \to \infty ,$$

where d_H is the Haussdorf distance between two sets in $\mathbb{R}^d$.

Remark Instead of starting with the symmetric random X_t walk and N_T , on cans also take Brownian motion and replace N_T by the so called volume of the Wiener sausage $\lambda\Big(\bigcup_{s \leq T} B_\delta(X_s)\Big)$, where $\delta > 0$ is arbitrary and does not matter at all.

(5.2) has for d=1 and the Wiener sausage been proved by Schmock [SM] .

The confinement property (5.3) has been proved for $d=2$ independently in [B5] and by Sznitman [SZ] (for the Wiener sausage). The last paper contains also (5.2) . The general case $d \geq 3$ contains some challenging problems. I have no doubt that the result is true in any dimension, but a complete proof is lacking.

There are a number of variations and other open problems in this field. I just mention one:

In our discrete model of a continuous time random walk on the d-dimensional lattice, we can write $N_T = \sum_x 1_{(0,\infty)}(l_T(x))$, where $l_T(x) = \int_0^T 1_x(X_s)ds$. The question is, what happens if we replace $1_{(0,\infty)}$ by another concave function $\varphi : [0,\infty) \rightarrow [0,\infty)$, satisfying $\varphi(0) = 0$, $\varphi'(0) = \infty$, $\lim_{t \rightarrow \infty}\varphi(t)/t = 0$, e.g. $\varphi_\alpha(t) = t^\alpha$, $0 < \alpha < 1$. This is motivated in part by a paper of Greven and den Hollander [GH1] on branching random walks in random environments. It turns out that for φ_α , the appropriate rescaling is given by $\tau = T^{(2+(d-2)(1-\alpha))/(2+d(1-\alpha))}$, i.e.

$$\lim_{T \rightarrow \infty} \frac{1}{\tau} \log z_T = -\inf\left\{ \int g^{2\alpha}dx + \tfrac{1}{2}\int |\nabla g|^2 dx : \int g^2 dx = 1\right\} .$$

This will be published in a forthcoming paper. The variational problem has solutions of compact support.

There should also be a corresponding confinement property, but this is open, too.

6. Polaron type models

The partition function of the so called Fröhlich polaron is given by

$$z_T(\alpha) = E\left(\exp\left(\frac{\alpha}{2}\int_0^T dt \int_0^T ds \, e^{-|t-s|} \frac{1}{|X_t - X_s|} \right) \right),$$

where X_t , $t > 0$, is a 3-dimensional Brownian motion. For the physical background, see Feynman [F]. The large deviation theory can be used to give an evaluation of this partition function for large α and T (see Donsker & Varadhan [DV2]). In order to do this, one first uses rescaling. Substituting $X_t = (1/\alpha)X_{t\alpha^2}$ we get, with $\lambda=1/\alpha^2$ and $T'=T\alpha^2$:

$$z_{T',\lambda} = z_T(1/\lambda) = E\left(\exp\left(\frac{\lambda}{2} \int_0^{T'} dt \int_0^{T'} ds \, e^{-\lambda|t-s|} \frac{1}{|X_t' - X_s'|} \right) \right).$$

We are dropping now the dashes in T' and X_t' .We are interested in the behaviour for first $T \rightarrow \infty$ and then $\lambda \rightarrow 0$. For fixed t not too close to the boundary of the interval $[0,T]$, the function $[0,T] \ni s \mapsto (\lambda/2)e^{-\lambda|t-s|}$ is nearly a probability density function. If $\lambda \sim 0$, then this density is very spread out. One therefore compares $z_{T,\lambda}$ with

$$\tilde{z}_T = \mathbb{E}\left(\exp\left(\frac{1}{T} \int_0^T dt \int_0^T ds \, \frac{1}{|X_t - X_s|} \right) \right) .$$

The expression in the exponent is of the form considered in Example (3.2) , with $V(x,y) = 1/|x-y|$, and L_T the empirical measure of the Brownian motion. Of course, this V is not bounded, which causes some additional technical problems. Anyway, if one believes that $z_{T,\lambda}$ should be close to $\tilde{z}_T$ for large T and then λ close to 0 , the following result of Donsker and Varadhan looks plausible:

Theorem 6.1 ([DV2])

$$\lim_{\lambda \to 0} \lim_{T \to \infty} \frac{1}{T} \log z_{T,\lambda} = \lim_{T \to \infty} \frac{1}{T} \log \tilde{z}_T$$
$$= \sup\left\{ \int \int \frac{1}{|x-y|} g^2(x) \, g^2(y) dx \, dy - \frac{1}{2} \int |\nabla g(x)|^2 dx : \int g^2 dx = 1 \right\}$$

The solutions of the variational problem are unique, up to translation, as has been proved by E. Lieb [L] , i.e. the set of functions g maximising the right hand side can be described by $\{ g_x = g_o(. - x) : x \in \mathbb{R}^3 \}$, where g_o is a rotationally symmetric function.

We introduce now the path measures

(6.2) $\qquad d\hat{P}_{T,\lambda} = \exp\left(\frac{\lambda}{2} \int_0^T dt \int_0^T ds \, e^{-\lambda|t-s|} \frac{1}{|X_t - X_s|} \right) dP \Big/ z_{T,\lambda} ,$

(6.3) $\qquad d\tilde{P}_T = \exp\left(\frac{1}{T} \int_0^T dt \int_0^T ds \, \frac{1}{|X_t - X_s|} \right) dP \Big/ \tilde{z}_T .$

From the above Theorem and the results of Section 3, it would look plausible that

(6.4) $\qquad \lim_{\lambda \to 0} \lim_{T \to \infty} \hat{P}_{T,\lambda} = \lim_{T \to \infty} \tilde{P}_T = \int_{\mathbb{R}^3} dx \, g_o(x) \, Q^x ,$

where Q^x is the appropriate h-path transformation corresponding to g_x , i.e. just the diffusion with generator $\frac{1}{2}\Delta - \nabla(\log g_x)\nabla$. Due to a number of technical problems with the Fröhlich polaron, there are, up to now, no proofs. However, I have no doubts, that the second equality in (6.4) is correct, while the first is not. I will try to give an explanation of this.

It is better to look at a slightly more general situation: X_t , $t \geq 0$, should be a nice Markov process on a state space E and $V : E \times E \to \mathbb{R}$ a suitable function. As the only results which have been proved, are two very special examples given below, I will not be very precise here. Our two path measures are then the ones where in (6.2) and (6.3) , one replaces $|X_t - X_s|^{-1}$ by $V(X_s, X_t)$. We do a further rescaling,

$$(6.5) \quad \frac{\lambda}{2} \int_0^T dt \int_0^T ds \, e^{-\lambda|t-s|} \, V(X_s, X_t) = \frac{1}{2\lambda} \int_0^{\lambda T} dt \int_0^{\lambda T} ds \, e^{-|t-s|} \, V(X_{t/\lambda}, X_{s/\lambda}) \, .$$

In the end, we want to let first $T \to \infty$ and then $\lambda \to 0$, but for the moment, we do both things together, by introducing $R = \lambda T$, which is kept fixed and letting $\lambda \to 0$.

We first look what happens with the corresponding partition function

$$z_{R/\lambda, \lambda} = \mathbb{E}\left(\exp\left(\frac{1}{2\lambda} \int_0^R dt \int_0^R ds \, e^{-|t-s|} \, V(X_{t/\lambda}, X_{s/\lambda}) \right) \right) \, .$$

Chopping the time interval $[0,R]$ into small pieces $[\frac{iR}{N} , \frac{(i+1)R}{N}]$, $i=0, \ldots ,$ N-1, letting $\lambda \to 0$ and finally $N \to \infty$, it is not difficult to see, that at least under the conditions of Section 3 , we have

$$(6.6) \quad \lim_{\lambda \to 0} \lambda \log z_{R/\lambda, \lambda}$$

$$= \sup\left\{ \frac{1}{2} \int_0^R dt \int_0^R ds \, e^{-|t-s|} \, V(\mu_t, \mu_s) - \int_0^R I(\mu_t) dt : \mu_. \in C([0,R] \to \mathcal{M}_1^+(E)) \right\},$$

where $V(\mu, \nu) = \int \int V(x,y) \mu(dx) \nu(dy)$. If R is large, then the main contribution to the sup on the right hand side of (6.6) is coming from functions $\mu_. \in C([0,R] \to \mathcal{M}_1^+(E))$, which are constant most of the time and therefore, this sup is $\approx R \sup_\mu(V(\mu,\mu) - I(\mu))$ which is the Donsker-Varadhan result. However, for the path measure, an exact discussion of the maximisers of this time inhomogeneous variational problem is necessary. Let us denote by $K_{inhom}(R)$ the set of maximisers on the right hand side of (6.6). Up to now, we don't have any general results, but we believe, that the picture is the following: For $R \to \infty$, the set $K_{inhom}(R)$ converges to a set $K_{inhom}(\infty)$ of continuous mappings $\mu_. : [0,\infty) \to \mathcal{M}_1^+(E)$, which have the property that for $t \to \infty$, μ_t approaches an element $\mu \in K = \{\nu \in \mathcal{M}_1^+(E) :$

$V(\nu,\nu) - I(\nu)$ is maximal$\}$. If $\overline{K} = \{ \int_0^\infty e^{-t} \mu_t \, dt : \mu_t \in K_{inhom}(\infty) \}$, then the possible limits of $\hat{P}_{T,\lambda}$, with first $T \to \infty$ and then $\lambda \to 0$ are mixtures of the measures $Q^{\bar{\mu}}$, $\bar{\mu} \in \overline{K}$. To prove this in any generality, one would need a quite detailed analysis of the inhomogeneous variational problem and its connection with the homogeneous one. Especially quite strong statements about certain uniform approximations for $R \to \infty$ is needed. That this is not quite easy is also reflected by the fact that the analysis has certainly to go beyond that of the Donsker-Varadhan result, which essentially is the statement that $(1/R) \times$ the right hand side of (6.6) converges to $\sup_\mu (V(\mu,\mu) - I(\mu))$.

In [BDS] we have been able to discuss two special examples, which fully confirm the above picture:

Example 6.7 X_t , $t \geq 0$, is an ordinary symmetric jump process on $E = \{0,1\}$. V is given by $V(x,y) = \tau > 0$ if $x \neq y$ and $V(x,x) = 0$. This is certainly the simplest example one could think of. The mean field model, i.e. $\tilde{P}_T$ exhibits a phase transition: If $\tau \leq 1$, then K contains just one element, obviously then the measure $(1/2,1/2)$. For $\tau > 1$, there are two elements : (p_-,p_+) and (p_+,p_-) with $p_\pm = (1/2)(1 \pm (1 - 1/\tau^2)^{1/2})$. Then it follows from the results in Section 3 that

$$\lim_{T \to \infty} \tilde{P}_T = \frac{1}{1 + \gamma} Q^\gamma + \frac{\gamma}{1 + \gamma} Q^{1/\gamma} ,$$

where Q^γ is the law of the Markov chain with Q-matrix $\begin{bmatrix} -\gamma & \gamma \\ 1/\gamma & -1/\gamma \end{bmatrix}$, and

where $\gamma = 1$ if $\tau < 1$ and $\gamma = \tau + \sqrt{\tau^2 - 1}$ if $\tau > 1$.

In the polaron model, the inhomogeneous variational problem leads, for the determination of $\mu_t = (p_t, 1 - p_t)$, after some transformations, to the differential equation

$$y''(t) = y(t) - 2\tau \frac{y(t)}{\sqrt{4 + y(t)^2}} .$$

Finally, one gets

$$\lim_{\lambda \to 0} \lim_{T \to \infty} \hat{P}_{T,\lambda} = \frac{1}{1 + \overline{\gamma}} Q^{\overline{\gamma}} + \frac{\overline{\gamma}}{1 + \overline{\gamma}} Q^{1/\overline{\gamma}} ,$$

where $\bar{\gamma}$ corresponds to an adjusted $\tau : \bar{\tau} = (\tau + 1/\tau)/2$ for $\tau > 1$ and $\bar{\tau}$ $= \tau$ for $\tau \leq 1$.

It should be noted that already for a finite state Markov chain $|E| = m$, the inhomogeneous variational problem leads to a system of $m - 1$ coupled nonlinear differential equations and for the true polaron to a nonlinear partial differential equation. We are not yet able to discuss this in any generality.

Example 6.8 Here X_t , is a one-dimensional Brownian motion and $V(x,y) =$ $-\tau^2(x-y)^2$ for a parameter $\tau > 0$. Remark that even the mean field model does not fall into Section 3, because the Brownian motion does not satisfy the ergodicity assumption used there. This example is much simplified by the fact that everything stays in the realm of Gaussian measures, so can be analysed by investigating the covariance operators. The discussion of the polaron case is not without subtleties, but the outcome is the one expected by the analysis in Section 3 and the discussion given above.

The variational problems can be solved explicitly. K is given by $\{N(x,4\tau^2) :$ $x \in \mathbb{R}\}$, where $N(a,\sigma^2)$ denotes the normal distribution with mean a and variance σ^2. $K_{inhom}(R)$ is given by

$$K_{inhom}(R) = \left\{ t \mapsto N(x, 2\tau^2(2 - e^{-t} - e^{-(R-t)})) : x \in \mathbb{R} \right\} .$$

It is easily checked that our h-path transformed measures are in any case Ornstein-Uhlenbeck processes, and therefore our limits are mixtures of Ornstein-Uhlenbeck processes. What comes out is

$$\lim_{T \to \infty} \tilde{\mathbb{P}}_T = \mathbb{S}^\tau \quad , \quad \lim_{\lambda \to 0} \lim_{T \to \infty} \widehat{\mathbb{P}}_{T,\lambda} = \mathbb{S}^{\tau/\sqrt{2}} \quad ,$$

where

$$\mathbb{S}^\tau = \int_{\mathbb{R}} \mathbb{Q}^{c,\tau} \left(\frac{\tau}{4\pi} \right)^{1/4} h_{c,\tau}(0) \, dc \quad ,$$

$\mathbb{Q}^{c,\tau}$ being the Ornstein-Uhlenbeck process given by the stochastic differential equation

$$dX_t = d\beta_t - \tau(X_t - c)dt \quad \text{and} \quad h_{c,\tau} = (\tau/\pi)^{1/4} \exp(-\tau(x - c)^2/2) .$$

References

[B1] Bolthausen, E. : Laplace approximations for sums of i.i.d. random vectors. Probab. Theory and Rel. Fields **72** (1986) 305-318

[B2] Bolthausen, E. : Laplace approximations for sums of i.i.d. random vectors: Part II : Degenerate maxima and manifolds of maxima. Probab. Theory and Rel. Fields **76** (1987) 167-206

[B3] Bolthausen, E. : Markov process large deviation in the τ-topology. Stoch. Proc. Appl. **25** (1987) 95-108

[B4] Bolthausen, E. : On self-repellent one dimensional random walks , Probab. Theory and Rel. Fields **86** (1990) 423-441

[B5] Bolthausen, E. : Localization of a two-dimensional random walk with an attractive path interaction , to appear in Ann. Prob.

[BDS] Bolthausen, E. , Deuschel, J.D. and Schmock U. : Convergence of path measures arising from a mean field or polaron type interaction , to appear in Prob. Theory and Related Fields

[BDT] Bolthausen, E. , Deuschel, J.D. and Tamura Y. : Precise estimate for large deviations for nonsymmetric Markov processes , in preparation (1992)

[BM] Brydges, D. and Maya I.M. : An application of Berezin integration to large deviation, J. of Theoretical Probability **4** (1991) 371-390

[BS] Bolthausen, E. and Schmock, U. : On the maximum entropy principle for uniformly ergodic Markov chains , Stochastic Proc. Appl. **33** (1989) 1-27

[CCC] Choi, B.S. , Cover, T.M. and Csiszar, I. : Conditioned limit theorems under Markov conditioning. IEEE Trans. Inform. Theory IT-35 (1987) 788-801

[DS] Deuschel, J.D. and Stroock, D.W. : Large Deviations. San Diego: Academic Press 1989

[DSZ] Deuschel, J.D. , Stroock, D.W. and Zessin H. : Microcanonical distribution for lattice gases. Comm. Math. Phys. **139** (1991) 83-101

[DV1] Donsker, M.D. and Varadhan, S.R.S. : On the number of distinct sites visited by a random walk. Comm. Pure Appl. Math. **32** (1979) 721-747

[DV2] Donsker, M.D. and Varadhan, S.R.S. : Asymptotics for the polaron , Comm. Pure Appl. Math. **36** (1983) 505-528

[F]   Feynman, R. : Statistical Mechanics. W.A. Benjamin 1972

[G]   Georgii, H.O. : Large deviations and maximum entropy principles for interacting random fields on $\mathbb{Z}^d$, to appear in Ann. Prob.

[GH1] Greven, A. and den Hollander, F. : Population growth in random media I, II . J. Stat. Phys. **65** (1991) 1113-1154

[GH2] Greven, A. and den Hollander, F. : A variational characterization of the speed of a one-dimensional self repellent random walk. Preprint (1991)

[HS]   Hara and Slade, G. : Self avoiding walk in 5 or more dimension. Preprint (1991)

[KT]   Kusuoka, S. and Tamura, Y. : Precise estimate for large deviation of Donsker-Varadhan type. J. of The Faculty of Sci. , Univ. Tokyo, Sec. IA, **38** (1991) 533-565

[L]   Lieb, E. : Existence and uniqueness of the minimising solution of Choquard's non-linear equation. Studies in Appl. Math. **57** (1977) 93-105

[S]   Schaefer, H.H.: Banach Lattices and Positive Operators. Springer 1974

[SM]   Schmock, U. : Convergence of one dimensional Wiener sausage path measure to a mixture of Brownian taboo processes. Stochastics **29** (1989) 203-220

[SZ]   Sznitman, A.-S. : On the confinement property of two-dimensional Brownian motion among poissonian obstacles. Comm. Pure Appl. Math. **44** (1991) 1137-1170

Erwin Bolthausen   e-mail: K563720@CZHRZU1A.BITNET
Institut für Angewandte Mathematik
Universität Zürich
Rämistrasse 74
CH-8001 Zürich

Contemporary Mathematics
Volume **149**, 1993

Some remarks on products of random affine maps on $(R^+)^d$

ARUNAVA MUKHERJEA

ABSTRACT. Let X_0, X_1, ... be a sequence of i.i.d. random variables taking values in a family of affine maps from $(R^+)^d$ into itself. Let $Z_n \equiv X_0 X_1 \ldots X_n$ and $W_n \equiv X_n X_{n-1} \ldots X_0$, where the multiplication is simply the composition of maps. Here we identify the X_i's with appropriate random matrices and consider problems on the rate of growth of the sequence $\|W_n u\|$, $u \in (R^+)^d$, and also on the identification of the recurrent sets for the random walks (Z_n) and (W_n).

1. Introduction

Doeblin, as is well-known, made significant contributions towards our understanding of recurrence/transience for Markov chains. The 1937 ergodic theorem of Doeblin and Fortet ([**DF**], see also Iosifescu [**I**]) has also implications for certain Markov chains in the context of fractals/attractors. In this paper, we will discuss asymptotic behavior in these contexts for products (with respect to composition) of random affine maps on $(R^+)^d$. Such affine maps have been often used in the study of fractals/attractors, see [**B**].

Consider a family Q of affine maps from $(R^+)^d$ into itself with pointwise convergence topology such that each T in Q is of the form

(1) $$T(x) = A(T)x + B(T),$$

where $A(T)$ is a $d \times d$ nonnegative matrix, $B(T)$ and x are $d \times 1$ nonnegative vectors. Let X_0, $X_1, \ldots$ be a sequence of i.i.d. Q-valued random variables with common distribution μ. Write:

1991 *Mathematics Subject Classification.* Primary 60B10, 60J10.

Key words and phrases. Non-negative matrices, convolution of probability measures, completely simple semigroup.

This paper is in final form and no version of it will be submitted for publication elsewhere

$$W_n \equiv X_n X_{n-1} \ldots X_0, \ Z_n \equiv X_0 X_1 \ldots X_n,$$

where the multiplication is simply the composition of maps. For every x in $(R^+)^d$, the attractor $\mathcal{A}(x)$ is then defined as the set of all $y \in (R^+)^d$ such that $\Pr(W_n x \in N(y) \, i.o.) > 0$ for open $N(y)$ containing y. It is shown in Mukherjea (1992) that if x is strictly positive (that is, if each $x_i > 0$, where $x = (x_1, x_2, \ldots, x_d)$), then $\mathcal{A}(x) = \{T(x) : T \in R(W)\}$, $R(W)$ being all T as in (1) and $\Pr(W_n \in N(T) \, i.o.) > 0$ for open $N(T)$ containing T. In this paper, we will make certain remarks/assertions concerning: 1) the rate of growth of $\|X_n X_{n-1} \ldots X_0 u\|$ for u in $(R^+)^d$; and 2) the set of recurrent states of the random walks (Z_n) and (W_n). In Section 2, the main result is contained in (19) and (20); in Section 3, (29) contains the main result.

2. The rate of growth of $\|W_n u\|$

In this section and the rest of this paper, we will identify maps T of the form (1) with $(d+1) \times (d+1)$ nonnegative matrices (with respect to matrix multiplications and usual topology) via the topological isomorphism

$$(2) \qquad\qquad T \longmapsto \begin{pmatrix} A(T) & B(T) \\ 0 & 1 \end{pmatrix} \equiv p(T)$$

where $0 \equiv (0, 0, \ldots, 0)$ and 1 is the number 1. Then for x in $(R^+)^d$, the element $T(x)$ in $(R^+)^d$ is identified with the element $p(T) \cdot \begin{pmatrix} x \\ 1 \end{pmatrix}$.

Let us consider the compact space

$$Y = \left\{ y : \text{each } y_i \geq 0, \ y_1^2 + y_2^2 + \ldots + y_{d+1}^2 = 1 \right\},$$

with usual topology. Let A be a $(d+1) \times (d+1)$ nonnegative matrix whose last row is $(0, 0, \ldots 0, 1)$. Then we define the action

$$A.y = \frac{Ay}{\|Ay\|} \ ,$$

where $(Ay)_i = \Sigma \ A_{ij} y_j$ and for $y = (y_1, y_2, \ldots, y_{d+1})$, $y \in Y$, $\|y\| = \sqrt{y_1^2 + \ldots + y_{d+1}^2}$. Then this defines a continuous action which is well-defined whenever either A has no zero columns or $y_{d+1} > 0$. Notice that for matrices A, B and any element $y \in Y$,

$$A.(B.y) = (AB).y,$$

whenever both sides are well-defined. Now let S be the closed multiplicative semigroup generated by $S(\mu)$, the support of μ. If $\lambda_1, \lambda_2 \in P(S) \equiv$ the set of probability measures on S and $\beta \in P(Y)$ with $\beta\{y \in Y : y_{d+1} = 0\} = 0$,

(3)
$$\lambda_1 * (\lambda_2 * \beta) = (\lambda_1 * \lambda_2) * \beta,$$

where for $\lambda \in P(S)$, $\lambda * \beta$ is defined by

$$\lambda * \beta (\mathcal{B}) = \int \lambda \{A : A.y \in \mathcal{B}\} \ \beta(dy).$$

Note that (3) holds for any $\beta \in P(Y)$ if $S(\lambda_1)$ and $S(\lambda_2)$ do not contain any matrix which has a zero column.

Let us now assume that there exist matrices $p(T_i)$, $1 \leq i \leq r$, in $S(\mu)$ such that

(4)
$$\|A(T_1) A(T_2) \dots A(T_r)\| < 1.$$

Then, the set $m(S) \equiv \{x \in S : x \text{ has minimal rank}\}$ consists of rank one matrices.

Let us also assume that

(5)
$$\{\mu^n : n \geq 1\} \text{ is tight.}$$

Let X_0, X_1, $\dots$ be a sequence of $(d+1) \times (d+1)$ i.i.d. matrices of the form (2) with distribution μ. Following [**FK**], where invertible real matrices have been considered, we now define a new process (Z_n^*) on the space $M = S(\mu) \times Y$ as follows:

Let $u \in Y$ such that $u_{d+1} > 0$ and $Z_0^* = (X_0, u)$. Let $Z_1^* = (X_1, X_0.u)$, $Z_2^* = (X_2, X_1 X_0.u)$, $\dots$, $Z_{n+1}^* = (X_{n+1}, X_n X_{n-1} \dots X_0.u)$. Then (Z_n^*) is a Markov process on M with transition function given by

(6)
$$\begin{aligned} &P\left(Z_{n+1}^* \in \mathcal{D} \mid Z_n^* = (s, y)\right) \\ &= \mu\{t \in S(\mu) : (t, s.y) \in \mathcal{D}\}. \end{aligned}$$

Now suppose that λ is an invariant probability measure for (Z_n^*) on M so that

(7)
$$\lambda(\mathcal{D}) = \int P((s, y), \mathcal{D}) \ \lambda(d(s, y)).$$

Let f be any bounded continuous function on Y. Define g on M by $g(s, y) = f(s.y)$. It follows from (7) that

(8)
$$\begin{aligned} &\int f(s.y) \lambda(d(s, y)) \\ &= \int g(s, y) \lambda(d(s, y)) \\ &= \int \left[\int g(t, s.y) \mu(dt)\right] \lambda(d(s, y)) \\ &= \int \int f(ts.y) \mu(dt) \lambda(d(s, y)). \end{aligned}$$

Let us define the probability measure λ_0 on Y by

(9) $\lambda_0(\mathcal{B}) = \lambda\{(s, y) \in M : s.y \in \mathcal{B}\}$.

Then we have from (8) and (9) that

$$\begin{aligned}
\int f d\lambda_0 &= \int f(s.y)\,\lambda(d(s, y)) \\
&= \int \int f(ts.y)\,\mu(dt)\,\lambda(d(s, y)) \\
&= \int \left[\int f(t.(s.y))\,\mu(dt)\right]\lambda(d(s, y)) \\
&= \int \left[\int f(t.y)\,\mu(dt)\right]\lambda_0(dy).
\end{aligned}$$

It follows that λ_0, defined in (7), is an invariant measure for the process $(X_n X_{n-1} \ldots X_0.u)$. Furthermore, notice that for $\mathcal{D} = \mathcal{A} \times \mathcal{B} \subset M$, we have from (7),

$$\lambda(\mathcal{D}) = \int \mu(\mathcal{A}).I_{\mathcal{B}}(s.y)\,\lambda(d(s, y)) = \mu(\mathcal{A})\,\lambda_0(\mathcal{B}),$$

so that λ is uniquely determined by μ and λ_0. It also shows that a unique invariant probability for the process $(X_n X_{n-1} \ldots X_0.u)$ on Y always leads to a unique invariant probability for the process (Z_n^*) on M.

Let us now assume that $S(\mu)$ is compact and that no matrix in $S(\mu)$ has a zero column. Then M is compact. Recall that $u_{d+1} > 0$. Then the process Z_{n+1}^*, given by $Z_{n+1}^* = (X_{n+1}, X_n \ldots X_0.u)$, takes values, with probability 1, in $S(\mu) \times \{y \in Y : y_{d+1} > 0\}$. Notice that then a probability measure λ satisfying (7) must satisfy

$$\lambda\{(s, y) \in M : y_{d+1} = 0\} = 0$$

and consequently, the probability measure λ_0 on Y defined by (7) must satisfy

(10) $\lambda_0\{y \in Y : y_{d+1} = 0\} = 0$.

Now let ν be the weak limit of $\frac{1}{n}\sum_{k=1}^{n}\mu^k$. Then $S(\nu) = m(S)$ and for any $\beta \in P(Y)$ with $\beta\{y : y_{d+1} = 0\} = 0$, we have:

$$\mu * (\nu * \beta) = (\mu * \nu) * \beta = \nu * \beta;$$

furthermore,

$$\begin{aligned}
\mu * \beta = \beta &\implies \mu^n * \beta = \beta \text{ for } n \geq 1 \\
&\implies \left(\frac{1}{n}\sum_{k=1}^{n}\mu^k\right) * \beta = \beta \text{ for } n \geq 1 \\
&\implies \nu * \beta = \beta.
\end{aligned}$$

Let us now make the following observation:

If $\beta \in P(Y)$ such that $\beta\{y \in Y : y_{d+1} = 0\}$ is zero, then for $\mathcal{A} \subset Y$,

$$\nu * \beta(\mathcal{A}) = \int \nu\{A \in m(S) : A.y \in \mathcal{A}\}\,\beta(dy)$$

and therefore, for all such β, $\nu * \beta$ is the same since for $y \in Y$ with $y_{d+1} > 0$,

$$A.y = (a_1, \ldots a_d, 1)\big/\sqrt{1 + a_1^2 + \ldots + a_d^2}\,,$$

when the $(d+1)$th column of the matrix A in $m(S)$ is $(a_1, a_2, \ldots, a_d, 1)$. All these mean that there is exactly <u>one</u> invariant measure β in $P(Y)$ such that

$$\mu * \beta = \beta \text{ and } \beta\{y \in Y : y_{d+1} = 0\} = 0.$$

Consequently there is a unique invariant probability λ for the process (Z_n^*). By Theorem 1.4 in [**FK**], for any continuous function g on M, we have:

$$(11) \qquad \frac{1}{n+1}\sum_{k=0}^{n} g(Z_k^*) \longrightarrow \int g\,d\lambda \text{ a.s.}$$

Define on M a function g by

$$(12) \qquad g(s, y) = \log\left(\frac{\|y\|}{\|sy\|}\right).$$

Notice that g is continuous, and also,

$$(13) \qquad \begin{aligned} g(s_2, s_1.y) &= \log\left(\frac{\|s_1.y\|}{\|s_2(s_1.y)\|}\right) \\ &= \log\left(\frac{\|s_1 y\|}{\|(s_2 s_1)y\|}\right). \end{aligned}$$

It follows from (11), (12) and (13) that for any $u \in Y$ with $u_{d+1} > 0$,

$$(14) \qquad \frac{1}{n+1}\log\|X_n X_{n-1}\ldots X_0 u\| \longrightarrow \int \log\|sy\|\,d\lambda(s, y)$$

almost surely.

We now drop the assumption of compactness on $S(\mu)$, but assume that

$$(15) \qquad \mu\{\text{the matrices which have a zero column}\} = 0.$$

Then it is easily verified that the statement in (15) remains true when μ there is replaced by μ^k, where k is any positive integer.

Let $M^* = S(\mu)^* \times Y$, where $S(\mu)^*$ is the one point compactification of $S(\mu)$. Consider the function g as defined in (12). For each finite positive number m, let us define g_m on M^* as follows:

$g_m = g$ for $y \in Y$ and $s \in S(\mu)$ satisfying
$$\|s\| \leq m \text{ and } k(s) \geq \tfrac{1}{m},$$

where

$$k(s) = \min \left\{ \sum_{i=1}^{d} s_{ij} : 1 \leq j \leq d \right\} ;$$

let g_m be extended continuously to all of M^* and still be bounded by $\log m + \log(d+1)$, see (16) below. Let us now define the sets A_m, B_m and C_m by

$$A_m = \{ s \in S(\mu) : \|s\| > m \},$$
$$C_m = \{ s \in S(\mu) : k(s) < \tfrac{1}{m} \} \text{ and } B_m = A_m \cup C_m.$$

Note that for any $s \in S$ and $y \in Y$,

(16) $$\frac{1}{d+1} \cdot k(s) \leq \|sy\| \leq \|s\|$$

and therefore,

$$|\log \|sy\|| \leq \log(d+1) + |\log k(s)| + |\log \|s\||.$$

It follows that we have, almost surely,

$$\left| \tfrac{1}{n+1} \sum_{k=0}^{n} g(Z_k) - \tfrac{1}{n+1} \sum_{k=0}^{n} g_m(Z_k) \right|$$
$$\leq \tfrac{1}{n+1} \sum_{k=0}^{n} [\log(d+1) +$$
$$|\log k(X_k)| + |\log \|X_k\|| +$$
$$\log m] I_{B_m}(X_k),$$

which converges as $n \longrightarrow \infty$, by the ergodic theorem to

(17) $$\int_{B_m} |\log \|s\|| \, \mu(ds) + \int_{B_m} |\log k(s)| \, \mu(ds)$$
$$+ (\log(d+1) + \log m) \mu(B_m).$$

Let us now assume that

(18) $$\int (|\log \|s\|| + |\log k(s)|) \, \mu(ds) < \infty.$$

Then since $\lim\limits_{m \to \infty} \mu(B_m) = 0$, the expression (17) goes to zero as $m \to \infty$. Also note that because of (16),

$$\left| \int g \, d\lambda - \int g_m \, d\lambda \right| \longrightarrow 0, \text{ as } m \longrightarrow \infty.$$

By Theorem 1.4, [**FK**],

$$\lim_{n \to \infty} \frac{1}{n+1} \sum_{k=0}^{n} g_m(Z_k) = \int g_m \, d\lambda \quad \text{a.s.}$$

Thus, we have proven that under the assumptions of (4), (5), (15) and (18), for any u such that $u \in (R^+)^{d+1}$ and $u_{d+1} > 0$,

$$(19) \qquad \lim_{n \to \infty} \frac{1}{n+1} \log \|W_n u\| \text{ exists a.s.}$$

where $W_n \equiv X_n X_{n-1} \ldots X_0 u$, and this limit is, almost surely, a constant, independent of u. This limit is 0 a.s. because of the following reason. Because of (5), every element in $m(S)$ is recurrent (actually positive recurrent), so that for any $u > 0$ and any $s \in m(S) \subset R(W)$, the element of $su \in \mathcal{A}(u)$ (defined earlier) and consequently for any open set N containing su,

$$\Pr(W_n u \in N \text{ i.o.}) > 0$$

so that there is a positive probability that $\liminf_{n \to \infty} \|W_n u\|$ is bounded.

Let us also point out that it follows from our arguments that (19) remains true under (4), (5) and the assumption of $S(\mu)$ being compact and not containing any matrix which has a zero column, replacing (18). Finally, we mention that if over and above (4), (5), (15) and (18), we also assume that $S(\mu)$ contains at least one matrix $p(T)$, see (2), where $B(T) \neq 0$, then for any $v \in (R^+)^d$,

$$(20) \qquad \lim_{n \to \infty} \frac{1}{n} \log \|W_{1n} v + W_{2n}\| = 0 \text{ a.s.}$$

where $W_n = \begin{pmatrix} W_{1n} & W_{2n} \\ 0 & 1 \end{pmatrix}$.

3. Identification of the recurrent set

As in Section 2, the i.i.d. sequence (X_n) of random affine maps of the form (1) will be considered here as $(d+1) \times (d+1)$ i.i.d. random nonnegative matrices of the form (2) with distribution μ. Let S be the closed (with usual topology) multiplicative semigroup generated by $S(\mu)$. Write again

$$W_n \equiv X_n X_{n-1} \ldots X_0, \quad Z_n \equiv X_0 X_1 \ldots X_n$$

so that both W_n and Z_n have distribution μ^{n+1}. Let us also use the notation

$$A^{-1}B = U\{x^{-1}B : x \in A\},$$

where $A \subset S$, $B \subset S$ and $x^{-1}B = \{y \in S : xy \in B\}$. Define the set $\mathcal{I}$ by

$$\mathcal{I} = \{x \in S : x_{i,d+1} > 0 \text{ for } 1 \leq i \leq d\} \text{ the } (d+1)$$

th column of x is strictly positive.

Let $J = J_c \cup J_r$, where J_r (resp. J_c) is the set of all matrices in S which have at least one zero row (resp. zero column). By A^c, we denote the complement of A. By $\overline{A}$, we denote the closure of A. Notice that J^c, J_r^c and J_c^c are all, when nonempty, subsemigroups of S. Define R, $R(Z)$ and $R(W)$, subsets of S, by

$$R = \left\{ x : \sum_{n=1}^{\infty} \mu^n (N) = \infty \text{ if } x \in N, N \text{ open} \right\},$$
$$R(Z) = \{x : \Pr(Z_n \in N \text{ i.o.}) > 0 \text{ if } x \in N, N \text{ open}\},$$
$$R(W) = \{x : \Pr(W_n \in N \text{ i.o.}) > 0 \text{ if } x \in N, N \text{ open}\}.$$

Notice that $R(Z) \cup R(W) \subset R$, R is an ideal of S, $R(Z)$ is a left ideal of S and $R(W)$ is a right ideal of S, when they are nonempty. The set $m(A)$ is the set of all matrices in A with the minimal rank, that is, $m(A) = \{y \in A \mid \text{rank } y \leq \text{rank } x \text{ for all } x \in A\}$. It is proven in [M] that when $R \cap J_r^c \neq \phi$,

$$(21) \qquad R(Z) \cap J_r^c \subset R \cap J_r^c = R(W) \cap J_r^c = m(S \cap J_r^c),$$

which is a completely simple minimal ideal of $S \cap J_r^c$ (see [MT] for definition); and when $R \cap J_c^c \neq \phi$,

$$(22) \qquad R(W) \cap J_c^c \subset R \cap J_c^c = R(Z) \cap J_c^c = m(S \cap J_c^c),$$

which is a completely simple minimal ideal of $S \cap J_c^c$. Our aim here is to identify the set $R(Z)$, when nonempty, under the following mild assumptions:

$$(23) \qquad \begin{array}{c} \mathcal{I} \neq \phi, \ \mu(J) = 0; \\ R(Z) \neq \phi. \end{array}$$

Elsewhere, under different assumptions, we will determine $R(W)$ and R.

It follows from (23) (see [M]) that

$$(24) \qquad \lim_{n \to \infty} \mu^n (\mathcal{I}) = 1,$$

since $\mathcal{I}$ is an open ideal of J_r^c. Notice that the special element s^* (the matrix where the only non-zero element is the 1 on its last row) belongs to $R(Z)$ whenever $s^* \in S$ and $R(Z) \neq \phi$, since $s^*.R(Z) \subset R(Z)$. Now suppose that $R(Z) \neq \{s^*\}$. Let $z \in R(Z) - \{s^*\}$ and K be open with compact closure such that $z \in K$. We claim:

$$(25) \qquad \overline{K} \cap (\mathcal{I} - J_c) R(Z) \neq \phi.$$

If (25) is false, then for any $x \in \mathcal{I} - J_c$, $x^{-1} \overline{K}$ is compact and $\Pr(Z_n \in x^{-1}\overline{K} \text{ i.o.})$ $= 0$. Let A be any compact subset of $\mathcal{I} - J_c$. Let $z \in V$, V open, $\overline{V}$ compact,

$\overline{V} \subset K$. For any $y \in V$, $x \in A$, there exist open sets N_1, N_2, $x \in N_1$, $y \in N_2$ such that $N_1^{-1} N_2 \subset x^{-1} K$. (See [M]). Then it follows that there exists open N, $y \in N$ such that

$$(26) \qquad\qquad \Pr\left(Z_n \in A^{-1} N \text{ i.o.}\right) = 0,$$

implying that

$$(27) \qquad\qquad \Pr\left(Z_n \in A^{-1} \overline{V} \text{ i.o.}\right) = 0.$$

Since $z \in R(Z)$, for some $\delta > 0$,

$$\Pr\left(Z_n \in V \text{ i.o.}\right) = \delta.$$

Because of (24), we could choose A and m such that $\mu^m(A) > 1 - \delta$, which means that

$$0 < \Pr\left(Z_m \in A, Z_n \in V \text{ i.o.}\right)$$
$$\leq \Pr\left(Z_n \in A^{-1} V \text{ i.o.}\right),$$

which is a contradiction, proving (25).

Thus, there exists $u \in \mathcal{I} - J_c$ and $w \in R(Z)$ such that $uw \in \overline{K}$. It is easily verified that $uw \in R(Z) \cap J_r^c$. This proves that

$$R(Z) - \{s^*\} \subset cl\left[R(Z) \cap J_r^c\right],$$

which implies (because of (21)) that

$$(28) \qquad\qquad R(Z) \subset cl\left[m\left(S \cap J_r^c\right)\right] \cup \{s^*\}.$$

Notice that $m(S) \cap J_r^c \neq \phi$, since $\mathcal{I}.m(S) \subset \mathcal{I}$. Then it can be shown that $m(S)$ is a completely simple minimal ideal of S iff $m(S) \cap J_r^c (\equiv m(S \cap J_r^c))$ is a completely simple minimal ideal of $S \cap J_r^c$. It follows from (28) that

$$(29) \qquad \left.\begin{array}{c} R(Z) \subset m(S), \\ m(S) \text{ is a completely simple minimal ideal of } S, \\ m(S) \subset R, R(Z) = XGY_1, \end{array}\right\}$$

where $X \equiv E(Se)$, $G = eSe$, Y_1 is some subset of $Y \equiv E(eS)$, e is an idempotent in $m(S)$. [Here, $E(B)$ is the set of all idempotent elements in B.]

With extra effort, we can actually prove much more (and this will appear in a separate publication elsewhere) as we describe below.

If (23) holds, then $R(Z) = m(S)$ and every point in $m(S)$ is positive recurrent with respect to (Z_n) as well as (W_n); moreover, then, there is a μ-invariant probability measure ν on S such that $S(\nu) = m(S)$ and $\mu * \nu = \nu * \mu = \mu$, and then the attractor $\mathcal{A}(x)$ for $x \in (R^+)^d$ and $x > 0$, defined in Section 1, happens to be the set

$$\left\{ y \in (R^+)^d : \begin{pmatrix} y \\ 1 \end{pmatrix} = A \begin{pmatrix} x \\ 1 \end{pmatrix} \text{ for } A \in S(\nu) \right\}$$

except for a set of d-dimensional Lebesgue measure zero, so that $\mathcal{A}(x)$ is, in this sense, the support of the μ-invariant probability $\nu * \delta_x$.

Finally, it is relevant to mention that for $d = 1$, writing $X_n = \begin{pmatrix} \xi_n & \eta_n \\ 0 & 1 \end{pmatrix}$, one can show (using [G]) that if

$$\text{(30)} \qquad \begin{aligned} P(\xi_0 = 0 \text{ or } \eta_0 = 0) = 0, \ -\infty \le E(\log \xi_0) < 0, \\ E \log \max\{\eta_0, 1\} < \infty, \end{aligned}$$

then $R(Z) = m(S)$ and (μ^n) is tight. Conversely, if $R(Z) \ne \phi$ and $E(\log \xi_0) > -\infty$, then the last inequality in (30) must hold.

References

[B] M.F. Barnsley and S. Demko, *Iterated function systems and the global construction of fractals*, Proc. Roy. Soc. London Ser. A, **399**, (1985), 243-275.

[DF] W. Doeblin et R. Fortet, *Sur les chaines a liasons completes*, Bull. Soc. Math. France, **65**, (1937), 132-148.

[FK] H. Furstenberg and Y. Kifer, *Random matrix products and measures on projective spaces*, Israel J. Math., **46**, (1983), 12-32.

[G] A.K. Grintsevichyus, *On the continuity of the distribution of a sum of dependent variables connected with independent walks on lines*, Theory Probab. Appl., **19**, (1974), 163-168.

[I] M. Iosifescu, [See the contribution in this volume.]

[K] Y. Kifer, *Ergodic Theory of Random Transformations*, Birkhauser, Boston, 1986.

[M] A. Mukherjea, *Recurrent random walks in nonnegative matrices: attractors of certain iterated function systems*, Probab. Theory Related Fields, **91**, (1992), 297-306.

[MT] A. Mukherjea and N.A. Tserpes, *. Measures in topological semigroups: convolution products and random walks*, Lecture Notes in Math., vol 547, Springer-Verlag (Berlin-Heidelberg-New York), 1976.

DEPARTMENT OF MATHEMATICS, UNIVERSITY OF SOUTH FLORIDA, TAMPA, FLORIDA 33620

Contemporary Mathematics
Volume **149**, 1993

A MULTIVARIATE LOOK AT E. SPARRE ANDERSEN'S EQUIVALENCE PRINCIPLE

P. E. NÜESCH

ABSTRACT. For the fluctuation theory of sums of symmetrically dependent random variables two approaches are used: an analytical one due to E. Sparre Andersen and a combinatorical one due to W. Feller and F. Spitzer. Our approach using multivariate statistics is situated between the analytical and the combinatorical one. The proofs rely on Farkas' lemma for separating hyperplanes and on orthant probabilities of positive vectors with special variance-covariance matrices.

1. INTRODUCTION

Ten years after the premature death of W. Doeblin a series of papers by E. Sparre Andersen on fluctuations of sums of random variables [1] [2] [3] [4] [5] started to appear. According to Feller [12] [p. 418, 420] they created a sensation and were greeted in probabilistic circles "with incredulity". Paul Lévy in his biographical survey of Doeblin's work puts him among the olympians in the field of sums of random variables right at the end of the line Bernoulli, de Moivre, Laplace, Cauchy, Kolmogorov. (Today, fifty years later, Doeblin's name is probably more intimately connected with the beginning of the theory of Markov chains than particular contributions to the classical problems of sums of random variables). E. Sparre Andersen's results are not so much a direct continuation than a continuation in spirit of Doeblin's work. (Spitzer [16] refers explicitly to Doeblin's paper of 1940, the number 13 in P. Lévy's list). Although greeted as sensational they leave in today's treatises of probability theory practically no trace. One reason is that a combinatorial approach first promoted by Feller [11] and Spitzer [16] reduced the "extraordinarily intricate and complex" proofs of Sparre Andersen to "extreme simplicity".

One reproach is that he continued to use analytical methods of proofs, and therefore did not recognize the validity of the results for symmetrically dependent variables but kept assuming independence. Only in his paper presented to the fifth Berkeley Symposium [7] did he use an algebraic approach which permitted the generalization to symmetrically dependent variables. And by 1965 the "equivalence principle" of [6] had been replaced by the more modest "fluctuations of sums of random variables" a term already used in [4] and [5]. The "equivalence principle"

1991 *Mathematics Subject Classification.* Primary 60E07; Secondary 62H10.

Key words and phrases. Fluctuations of sums of random variables, contrasts, disjoint cones, order restricted inference.

This paper is in final form and no version of it will be submitted for publication elsewhere

was to my knowledge again used only in Jacobs [14]. Basically, it says that some events are equivalent, like e.g. the number of permutations with exactly r strictly positive partial sums equals the number of permutations in which the first maximum among these partial sums occurs at the place r. Feller [12] [Lemma 2, p. 420], Jacobs [14] [p. 61, Theorem 2.1. i)].

In the following we approach the problem of equivalence by methods of multivariate statistical analysis. These will rely on the one hand heavily on symmetry properties of contrasts and are thus close to the above mentioned combinatorial or distribution free approach, on the other hand the distribution of the original set of variables creeps in since the method applies only to those distributions which permit through linear transformation the elimination of the covariance between contrasts. The mutual dependence, however, can be easily handled since the method makes use of orthant probabilities, which by nature are scale invariant and the correlation structure of the contrasts makes the original dependence disappear.

This approach is used in multivariate statistical analysis in what is called order restricted inference. A key paper connecting Sparre Andersen's work to this relatively recent branch of multivariate analysis is Brunk [9]. The idea of cyclic shift transformations used there will be instrumental in our approach.

2. Method of Proof

The original $p+1$ variables respectively their partial sums are reduced to a set of p appropriate contrasts in such a way that the inclusion between the non-reduced version becomes non-negativity for the reduced one. The probability to calculate is a so-called orthant probability. The equivalence of the two events is reduced to the equality of their respective orthant probabilities. This method is illustrated for the following two problems:

Let $X_1, \ldots, X_p, X_{p+1}$ be a set of identically distributed, equally correlated random variables whose joint distribution is elliptically contoured. Denote

$$S_i = \sum_{k=1}^{i} X_k , \quad \bar{X}_i = \frac{1}{i} S_i .$$

The following two results hold

(2.1) (I) $P\left[\cap_{i=1}^{p} \{X_i > X_{p+1}\}\right] = \dfrac{1}{p+1}$

       (II) $P\left[\cap_{i=1}^{p} \{\bar{X}_i > \bar{X}_{p+1}\}\right] = \dfrac{1}{p+1} .$

While in (I) it is immaterial which one of the $(p+1)$ variables X_i is to be exceeded by all others, in result (II) it has to be $\bar{X}_{p+1}$. Clearly, the inequality signs in both results can be reversed. Result (I) is a slight extension of the well-known result that the probability of a specified one of $(p+1)$ independent identically distributed variables being smaller (or bigger) than all the others is $(p+1)^{-1}$ (see e.g. Stuart [17] [p. 376]). Result (II) has been proven by Andersen [5] [Theorem D, p. 197].

Instead of attacking the problem by showing the equivalence of the two events in brackets on the left hand side of (2.1), our proof in section 3 will show that each event has the same probability. The natural contrasts to use are

$$(2.2) \qquad \begin{aligned} Y_i &= X_i - X_{p+1} \quad \text{for (I)} \quad \text{and} \\ Z_i &= \bar{X}_i - \bar{X}_{p+1} \quad \text{for (II)} \quad i = 1, \ldots, p \end{aligned}$$

and the respective orthant probabilities will be

$$(2.3) \qquad \begin{aligned} P(\mathbf{Y} > \mathbf{0}) \quad \text{for (I)} \quad \text{and} \\ P(\mathbf{Z} > \mathbf{0}) \quad \text{for (II)} \end{aligned}$$

with $(\mathbf{Y})_i = Y_i$ and $(\mathbf{Z})_i = Z_i$. [The multivariate inclusion for vectors means that it holds for each component].

The vectors have elliptically contoured $p-$variate distributions. So the orthant probabilities (2.3) will depend only on the second moments, the variance-covariance matrices.

A square root $\mathbf{U}$ of the inverse of the positive definite matrix $\mathbf{\Sigma_Y}$ leads to the transformation

$$\mathbf{Y} = \mathbf{U\,X}$$

with a spherically distributed $\mathbf{X}$. $p+1$ square roots are constructed which transform the positive orthant on non-overlapping cones that span the entire E^p-space. These cones, being orthogonal transforms of each other, contain the same probability mass which gives for one cone the mass $(p+1)^{-1}$, which is the content of $\{\mathbf{Y} > \mathbf{0}\}$.

Both problems can be treated entirely analogously. Because of the equivalence it is not surprising that the two developments will coincide somewhere along the way.

3. Proof of the Equivalence

Orthant probabilities are scale invariant. We have therefore to deal with the correlation matrix only. The first and second moments of $\mathbf{Y}$ are

$$(3.1) \qquad \begin{aligned} E(\mathbf{Y}) &= \mathbf{0} \\ \mathbf{\Sigma}_Y &= \frac{1}{2}(\mathbf{I} + \mathbf{E}) \end{aligned}$$

with $\mathbf{I}$ the identity matrix and $\mathbf{E} = \mathbf{1}\,\mathbf{1}^T$ with $\mathbf{1}^T = (1, \ldots, 1)$.

The Y-contrasts are correlated with correlation coefficient $\frac{1}{2}$. The analogous construction for (II) gives

$$(3.2) \qquad E(\mathbf{Z}) = \mathbf{0}$$

$$\mathbf{\Sigma}_Z = (s_{ij}) \text{ where } s_{ij} = \mathrm{Corr}(Z_i, Z_j) = \sqrt{\frac{i(p-j+1)}{j(p-i+1)}}$$

with the identical content for the same orthant, $P(\mathbf{Z} > \mathbf{0}) = \frac{1}{p+1}$.

Observe that for both problems the first two moments of the original set $\{X_i\}$ of variables disappear.

For problem (I)

$$\Sigma_Y^{-1} = 2\left(\mathbf{I} - \frac{1}{p+1}\mathbf{E}\right)$$

$$= \frac{2p}{p+1}\left[\mathbf{I} + \frac{1}{p}(\mathbf{I} - \mathbf{E})\right].$$

The correlation structure of the inverse distribution is again an equal correlation, with $\rho_{ij} = -\frac{1}{p}$ for all i, j. Let

$$(3.3) \qquad\qquad \mathbf{A} = \mathbf{I} + \frac{1}{p}(\mathbf{I} - \mathbf{E})$$

and $\mathbf{U}$ be the upper triangular matrix of the Choleski decomposition $\mathbf{A} = \mathbf{U}^T\mathbf{U}$. Without explicity decomposing the matrix $\mathbf{A}$, one establishes for the column vectors $\mathbf{u}_i$ $(i = 1, \ldots, p)$ of $\mathbf{U}$ the following properties :

$$(3.4) \qquad\qquad \bullet \text{ all } \mathbf{u}_i\text{'s are unit vectors, since } \mathbf{u}_i^T\mathbf{u}_i = a_{ii} = 1$$

$$\bullet \text{ all angles } \angle(\mathbf{u}_i, \mathbf{u}_j) \text{ are the same, since } \mathbf{u}_i^T\mathbf{u}_j = a_{ij} = -\frac{1}{p}.$$

Define

$$(3.5) \qquad\qquad \mathbf{u}_{p+1} = -\sum_{i=1}^{p}\mathbf{u}_i.$$

This vector $\mathbf{u}_{p+1}$ shows the same two properties (3.4) :

$$\mathbf{u}_{p+1}^T\mathbf{u}_{p+1} = (\textstyle\sum \mathbf{u}_i^T)(\sum \mathbf{u}_j) = p(p-1)a_{ij} + p\,a_{ii} = 1$$

$$\mathbf{u}_{p+1}^T\mathbf{u}_j = -(\textstyle\sum \mathbf{u}_i)\mathbf{u}_j = -[(p-1)a_{ij} + a_{ii}] = -\frac{1}{p}.$$

This fact leads to the definition of $p+1$ matrices $\mathbf{U}_i$ obtained by cyclical shift from each other.
Let

$$(3.6) \qquad\qquad \mathbf{U}_i = \{\mathbf{u}_{i+1}, \ldots, \mathbf{u}_p, \mathbf{u}_{p+1}, \mathbf{u}_1, \ldots, \mathbf{u}_{i-1}\} \quad i = 1, \ldots, p+1.$$

Note that the shift $\mathbf{T}$ with $\mathbf{T}\mathbf{U}_i = \mathbf{U}_j$ is orthogonal. This leads to the matrix $\mathbf{U}_i^T\mathbf{U}_j$, which is the cyclically modified matrix $\mathbf{A}$, whose principal diagonal has been shifted to $j - i$, mod $(p+1)$.

In problem (II) the correlation of the inverse distribution associated to the matrix is tridiagonal, with $\rho_{ij} = -\frac{1}{2}$ for $|i-j| = 1$, the other contrasts being uncorrelated. [The inverse of a symmetric Green matrix $g_{ij} = a_i\,b_j$, $j \geq i$ is tridiagonal]. (See e.g. Karlin [15] [p. 114], Gantmacher-Krein [13] [p. 95]).

In the case of $\rho_{ij} = -\frac{1}{2}$ for adjacent contrasts, one obtains $a_i = \frac{2}{p+1}i$, $b_j = p - j + 1$. The correlation matrix associated to this Green matrix $\mathbf{G} = \big(g_{ij} = a_i\,b_j\big)$ has off diagonal elements

$$\frac{g_{ij}}{\sqrt{g_{ii}\,g_{jj}}} = \sqrt{\frac{i(p-j+1)}{j(p-i+1)}}$$

$$= s_{ij},$$

as required in (3.2). We have

$$(3.7) \qquad \mathbf{B} = (b_{ij}) \text{ with } \begin{cases} b_{ii} = 1 \\ b_{ij} = -\frac{1}{2} & |i - j| = 1 \\ b_{ij} = 0 & |i - j| > 1. \end{cases}$$

The Choleski decomposition $\mathbf{B} = \mathbf{V}^T \mathbf{V}$ gives column vectors $\mathbf{v}_i$ with the properties that

(3.8)
- all $\mathbf{v}_i$'s are unit vectors
- angles between adjacent $\mathbf{v}_i$'s are identically equal to $\frac{2\pi}{3}$,

non adjacent $\mathbf{v}_i$'s are orthogonal.

The definition

$$(3.9) \qquad \mathbf{v}_{p+1} = -\sum_{i=1}^{p} \mathbf{v}_i$$

leads also in this case to a new vector with these same properties with respect to the other $\mathbf{v}_i$'s.
The matrix

$$(3.10) \qquad \mathbf{V}_i = \{\mathbf{v}_{i+1}, \ldots, \mathbf{v}_p, \mathbf{v}_{p+1}, \mathbf{v}_1, \ldots, \mathbf{v}_{i-1}\} \quad i = 1, \ldots, p+1$$

gives rise to $\mathbf{V}_i^T \mathbf{V}_j$, the cyclically modified matrix $\mathbf{B}$. Again, the shift $\mathbf{T}$ with $\mathbf{T}\mathbf{V}_i = \mathbf{V}_j$ is orthogonal. Also here the principal diagonal of $\mathbf{B}$ has been shifted to $j - i$, mod $(p + 1)$. Straightforward calculations give

$$(3.11) \quad \mathbf{N} = (\nu_{hk}) = \begin{cases} 1 & \text{if} & h - k = j - 1 \bmod (p+1) \\ -1 & \text{if} & k = p + 1 - (j - i) \bmod (p+1) \\ 0 & \text{otherwise} \end{cases}$$

and the

Theorem 3.1.
$$\mathbf{N} = \mathbf{A}^{-1} \mathbf{U}_i^T \mathbf{U}_j = \mathbf{B}^{-1} \mathbf{V}_i^T \mathbf{V}_j = \mathbf{M}.$$

It is at this point where the two approaches merge. The second part of the proof will be shown for the matrix $\mathbf{A}$ of problem (I) only. It applies without modification to $\mathbf{B}$ of problem (II). For theorem (3.3) and theorem (3.4) we use an idea proposed by Anis-Lloyd [8]. Here the following lemma will be used.

Lemma 3.2. *(Farkas [10])*
Exactly one of the following alternatives holds: Either the system of equations $\mathbf{C}\mathbf{x} = \mathbf{b}$ *has a non-negative solution, or the inequalities* $\boldsymbol{\pi}\mathbf{C} \geq \mathbf{0}$, $\boldsymbol{\pi}\mathbf{b} < 0$ *have a solution.*

Geometrically this means that either $\mathbf{b}$ lies in the cone spanned by the columns of $\mathbf{C}$, or there is a hyperplane $\boldsymbol{\pi}\mathbf{x} = 0$ separating $\mathbf{b}$ and the cone.

Theorem 3.3. *The cones* $\mathcal{X}_i = \{\mathbf{y} \mid \mathbf{y} = \mathbf{U}_i \mathbf{x}, \mathbf{x} \geq \mathbf{0}\}$ *of* E^p *(*$i = 1, 2, \ldots,$ $p + 1$*) are non-overlapping.*

Proof. Suppose $\mathbf{y}^0$ belongs to both regions $\mathcal{X}_i$ and $\mathcal{X}_j$

(3.12)
$$
\begin{aligned}
\mathbf{y}^0 &= \mathbf{U}_i\,\mathbf{x}^i = \mathbf{U}_j\,\mathbf{x}^j \\
&\quad\ \mathbf{U}_i^T\,\mathbf{U}_i\,\mathbf{x}^i = \mathbf{U}_i^T\,\mathbf{U}_j\,\mathbf{x}^j \\
&\quad\ \mathbf{A}^{-1}\mathbf{U}_i^T\,\mathbf{U}_i\,\mathbf{x}^i = \mathbf{A}^{-1}\,\mathbf{U}_i^T\,\mathbf{U}_j\,\mathbf{x}^j \\
&\quad\ \mathbf{x}^i = \mathbf{N}\,\mathbf{x}^j\ .
\end{aligned}
$$

For the choice

$$
\begin{aligned}
\boldsymbol{\pi} &= -\mathbf{e}_{p+1-(j-i)mod(p+1)}^T \\
\mathbf{b} &= \mathbf{x}^i \\
\mathbf{C} &= \mathbf{N}
\end{aligned}
$$

one gets

$$
\begin{aligned}
\boldsymbol{\pi}\,\mathbf{b} &< 0 \\
\boldsymbol{\pi}\,\mathbf{C} &= \mathbf{1}^T > \mathbf{0}^T
\end{aligned}
$$

since column $p + 1 - (j - i)$ mod $(p + 1)$ of $\mathbf{C}$ is -1. Thus there does not exist a non-negative solution of system (3.12).

Since the $\boldsymbol{\pi}$ chosen is the row $(j - i)$ of $\mathbf{N}$, it follows that

$$
\mathbf{x}_{j-i}^i = -\mathbf{x}_{p+1-(j-i)mod(p+1)}^j\ .
$$

Thus, according to the non-negativity these components are zero, all other components of $\mathbf{x}^j$ are positive. The common boundary of $\mathcal{X}_i$ and $\mathcal{X}_j$ is therefore a $(p - 1)$ dimensional hyperplane. $\square$

Theorem 3.4. *The regions $\mathcal{X}_i$ are exhaustive.*

Proof. Assume there exists a $\mathbf{y}^0$ that does not lie in any $\mathcal{X}_i$. This means there exists a hyperplane $\boldsymbol{\pi}\,\mathbf{x} = 0$ separating $\mathbf{y}^0$ from the u_i's with

$$
\boldsymbol{\pi}\,\mathbf{y}^0 > 0 \qquad \boldsymbol{\pi}\,\mathbf{u}_i = \alpha_i < 0 \qquad i = 1,\ldots,p+1
$$
$$
0 > \alpha_{p+1} = \boldsymbol{\pi}\,\mathbf{u}_{p+1} = \boldsymbol{\pi}\left(-\sum_{i=1}^{p}\mathbf{u}_i\right) = -\sum(\boldsymbol{\pi}\,\mathbf{u}_i)
$$
$$
= -\sum \alpha_i > 0\ .
$$

The assumption leads to a contradiction and is therefore false.

Matrix $\mathbf{N}$ of problem (I) equals matrix $\mathbf{M}$ of problem (II) (Theorem (3.1)), and vector $\mathbf{v}_{p+1}$ (3.9) and the matrices $\mathbf{V}_i$ (3.10) are defined as $\mathbf{u}_{p+1}$ (3.5) and $\mathbf{U}_i$ (3.6). Theorems (3.3) and (3.4) are valid for regions $\mathcal{X}_i$ which are $\mathbf{V}_i$-maps of the positive orthant. This proves the equivalence. $\square$

REFERENCES

1. Andersen, E.S. : *On the number of positive sums of random variables*, Skand. Aktuarietidsskrift **32** (1949) 27-36.

2. Andersen, E.S. : *On the frequency of positive partial sums of a series of random variables*, Mat. Tidsskrift **B** (1950) 33-35.

3. Andersen, E.S. : *On sums of symmetrically dependent random variables*, Skand. Aktuarietidsskrift **36** (1953) 123-138.

4. Andersen, E.S. : *On the fluctuations of sums of random variables I*, Math. Scand. **1** (1953) 263-285.

5. Andersen, E.S. : *On the fluctuations of sums of random variables II*, Math. Scand. **2** (1954) 195-223.

6. Andersen, E.S. : *The equivalence principle in the theory of fluctuations of sums of random variables*, Colloq. Comb. Meth. In Prob. Th., 12-16 (1962) Aarhus.

7. Andersen, E.S. : *An algebraic treatment of fluctuations of sums of random variables*, Proc. Fifth Berkeley Sympos. Math. Statist. and Prob. (1965) Vol II : Contributions to prob. theory, Part I, Univ. of California Press (1967) Berkeley, 423-429.

8. Anis, A.A. and Lloyd, E.H. : *On the range of partial sums of a finite number of independent variates*, Biometrika **40** (1953) 35-42.

9. Brunk, H.D. : *A generalization of Spitzer's combinatorial lemma*, Z. Wahrscheinlichkeitstheorie 2 (1964) 395-405.

10. Farkas, J. : *Theorie der einfachen Ungleichungen*, J. Reine Angew. Math. **124** (1902) 1-27.

11. Feller, W. : *On combinatorial methods in fluctuation theory*, Harald Cramr Volume (ed. U. Grenander) (1959) New York, 75-91.

12. Feller, W. : *An introduction to probability theory and its applications*, vol. II, Wiley (1966) New York (2nd edition).

13. Gantmacher, F.R. und Krein, M.G. : *Oszillationsmatrizen, Oszillationskerne und kleine Schwingungen mechanischer Systeme*, Akademie-Verlag (1960) Berlin.

14. Jacobs, K. : *Das kombinatorische Aequivalenzprinzip und das arcsin-Gesetz von E. Sparre Andersen*, Selecta Mathematica I (1969) Springer-Verlag.

15. Karlin, S. : *Total positivity*, Volume I, Stanford University Press (1968) Stanford, California.

16. Spitzer, F. : *A combinatorial lemma and its application to probability theory*, Trans. Amer. Math. Soc. **82** (1956) 323-339.

17. Stuart, A. : *Equally correlated variates and the multinormal integral*, J. Roy, Statist. Soc. Ser. B, **20** (1958) 373-378.

DEPARTMENT OF MATHEMATICS
FEDERAL INSTITUTE OF TECHNOLOGY
CH-1015 LAUSANNE SWITZERLAND

Contemporary Mathematics
Volume **149**, 1993

REGENERATION FOR CHAINS OF INFINITE ORDER AND RANDOM MAPS

PETER NEY AND ESA NUMMELIN

ABSTRACT. Sufficient conditions are given for the existence of a regeneration structure for chains of infinite order. A relation between these chains and a class of random maps is established.

1. INTRODUCTION

There are some natural connections between (i) chains of infinite order, (ii) a class of processes defined in terms or random maps, and (iii) Gibbs measures. In this note we will discuss these relationships.

In [7], we proved the existence of a regeneration structure for infinite order chains, under a hypothesis on the decay of memory in the chain. This hypothesis translates into a continuity condition on the random maps and Gibbs measures, and thus the regeneration structure for the chains translates into such a structure for the maps and Gibbs measures.

Regeneration is a useful tool for proving limit theorems such as laws of large numbers, central limit theorems, and in some cases large deviation theorems (see e.g. [8], [9]).

Constructions of the type discussed here have appeared in Berbee [1] and Lalley [5]. The latter paper points out the relation of the chains to Gibbs measures. The relation of chains of infinite order to random maps in the case of the map $f(x) = 2x \bmod 1$ is treated in Harris [2]. An application of random maps to number theory is treated by Kalpazidou [4]. A general reference for the whole subject is Iosifescu and Grigorescu [3], which also contains an extensive bibliography.

2. CHAINS OF INFINITE ORDER

Let $\mathcal{S}$ be a countable state space. For $-\infty \le m \le n \le \infty$ we will write

$$x_m^n = (x_m, \ldots, x_n), \quad x_i \in \mathcal{S},$$

and call this a *word*.

1991 *Mathematics Subject Classification.* Primary 60–J10, 60K05.
Key words and phrases. Markov chains, random maps, regeneration.
Supported by Grant INT–8922371.
A detailed version of part of this paper will appear elsewhere.

Let $\mathcal{S}^n_m =$ the set of all such words. Let

$$(2.1) \qquad\qquad G = \{g(x^0_{-\infty}, x_1), \ x^0_{-\infty} \in \mathcal{S}^0_{-\infty}, \ x_1 \in \mathcal{S}\}$$

be a stochastic kernel, i.e. $g \geq 0$ is a real valued function on $\mathcal{S}^0_{-\infty} \times \mathcal{S}$ such that

$$(2.2) \qquad\qquad \sum_{x_1 \in \mathcal{S}} g(x^0_{-\infty}, x_1) = 1.$$

By iteration, g induces the kernels

$$G^N = \{g^N(x^0_{-\infty}, x^N_1), x^0_{-\infty} \in \mathcal{S}^0_{-\infty}, x^N_1 \in \mathcal{S}^N_1\}.$$

Thus, for any "initial measure" λ on $\mathcal{S}^0_{-\infty}$, g induces a sequence of random variables $X_1, X_2, \cdots$ taking values in $\mathcal{S}$. This sequence will *not* be Markovian of any finite order. It is called a *chain of infinite order*.

Associated with this chain is a Markov chain $\{X^n_{-\infty}; n = 1, 2, \cdots\}$ on $\mathcal{S}^0_{-\infty}$ with transition function

$$(2.3) \qquad\qquad P = \{p(x, y), x, y \in \mathcal{S}^0_{-\infty}\}$$

given by

$$(2.4) \qquad p(x, y) = \begin{cases} g(x^n_{-\infty}, x_{n+1}) & \text{if } x = x^n_{-\infty}, \ y = x^{n+1}_{-\infty}, \ x_i \in \mathcal{S}, \\ 0 & \text{otherwise.} \end{cases}$$

Remark. Regeneration structures for Markov chains under the existence of suitable non-singularity conditions are well known. The chain $\{X_n\}$ is however highly singular, and hence the usual splitting technique cannot be employed (see e.g. Nummelin [11]).

Thus far, any sequence of random variables $X_1, X_2, \cdots$ can be represented as generated by some kernel g, so if there is to be any interesting structure or limit theory other hypotheses must be imposed. We proceed to do so.

Let λ be a probability measure on $\mathcal{S}^0_{-\infty}$, and $P = \{p(x, y)\}$ be as in (2.4). The memory of the process $\{X_n\}$ along the sequence $\alpha = \alpha^0_{-\infty}$ is quantified by the sequence
$$(2.5)$$
$$\gamma_m(\alpha) = \inf \left\{ \frac{g^N(x^0_{-\infty}, z^N_1)}{g^N(y^0_{-\infty}, z^N_1)} : x^0_{-m+1} = y^0_{-m+1} = \alpha^0_{-m+1}, z^N_1 \in \mathcal{S}^N_1, N = 1, 2, \cdots \right\}.$$

Definition. Let the sequence $\{X_n\}$ be generated by the kernel g and initial measure λ. We will say that *the memory of* $\{X_n\}$ *decays along* $\alpha = \alpha^0_{-\infty}$ if

$$(2.6) \qquad\qquad \gamma_m(\alpha) \nearrow 1 \text{ as } m \nearrow \infty.$$

3. REGENERATION

A *regeneration structure* for $\{X_n, n = 1, 2, \cdots\}$ is a sequence of random times $\{T_i, i = 0, 1, \cdots\}$ and a measure ν on S^∞_1, such that

    (i) $\{T_{i+1} - T_i, i = 0, 1, \cdots\}$ are i.i.d.r.v.'s,

    (ii) $X^{T_0}_1, X^{T_1}_{T_0+1}, X^{T_2}_{T_1+1}, \cdots$ are independent, blocks, and

(iii) $\mathbb{P}_\lambda\{X_{n+1}^\infty \in \cdot \mid T_i = n, \mathcal{F}_0^n\} = \nu(\cdot)$ for all $i \geq 0$, $n \geq 1$, where $\mathcal{F}_0^n$ is the σ-field generated by X_0^n.

A point $\alpha = \alpha_{-\infty}^0$ will be called a *recurrence point* for (λ, P) if

$$\mathbb{P}_\lambda\{X_{n-m+1}^n = \alpha_{-m+1}^0 \quad \text{for infinitely many } n\} = 1$$

for all $m = 1, 2, \cdots$.

Memory hypothesis. Say that $\{X_n\}$ satisfies the memory hypothesis if the memory of $\{X_n\}$ decays along a recurrence point $\alpha = \alpha_{-\infty}^0$ in the sense of (2.6).

Theorem. *If $\{X_n\}$ satisfies the memory hypothesis then a regeneration structure for $\{X_n\}$ exists.*

This is proved in [7].

4. Iterates of Expanding Maps

Let f be a piecewise monotone map from the interval $[0, 1]$ into $[0, 1]$. Namely, let $\mathcal{S} = \{1, \cdots, d \leq \infty\}$ denote a finite or countably infinite set (the state space, or alphabet), and let $\{\mathcal{I}_k, k \in \mathcal{S}\}$ be a partition of $[0, 1]$ into intervals. Then we assume that f satisfies:

(4.1a) $$f : [0, 1] \to [0, 1],$$

(4.1b) $$f \quad \text{is monotone on} \quad I_k, \quad k = 1, \cdots d, \quad \text{and}$$

(4.1c) $$|f(x) = f(y)| > \gamma|x - y| \text{ for some } \gamma > 1 \\ \text{and for all} \quad x, y \in I_k, k = 1, \cdots, d.$$

Such maps have been extensively studied. For background see e.g. Lasotta and Mackey [6].

Let $\{f_n(\cdot) \ n = 1, 2, \cdots\}$ denote the iterates of f, namely

$$f_1(x) = f(x), \quad f_n(x) = f(f_{n-1}(x)), \quad n \geq 2.$$

Let ξ be a random variable taking values in $[0, 1]$, and $\mu = \mathcal{L}(\xi)$ be the probability measure of ξ. Let

(4.2) $$\xi_{-n} = f_n(\xi).$$

(The negative index will be convenient.) We are interested in the behaviour of $\mathcal{L}(\xi_{-n})$.

Define the itinerary of a point $x \in [0, 1]$ by

$$i(x) = \{i_k(x)\}_{-\infty}^0 = \{\cdots i_{-2}(x), i_{-1}(x), i_0(x)\},$$

where

$$i_{-n}(x) = k \text{ if } f_n(x) \in I_k.$$

For $a^0_{-\infty} = (\cdots a_{-2}, a_{-1}, a_0) \in \mathcal{S}^0_{-\infty}$, define

(4.3)
$$h(a^0_{-\infty}) = \{x : i(x) = a^0_{-\infty}\}.$$

One can show that $h(a^0_{-\infty})$ is a one point set or is empty. Let

$$\mathcal{I} = \text{the set of all trajectories}$$
$$= \{i(x) : x \in [0,1]\}.$$

Then $h^{-1} = i$, namely

$$[0,1] \; \overset{i}{\underset{h}{\rightleftarrows}} \; \mathcal{I},$$

and i and h are $1-1$.

Let σ denote the shift map on $\mathcal{I}$, namely

$$\sigma(a^0_{-\infty}) = a^{-1}_{-\infty}.$$

Then f applied to $x \in [0,1]$ corresponds to σ applied to $i(x)$. Thus we have the relations

(4.4)
$$
\begin{array}{ccccccc}
y & = & f(x) & \overset{i}{\longrightarrow} & i(y) & = & \sigma(a^0_{-\infty}) & = a^{-1}_{-\infty} \\
\psi \downarrow & & \uparrow f & & \uparrow \sigma & & & \\
h(a^0_{-\infty}) & = & x & \overset{i}{\longrightarrow} & i(x) & = & a^0_{-\infty} &
\end{array}
$$

Here

(4.5)
$$\psi_k(y) = (f \mid_{I_k})^{-1} = \text{the inverse of } f \text{ restricted to } I_k.$$

For each $y \in [0,1]$, let

(4.6)
$$\{q_i(y); i = 1, 2, \cdots\}$$

be a probability distribution on $\mathcal{S}$ ($q_i(y) \geq 0, \Sigma_i q_i(y) = 1$), and define a Markov chain $\{\xi_n\}_0^\infty$ on $[0,1]$ by the transition function

(4.7)
$$P\{\xi_{n+1} = \psi_i(y) \mid \xi_n = y\} = q_i(y).$$

(ξ_0 has arbitrary initial distribution.) Note that

(4.8)
$$f(\xi_{n+1}) = \xi_n, n = 0, 1, \cdots.$$

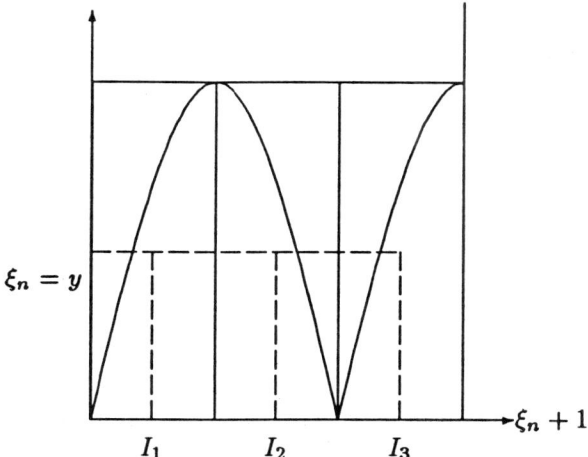

The figure illustrates the situation when $d = 3$.

The above construction leads to a two sided process taking values in the interval $[0, 1]$:

$$(4.9) \qquad \cdots \xi_{-2}, \xi_{-1}, \xi_0, \xi_1, \xi_2, \cdots,$$

where ξ_0 is an initial point or r.v., and

$$(4.10) \qquad \{\xi_n; n \geq 0\} \quad \text{is the M. C. defined by } (4.7),$$

while

$$(4.11) \qquad \xi_{-n} = f_n(\xi_0) \quad \text{for} \quad n = 1, 2, \cdots.$$

Thus starting with ξ_0 we iterate to the left by f_n and to the right by the (random) Markov chain. Now

$$(4.12) \qquad f(\xi_{n+1}) = \xi_n \quad \text{for all} \quad n = 0, \pm 1, \cdots.$$

Associated with $\{\xi_n\}_{-\infty}^{\infty}$ is its "label process" $\{X_n\}_{-\infty}^{\infty} \subset \mathcal{S}$, defined by

$$(4.13) \qquad X_n = k \quad \text{if} \quad \xi_n \in I_k, n = 0, \pm 1, \cdots.$$

This is not itself a M.C., but it is generated by the kernel

$$(4.14) \qquad P\{X_{n+1} = a \mid i(\xi_n) = a_{-\infty}^n\} = q_a[h(a_{-\infty}^n)]$$

and

$$(4.15) \qquad X_{-n} = i_{-n}(\xi_0) = k \quad \text{if} \quad f_n(\xi_0) \in I_k$$
$$\text{for} \quad n = 1, 2, \cdots.$$

TIME	$\cdots$	$-n$	$\cdots$	-2	-1	0	1	2	$\cdots$	n	$\cdots$
Iterates of f		$f_n(x)$	$\cdots$	$f_2(x)$	$f(x)$	x					
Itinerary of x		$i_{-n}(x)$	$\cdots$	$i_{-2}(x)$	$i_0(x)$						
Iterates of r.v. ξ_0		$\xi_{-n}=f_n(\xi_0)$	$\cdots$ iteration	$\xi_{-2}=f_2(x)$	$\xi_{-1}=f(\xi_0)$	ξ_0	ξ_1	ξ_2	$\cdots$	ξ_n The Markov chain	
Itinerary of ξ_0		$X_{-n}=i_{-n}(\xi_0)$		$X_{-2}=i_{-2}(\xi_0)$	$X_{-1}=i_{-1}(\xi_0)$	$X_0=\xi_0$	X_1	X_2		$\cdots$	
Regeneration										$X_{T_0}\cdots X_{T_1}$	

TABLE 4.1

See the table for the relationship between all the above quantities.

5. THE ITERATES OF f AND THEIR RELATION TO ∞-ORDER CHAINS

We can now write

$$(5.1) \qquad q_{a_1}[h(a_{-\infty}^0)] \doteq g(a_{-\infty}^0, a_1) \quad \text{for} \quad a_{-\infty}^0 \in \mathcal{S}_{-\infty}^0.$$

This is a probability kernel on $\mathcal{S}_{-\infty}^0 \times \mathcal{S}$ exactly as defined in section 2. Hence the labell process $\{X_n\}$ constructed from the M. C. $\{\xi_n\}$ in section 4, is equivalent (in distribution) to the ∞-order chain $\{X_n\}$ generated by (λ, g), as in section 2, namely

$$P\{X_1 = a \mid X_0 = a_0, X_{-1} = a_{-1}, \cdots\} = g(a_{-\infty}^0, a_1).$$

The important fact now is the following

Lemma. *Let f satisfy (4.1 a,b,c), and assume that q (defined in (4.7)) satisfies the uniform continuity condition*

$$|\log q_j(x) - \log q_j(y)| \le c|x - y|$$

for some $0 < c < \infty$, all $j \in \mathcal{S}$ and all $x, y \in [0, 1]$. Then the memory of the label sequences $\{X_n\}$ decays along all sequences $\alpha_{-\infty}^0 \in \mathcal{S}_{-\infty}^0$ (see 2.6).

This is proved in [10].

We can thus conclude from the theorem in section 3 that a regeneration structure exists for the label process $\{X_n\}$.

We now have i.i.d. blocks

$$(X_{T_i+1}, \cdots X_{T_{i+1}-1})i = 0, 1, \cdots.$$

By a standard construction of renewal theory, this block structure can be extended to the negative side, i.e. $i = -1, -2, \cdots$, by suitably defining the block that covers the origin, and then laying off the other blocks to the right and left. Call the resultant sequence $\{Y_n, n = 0, \pm 1, \cdots\}$. Then

$$(Y_{T_i+1}, \cdots, Y_{T_{i+1}-1}) \quad \text{are i.i.d. blocks}$$
$$\text{for} \quad i = 0, \pm 1, \pm 2, \cdots .$$

Let $\eta_0 = h(Y_{-\infty}^0), \cdots, \eta_n = f_n(\eta_0) = h(Y_{-\infty}^{-n})$, and $i_{-n}(\eta_0) = Y_{-n} =$ the itinerary of η_0 for $n = 0, 1, 2, \cdots$. Let $\tau_i = -T_{-i}$. Then

(i) $\mathcal{L}(f_{\tau_i}(\eta_0))$ does not depend on i,

(ii) The r.v. $f_{\tau_i}(\eta_0)$ is independent of $Y_{\tau_i+1}^\infty$, in particular of $Y_{\tau_i+1}^0$.

Thus the itinerary of η_0 under iteration by f is divided into independent blocks.

6. GIBBS MEASURES

We outline the idea why Gibbs measures admit a regeneration structure. See Lalley [5] and Bowen [12].

Let $\mathcal{F}$ be the family of all continuous functions $\varphi : \mathcal{S}_{-\infty}^0 \to \mathbb{R}$ for which

(6.1)
$$v_k = \mathrm{var}_k \varphi \doteq \sup\{|\varphi(x) - \varphi(y)| : x, y \in \mathcal{S}_{-\infty}^0, \ x_i = y_i, i = -k, \cdots, 0\} \leq b\alpha^k$$
$$\text{for some} \quad b < \infty, \alpha \in (0, 1).$$

Let $C(\mathcal{S}_{-\infty}^0) \doteq$ the continuous real valued functions f on $\mathcal{S}_{-\infty}^0$ such that $\mathrm{var}_k f \to 0$ as $k \to \infty$.

The Perron-Frobenius operator on $C(\mathcal{S}_{-\infty}^0)$ is defined by

(6.2)
$$(\mathcal{L}_\varphi f)(x_{-\infty}^0) = \sum_{y_1 \in \mathcal{S}} e^{\varphi(x_{-\infty}^0 y_1)} f(x_{-\infty}^0 y_1).$$

Let
$$k(x_{-\infty}^0, y_{-\infty}^1) = \begin{cases} \exp \varphi(x_{-\infty}^0 y_1) & \text{if } y_{-\infty}^1 = x_{-\infty}^0 y_1 \\ 0 & \text{otherwise.} \end{cases}$$

Then

(6.3)
$$\mathcal{L}_\varphi f(x_{-\infty}^0) = \sum_{y_{-\infty}^1 \in \mathcal{S}_{-\infty}^1} k(x_{-\infty}^0, y_{-\infty}^1) f(y_{-\infty}^1)$$
$$= \sum_{y_1 \in \mathcal{S}} f(y_{-\infty}^1) e^{\varphi(x_{-\infty}^0 y_1)}.$$

Write
$$e^{\varphi(x_{-\infty}^0 y_1)} = g_\varphi(x_{-\infty}^0, y_1).$$

(Note g_φ is *not* a stochastic kernel.)

Assume that all the sequences in $\mathcal{S}_{-\infty}^0$ are allowable and that $\mathcal{S}_{-\infty}^0$ is mixing in the sense of Bowen [12]. Then one has the

Ruelle-Perron Frobenius Theorem.

If $\varphi \in \mathcal{F}$ then there exist an eigenvalue λ, an eigenfunction $r(\,\cdot\,)$, and an eigenmeasure $\ell(\,\cdot\,)$ for $\mathcal{L}_\varphi$. Namely

$$(6.4) \qquad\qquad \mathcal{L}_\varphi r = \lambda r, \ \mathcal{L}_\varphi^* \ell = \lambda \ell, \ \ell r = 1.$$

Also $0 < c_1 < r(x) < c_2 < \infty$.

Now $k(\,\cdot\,,\,\cdot\,)$ is not a stochastic kernel but by Ruelle's theorem

$$(6.5) \qquad p(x_{-\infty}^0, y_{-\infty}^1) \doteq \frac{k(x_{-\infty}^0, y_{-\infty}^1) r(y_{-\infty}^1)}{\lambda r(x_{-\infty}^0)}$$

$$= \begin{cases} \dfrac{e^{\varphi(x_{-\infty}^0 y_1)} r(x_{-\infty}^0 y_1)}{\lambda r(x_{-\infty}^0)} = & \text{(def) } g(x_{-\infty}^0, y_1) \text{ if } x_{-\infty}^0 = y_{-\infty}^0 \\[2mm] 0 & \text{otherwise.} \end{cases}$$

is stochastic. The N-fold iterate is

$$(6.6) \qquad p^N(x_{-\infty}^0, x_{-\infty}^0 x_1^N) = g^N(x_{-\infty}^0, x_1^N) = \frac{e^{\sum_1^N \varphi(x_{-\infty}^i)} r(x_{-\infty}^N)}{\lambda^N r(x_{-\infty}^0)}.$$

In this way one generates a measure μ on $\mathcal{S}_1^\infty$ such that

$$(6.7) \qquad 0 < c' < \frac{\mu(x_1, \cdots, x_N)}{\exp\{-N \log \lambda + \sum_{i=1}^N \varphi(x_{-\infty}^i)\}} < c'' < \infty.$$

This is a Gibbs measure. Finally noting that

$$(6.8) \quad g^n(x_{-\infty}^0, x_1^n) = p^n(x_{-\infty}^0, x_{-\infty}^n) = g(x_{-\infty}^0, x_1) g(x_{-\infty}^1, x_2) \cdots g(x_{-\infty}^{n-1}, x_n)$$

and using (6.6), we see that

$$(6.9) \qquad \gamma_m^{-1} \doteq \sup \left\{ \frac{g^N(x_{-\infty}^0, x_1^N)}{g^N(y_{-\infty}^0, y_1^N)} : x_{-m+1}^N = y_{-m+1}^N, N = 1, 2, \cdots \right\}$$

$$\leq \exp\{|\log r(y_{-\infty}^0) - \log r(x_{-\infty}^0)|\}$$

$$+ \exp\{|\log r(y_{-\infty}^n) - \log r(x_{-\infty}^n)|\}$$

$$+ \exp \sum_{k=1}^n |\varphi(x_{-\infty}^k) - \varphi(y_{-\infty}^k)|.$$

Let $\rho_k = \mathrm{var}_q(r)$. Then

$$(6.10) \qquad\qquad \gamma_m^{-1} \leq \exp\{\rho_m + \rho_{m+n} + \sum_{k=1}^n \nu_{m+k}\}.$$

Hence by the condition (6.1) and the continuity of $r(\cdot)$, this means that

$$(6.11) \qquad\qquad\qquad \gamma_m \nearrow 1,$$

i.e. that $g(\,\cdot\,,\,\cdot\,)$ satisfies the memory hypothesis for infinite order chains (2.5). Hence there exists a regeneration structure for the measure induced by $\mathcal{L}_\varphi$.

REFERENCES

[1] H. Berbee, *Chains with infinite connections: Uniqueness and Markov representations*, Prob. Th. and Rel. Fields **76** (1987), 243–353.

[2] T. E. Harris, *On chains of infinite order*, Pacific J. of Math. (1955), 707–724.

[3] M. Iosifescu and S. Grigorescu, *Dependence with Complete Connections and its Applications*, Cambridge Univ. Press, Cambridge, 1990.

[4] S. Kalpazidou, *A Gaussian measure for certain continued fractions*, Proc. Am. Math. Soc. **96** (1986), 629–635.

[5] S. Lalley, *Regenerative representation for one dimensional Gibbs states*, Ann. Prob. **14** (1986), 1262–1271.

[6] A. Lasota and M. Mackey, *Probabilistic properties of deterministic systems*, Cambridge Univ. Press, Cambridge, 1985.

[7] P. Ney and E. Nummelin, *Regeneration for chains with infinite memory*, Prob. Th. and Rel. Fields (to appear).

[8] _____ , *Markov additive processes I: Eigenvalue properties and limit theorems*, Ann. of Prob. **15** (1987), 561–592.

[9] _____ , *Markov additive processes II: Large deviations*, Ann. of Prob. **15** (1987), 593–609.

[10] _____ , *Regeneration for random maps*, Technical report (1992).

[11] E. Nummelin, *General irreducible Markov chains and non-negative operators*, Cambridge Univ. Press, Cambridge, 1984.

[12] R. Bowen, *Equilibrium States and Ergodic Theory and Anosov Diffeomorphisms*, vol. 470, Springer Lecture Notes.

DEPARTMENT OF MATHEMATICS, UNIV. WISCONSIN, MADISON, WI 53706
E-mail address: ney@math.wisc.edu

DEPARTMENT OF MATHEMATICS, UNIV. HELSINKI, HELSINKI, FINLAND